THE URANTIA REVELATION

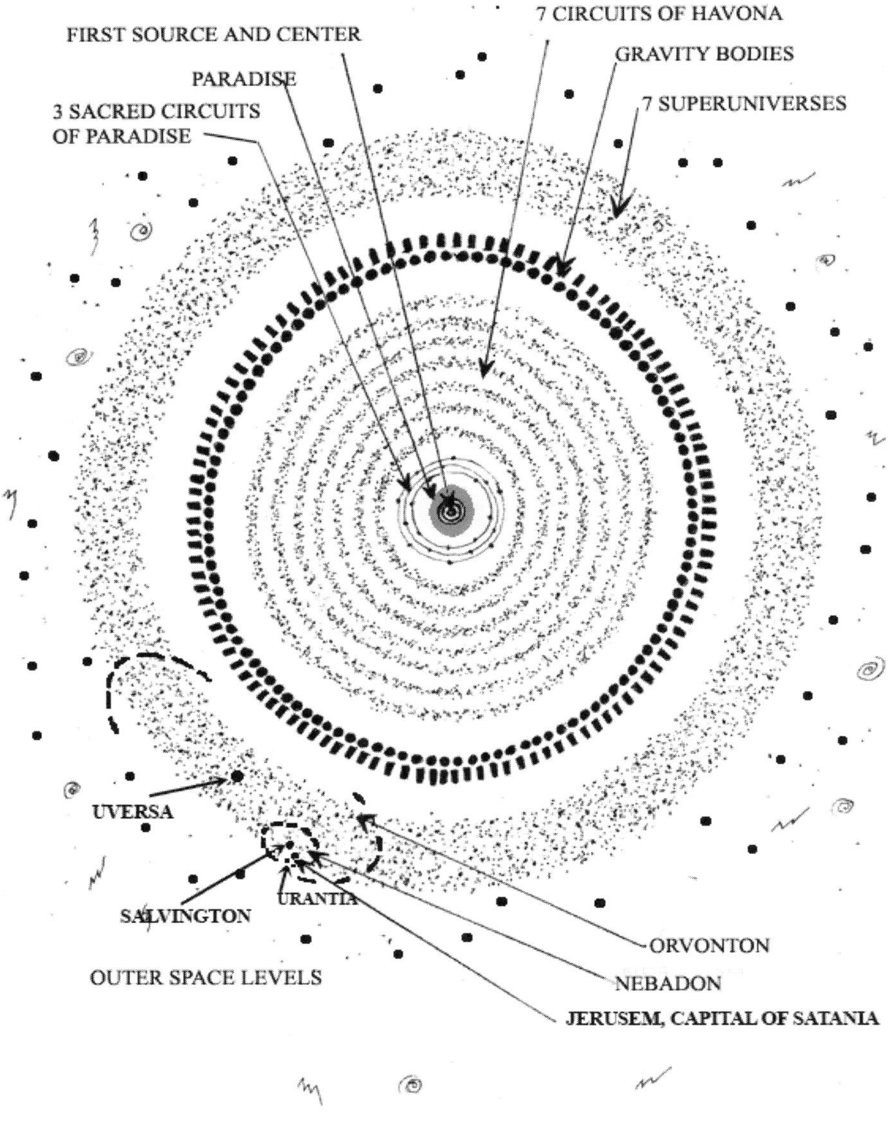

FIGURE 1. Urantia's Position within a Map of the Universe

THE URANTIA REVELATION

The Structure and Meaning of the Universe Explained

MALCOLM LOCKE

BASCOM HILL BOOKS
Minneapolis, Minnesota

Bascom Hill Books
212 Third Avenue North
Minneapolis, Minnesota 55401

Printed in the United States of America.

ISBN 978-1-935456-22-3

PETER IN THE COURT OF THE HIGH PRIEST

Peter is a careful man.
He peers unnoticed from behind
the lighted stone through the crowd
to where they carry him to death.

Peter almost spoke his name.
He even raised his hand.
But the word sank in his throat
when the fire's gleam from
the soldier's bronze,
the terror of the mob,
and the look on his friend's lips
made Peter, at last, understand.

To my friend
Best
Kathy

TABLE OF CONTENTS

ACKNOWLEDGEMENTS

I WISH TO THANK Donna Eagles, who created the graphics at the front of the book and most of the figures. She also broke the back of the complex glossary. I owe Donna a debt of gratitude for these services as well as enlightening conversation about *The Urantia Book*. I also wish to thank Elizabeth Cusack, who read through the manuscript and made many suggestions. My wife, Trisha, gave her customary careful attention to every word.

INTRODUCTION

Imagine that an interplanetary corps of nonhuman beings came to planet earth and dictated 1,153,000 words, over two thousand pages, of information about the history of the planet, the nature of God and reality, the purpose of evolution, the destiny and nature of man, and the life and teachings of Jesus of Nazareth. Apparently something like this actually did happen sometime in the early 1930s in Chicago, Illinois. You'd think this fact would be better known.

In reality *The Urantia Book* is little known, although in continuous publication since 1955, on Borders' shelf list, and since 2006 in the public domain and on the Internet in many languages (http://www.urantia.org/). *The Urantia Book* is downloadable for free so you can put it on your iPod and listen while at the gym. I've known enough people in my life who were seekers, looking for the truth about themselves and the world around them, trying to find real answers to big questions, trying to find out about the human past, and about how to lead their lives. Occasionally I've recommended that they have a look at *The Urantia Book*, or put it on their iPod, but they never do, or buy it but do not study it.

It takes about a year of daily, dedicated study to get through *The Urantia Book*. Although reading it probably changes you, *The Urantia Book* is for most people too big, too long, too unfamiliar, and much too strange. The book is massive and sometimes exceedingly abstract. You will need to deal with such sentences as "God the Ultimate will eventually and inevitably powerize and personalize as the Deity consequence of the unification of this experiential Ultimate Trinity in the expanding arena of the well-nigh limitless master universe" (Foreword). Or: "And then we go on to envisage the experiential Trinities as culminating in the experiential personalization of the Deity Absolute, while conceiving the Universal Absolute as constituting the universe and the extrauniverse phenomena of the manifest presence of the impersonal activities of the unified and coordinated Deity associations of supremacy, ultimacy, and infinity —the Trinity of Trinities" (paper 56). Romantic—but what does it mean? And when you do understand what it means, how can you know if it is true?

One can only learn so much from science, then factual knowledge fails. It can be popular to think that the world is a meaningless jumble from which no escape is possible, or to pray to the great god so-and-so who shed his blood so that you might go to heaven, or so that you might enjoy a bountiful harvest. *The Urantia Book* reveals a very different world, an immense cosmopolis filled with trillions of stars and billions of trillions of planets pullulating with irrepressible life, all moving toward a greater but unknown goal in the reaches of outer space. The universe is a gigantic bureaucracy with offices everywhere and hierarchies of every kind and plans and purposes and conspiracy and danger, and destiny, and failure, and success and reward. Planet Earth, whose real name is Urantia, is a life-experiment station on which higher nonhuman beings are developing newer forms of life-expression. Many strange things have happened on this planet. Everything is going to be okay, but you can certainly go wrong and it is bewildering to find out what is going on.

In this book I present a digest of what *The Urantia Book* says, avoiding as much as possible the context of my own experience, trying to keep myself out of it, working as a disinterested reporter, although I will sometimes put in historical dates familiar from modern scholarly research. Not to say everything *The Urantia Book* says—the book is for that—but the gist of it, the big picture, the main outlines of the story. As such this book should serve as an introduction and guide to *The Urantia Book*, or as a simplified and accessible substitute, a *vade mecum* for those wishing to know what was revealed in the fifth epochal revelation on Planet Earth. There is hardly anything more interesting.

ORIGIN OF *THE URANTIA BOOK*

The origins of *The Urantia Book* remain unclear, but some information is available. There are 196 papers on numerous topics that we might think of as cosmology, theology, philosophy, metaphysics, physics, geography, biology, anthropology, the history of religion, sociology, and in the case of Jesus of Nazareth, biography. Different nonhuman authors are cited by name or by an order of universe beings. For example, the first paper is attributed to "an Orvonton Divine Counselor, Chief of the Corps of Superuniverse Personalities assigned to portray on Urantia the truth concerning the Paradise Deities and the universe of universes." These different "authors" write in different styles and have different voices, sometimes highly diffuse and abstract, sometimes concise and telling, sometimes swollen and rhetorical.

Often the authors speak as scientists, noting that only certain things are known about, for example, the purpose of the ongoing organization of energies in outer space levels; but much remains mysterious. By no means do they claim omniscience, although their resources exceed anything in our experience. They can be pious: Of course it is hard to praise perfection without sounding trite. In general the style is expansive, often making the same point again and again from different angles, especially in the theological portions. The papers, like this book, are divided into four main sections. They do not proceed from the simple and finite to the complex and infinite, from human origins to divine destinies, but begin with the infinite, a revelation of the divine destiny:

Part I. The Universe, 46 papers
Part II. The Local Universe, 48 papers

Part III. The History of Urantia, 76 papers
Part IV. The Life of Jesus, 116 papers

Somehow the papers came out of a group that met in Chicago in the early 1930s. According to the story published in many places and of course on the Internet, in 1911 a neighbor approached a prominent Chicago physician and psychiatrist named William S. Sadler (1875–1969) and his wife Lena Sadler (1875–1939), a physician much interested in women's health and a relative of the William Kellogg who with his brother John Harvey Kellogg invented cornflakes and founded the famous cereal company. William Sadler had an extraordinary professional career, wrote forty-two books, and was influential in the American Medical Association. He was professionally interested in debunking so-called paranormal phenomena and had written a book on the topic, *The Mind at Mischief* (1929). Later Sadler recollected his experience with *The Urantia Book* in several papers including "How *The Urantia Book* Came Into Existence," taped on February 18, 1962 (available on the Web at www. urantiabook. org/indes_history.htm.)

The neighbor had come to the Sadlers for help because her husband had taken to falling into deep trances while asleep, so deep that she could not rouse him. His breathing was abnormal. She was concerned. The Sadlers visited the man while asleep. At some point he began to talk. The different voices claimed to be "student visitors." Of the sleeping subject Sadler wrote (in an appendix to *The Mind at Mischief*):

> While the sleep seems to be quite of a natural order, it is very profound, and so far we have never been able to awaken the subject when in this state; but the body is never rigid, and the heart action is never modified, although respiration is sometimes markedly interfered with. This man is utterly unconscious, wholly oblivious to what takes place, and unless told about it subsequently, never knows that he has been used as a sort of clearing house for the coming and going of alleged extra-planetary personalities. In fact, he is more or less indifferent to the whole proceeding, and shows a surprising lack of interest in these affairs as they occur from time to time … Much of the material secured through this subject is quite contrary to his habits of thought, to the way in which he has been taught, and to his entire philosophy. In fact, of much that we have secured, we have failed to find anything of its nature in existence. Its philosophic content is quite new, and we are unable to find where very much of it has ever found human expression.

Some have thought that the unnamed sleeper was Wilfred Custer Kellogg (1876–1956), a nephew of William Kellogg, whose niece was Lena Sadler, née Kellogg. Not until early 1925—that is, over ten years later—was a stenographer brought in to write down what the sleeping man was saying, taken down in "voluminous written documents."

At this time William Sadler and his wife, Lena, belonged to a discussion group called the Forum consisting of friends and former patients. The group met on Sundays to hold intellectual discussions. Sadler read something of what was written to this group, which showed considerable interest. At this time the voice of the speaking subject suggested that the preexisting Forum submit questions and that answers would be given by celestial beings "through the contact personality." A group of five, made up of the doctor, his wife, and three others selected the questions posed by the Forum and, evidently, presented them to the sleeper. We hear no more about a stenographer, and some have doubted that the written versions of the sleeper's voice were created in this way. Apparently some members of the contact commission, now all dead, claimed that the papers became physical objects in ways they did not understand.

At first the questions were casual, but after reproof from the sleeper's voice at the group's careless response, the group began to ask questions that would elicit a broad picture of universe reality. A paper would be produced, read before the Forum, discussed, and clarifications pursued. In this way the papers controlled by the contact group were revised in consultation with the sleeper's voice.

Someone transferred the handwritten originals to typewritten form. Between 1925 and 1935, ten years, the Forum group gathered 196 papers, but continued to revise them for the next twenty years, during which they continued to receive new communications. At last the Forum published the papers in October 1955, under international copyright, published by the Urantia Foundation on Diversey Parkway in Chicago. In 1983 the Urantia Foundation, connected in some way to the circle of Dr. Sadler, renewed its copyright of *The Urantia Book*, but in a court decision in Oklahoma in 2001 lost the copyright. At this time *The Urantia Book* passed into the public domain. Several editions are available in bookstores and on the Internet.

To help the reader of this book faced with many difficult and unfamiliar terms and categories, especially in the early chapters, I have provided a glossary at the end.

PART I

THEORETICAL CONSIDERATIONS

CHAPTER 1: GOD AND REALITY

IT WOULD BE WRONG TO CLAIM THAT, just looking around, everything is the way it is because God wants it that way. In a sense of course this must be true, because the universe is an expression of God's will. Nonetheless, that is not how the world works. Everything comes from God, but God, although one, is no simple thing. Jesus allowed himself to be crucified in accordance with the Father's will, but the Father did not wish him to die horribly in this way and, we might say, had nothing to do with it.

GOD

We cannot understand what God is. That would be absurd. The created cannot comprehend the creator. Furthermore, the word-symbols in human language are quite inadequate to represent divine realities, with which nothing in our experience is comparable. Even so we can say some things about God.

"God" is a word with many meanings, but ordinarily we take it to refer to a primal and infinite personality. If you recognize God, you recognize his personality, which exists. Through personality you can know and be known, love and be loved, and God is love. Nonetheless part of God is prepersonal and part is superpersonal; that is, before God had qualities as a personality, God had no qualities and God always transcends all qualities, because qualities, such as personality, are already a kind of limitation. God knows no limitation. That is the essence of God, his infinite unity, which one can perceive behind the appearance of things.

God is the oneness of all things, then, but for us God is a personality who can be known. Jesus called God the Father because God can be known as a human child knows his human father. Knowledge of the personality of God comes through revelation; although it can be experienced, it is scarcely discoverable. Personality reaches out to unify all realities. The infinite personality of the First Source and Center, the Universal Father, unifies all constituent elements of infinity and of the creation. Likewise can the personality of mortal humans, a direct bestowal of the Universal Father, unify the constituent elements of the individual mortal creature. Without the personality of God, one would be reduced to the philosophic dilemmas of materialism or pantheism. Personality does not require a body, and of course God has no body, not even the universe.

God is spirit, a Latin word that means "breath," something that is there, vivifying, but invisible. Like life, you can see the effects of breath but not the thing itself. God is like that, not hiding from his creatures, but dwelling in a realm whose light prevents approach. You can never look on the face of God and live, according to an old saying.

DIFFERENT LEVELS OF REALITY

All that we see is not all that there is. Reality falls into three levels:

(1) *finite* reality, where we live. Here there is time and space and all things have a beginning (although not necessarily an end).

(2) *absonite* reality, where things have no beginnings and no endings. Time and space are transcended. *Absoniters*, beings who live at this level of reality, do not begin. Somehow, they always were. We can accept this situation, but not understand it very well.

(3) *absolute* reality is even more elusive, because in absolute reality there is no time and space to transcend; all things are forever. Absolute reality characterizes the geographical center of the universe that *The Urantia Book* calls *Paradise*. It is a place. Paradise is an ancient Persian word meaning "enclosed garden." Material gravity proceeds from Paradise, the only motionless thing in the universe, and Paradise is the model for all the physical creation. Paradise has location but is not in space and not in time. The purpose of creatures with will (such as ourselves) is to traverse a very, very long path that ends, for the successful, in Paradise, the goal of mortal ascenders (such as ourselves). Paradise is the source of the physical universe, but a controller, not a creator. Paradise is unique, without parallel or category. It simply is and always was.

God, then, although one, appears in different ways to different levels of reality. The word "God" is a symbol representing the reality of the uncreated first origin of all things, but this reality is different to differing levels of being. For ourselves, by "God" we ordinarily mean the *Universal Father*. As a teacher, it was the mission of Jesus of Nazareth to reveal the nature and reality of the Universal Father. In our use of word symbols, "God" always has personality, whereas "Deity" does not necessarily; deity is divinity with or without personality. *Personality* is from the Roman word for "mask," as used in dramatic performances. The mask describes who the character is, so that one character can interact with another.

You can interact with the personality of God as the Universal Father, but as we have seen, God exists before the universe too, and within it, and at prepersonal levels, as God without personality. The notion of the prepersonal level of God is important to *The Urantia Book* because after the day of Pentecost in AD 30, somehow a speck, a spark, a portion of the prepersonal God comes to live within all normal human beings on this planet. Before this time this speck dwelled only within some human beings. This prepersonal infinite spark of God, sent by the Universal Father to indwell mortals, *The Urantia Book* calls the *Thought Adjuster* or sometimes the *Mystery Monitor*. The Thought Adjuster is the secret to human destiny. Even as you read these words, your own Thought Adjuster reaches out to you.

THE FIRST SOURCE AND CENTER

The *First Source and Center* is infinite and eternal, hence limited only by volition. God the Universal Father is the personality of the First Source and Center and through this personality maintains personal relations with all subordinate centers. All that exists depends directly or indirectly on the First Source and Center, including all that can be imagined.

Infinite and eternal, the First Source and Center escaped the bonds of infinity by the creation through will of subinfinite levels of reality. Ironically, the finite world is inevitable if the First Source and Center has no limitations, because the absence of the finite world is a kind of limitation. Nonetheless, events that humans can understand only as being sequential—first this, then that—in reality (so filled with shadows) happened at the same time, or never happened but always were. The universe never began, because it always was. Even if we cannot really understand this, we can still see why it must be so.

A way to refer to the First Source and Center is as the eternal *I AM*, the answer supposedly given to Moses from the burning bush when he asked, "Who are you?" The eternal I AM achieved personality when it became the Universal Father of the Eternal Son and the source of the Isle of Paradise, the geographical center of the universe. Coexistent with the differentiation between *Father* and *Son* was the appearance of the *Infinite Spirit* and the Central Universe of *Havona*, which spins around Paradise (see Figure 1 opposite title page).

The Infinite Spirit is also called the *Conjoint Actor*, because it works conjointly with the Father and the Son in various capacities. As the Father is *personality*, and the Son is *spirit*, the Infinite Spirit is *mind*. (It is confusing that the word "spirit" is used to describe the essence of the Son and to name the third person of Trinity.) Although we who live in time and space speak of the differentiation of the single I AM into the Trinity of Father, Son, Spirit, in reality there never was a time when the Trinity did not exist. The persons of the Trinity are coeternal with the I AM.

DIFFERING LEVELS OF HUMAN REALITY

For humans, reality exists on three different levels: the *body*, the electrochemical product of animal evolution; the *mind*, the thinking, perceiving, feeling, remembering awareness, both conscious and subconscious; the *spirit*, the level of the Thought Adjuster, a divine fragment that today (somehow) lives in the normal individual human. The tension between the personality, a gift of the Universal Father attached to the body, and the spirit, the Father himself hidden within, is designed to create, actually engender, a new entity, the *soul*, which does not die. By nature the mortal mind, which the personality manipulates, is linked to matter, but through choosing to do the Father's will the soul comes into being. How to know the Father's will is a problem, but somehow this is possible if such knowledge becomes a conscious goal. The prepersonal Thought Adjuster is destined to acquire personality through fusion with a human personality, ordinarily at some time after mortal death in association with the soul. Such fusion will create a new immortal being now with evolutionary experience, fairly compared to the emergence of a gorgeous butterfly from a lowly caterpillar.

The soul is neither material (as the body) nor spiritual (as the Thought Adjuster), but lives in an intermediate realm that *The Urantia Book* calls the morontial level. Adjuster-fusion usually takes place on the morontial level. *Morontia* designates a vast realm between the material and the spiritual, although it partakes of both (occultists have called this the *astral plane*). The soul is made of morontia, as our material bodies are made of cells. We might think of the morontial realm as consisting of a finer form of matter, but really it stands between the material and the spiritual.

The engendering of an immortal morontial soul is the purpose of human life but takes place independently of intellectual self-consciousness about what is happening. It just grows by itself, as one chooses to do the Father's will. As the soul comes into being during the mortal life, a new kind of consciousness evolves that enables the individual more capably to commune with the indwelling Adjuster and in exceptional cases Adjuster-fusion may even take place during the mortal life.

PERSONALITY

Personality is neither body, mind, nor spirit, nor is it the soul. Personality is the one unchanging reality in creature experience, a bestowal of the Universal Father that may (or may not) survive mortal death in accompaniment with the morontial soul born during mortal life. Although personalities belong to types, which those of wide experience can recognize, no two personalities in the entire universe are the same. Identical twins may have similar personalities, but as persons each is unique. When Thought Adjusters come forth from the Universal Father, they are all the same (though sometimes later modified), but personalities are diverse and original. Personality is the thing one comes to know, so you can recognize a personality regardless of the change in form or manifestation.

Creature personality has self-consciousness and a relatively free will, the power to choose, to make moral decisions. Hence morality cannot be advanced by law or by force, but only by the contagious influence of those who lead moral lives. Personality, which makes moral behavior possible, precedes and is independent of the bestowal of the Thought Adjuster.

The self-consciousness of personality makes possible the recognition of other beings endowed with will, the basis of social life. Personalities respond to the presence of other personalities. Self-consciousness is really a form of communal consciousness. The personality makes contact with the environment through mind. Therefore personality functions poorly in isolation from other personalities and longs for social interaction. "No man lives unto himself."

Personality is permanence in the face of constant change, the gift of the Universal Father. It can be given to any energy system that partakes of mind (which derives from the third person of Trinity) and spirit (which derives from the second person of Trinity). Personality is not entirely fettered by the chain of causation, but has real freedom, the power to cocreate. Personality alone is able to make a gift to God, who already has everything: the dedication of one's free will to the will of God. Personality is the one thing that can be added to spirit (the Son), proving the "primacy" of the Father, the giver of personality, over the Son.

Whereas the personality and the Adjuster are both changeless, the product of their interaction in the evolution of the soul is change, constant change. Personality is uniquely conscious of time and functions efficiently in the Local Universe of Nebadon, in the Superuniverse of Orvonton, and in the Central Universe of Havona. Life is a process that takes places between an organism and its environment. The personality imparts identity to the organism and enables the creation of continuous values amidst constant change.

DEATH

We think of death as a single thing, but in regard to the personality death is three things:

death of the soul. When the personality has unequivocally rejected survival and the decision has been recorded and verified on *Uversa*, capital of the Superuniverse of *Orvonton* in which we live, then the Adjuster is released from all obligation. That person is now absolutely dead, although the life mechanism, the body, may continue for awhile on its own momentum.

intellectual death. When the physical brain is so injured that the Adjutant Mind Spirits (about which more later) can no longer communicate with the individual, then the Adjuster departs and that person is dead, although the body, no longer human, may continue to live on. If the person had earlier chosen survival, then the personality may live on in association with the morontia soul.

physical death. When the rhythmic pulsations of the brain cease to function, the person is dead and the Adjuster instantaneously departs for Divinington (a special sphere that goes around Paradise: more later) with the memory transcription of the mortal career. The seraphic destiny guardian, either of the group to which the person belongs, or of the individual if it is someone of accomplishment (more later), takes possession of the morontia soul.

Both the Thought Adjuster with its memory transcription of the mortal career and the morontial soul under seraphic protection will be necessary to reassemble the individual's personality. After death, an Adjuster never returns to a planet of former indwelling. A soul without its Adjuster is in unconscious sleep, either long or short. After reassembly of the personality on the mansion worlds surrounding *Jerusem*, capital of our Local System of *Satania*, no communication whatever is permitted between such resurrected beings and dwellers of the planet where once they lived. However, on the morontia worlds you will recognize those whom knew you in earthly life, if they have survived, and they will recognize you.

TRINITY AND UNITY

We may think of the three persons of the Trinity as circuits that hold the universe together. All things with *personality* are within the circuit of the Universal Father; that which responds to the Eternal Son is within the circuit of *spirit* (the Son cannot bestow personality); the third principle, the Conjoint Actor (or Infinite Spirit), governs the circuit of *mind*, which in personality intervenes between spirit and matter: *Matter* is whatever responds to the gravity circuit of nether Paradise.

The following chart may illustrate various aspects of the three persons of Trinity:

	FATHER	SON	INFINITE SPIRIT
process:	thought	word	knowledge
essence:	personality	spirit	mind
quality:	love	mercy	service

In spite of such distinctions, God is nonetheless one, who as the Universal Father created all there is, controls all there is, and upholds all there is. The children of time wish to know the eternal God, the Universal Father, and to be perfect as the Father is perfect in eternity. Such perfection is possible. The myriads of planetary systems that fill the universe were made to support intelligent life that can know God. The search for God is the supreme adventure for inhabitants of these numberless planets swirling around unnumbered suns, including ourselves.

THOUGHT ADJUSTERS AGAIN

The term "Thought Adjuster" is appropriate because as a prepersonal but real "piece of God" it strives in secret ways to adjust our thoughts, to make our thoughts loving, selfless, and worshipful, so that union between the personality and Adjuster may take place. We could in fact not expand our concept of God without the presence of the Thought Adjusters within us, together with assistance from the Spirit of Truth (about which more later in the story about Jesus). Sometimes *The Urantia Book* calls this piece of God inherent in the individual the Mystery Monitor, evidently because it monitors our constant thoughts and because its presence is a mystery—separate, other, standing behind and beyond. The Mystery Monitor is the mystery of the finite and the infinite, the bounded and the eternal, the material and the spiritual somehow coming together.

It may be preposterous to suppose that the infinite God, creator of the universe, divides himself into little pieces in order to inhabit our human minds, but mysteries have a right to be preposterous. Because of this indwelling it is literally true that God shares directly all human afflictions and triumphs. Only by faith—belief without evidence—can the mortal, bound by time and space, apprehend the timeless and eternal, the Adjuster's presence. One cannot prove its existence through science or logical deduction, but only through experience, which is always personal.

God's knowledge of the finite is possible, but not so long as God retains his infinite knowledge of himself. In the temporal finite realms where we dwell, the infinite God can through the Thought Adjuster gain experience of finitude, and through the union of the mortal personality with the infinite God create a fresh immortal being with evolutionary experience. In our day-to-day lives, we may not be aware of communion with the Adjuster, but in the fruits of one's life is revealed that communion. "How will we know them? You will know them by their fruits."

THE CREATOR SONS

God *did* personally create the perfect Central Universe of Havona and the Isle of Paradise at its center, but he *did not* personally create the universes of time and space, in one of which we now dwell. He does, however, through the gravity circuits emanating from nether Paradise control physical and other features of such subordinate universes, which were created by members of the *Paradise Corps of Creator Sons*. The Creator Sons are a special order of beings who embody the aspect of the Eternal Son within the primordial Trinity and come into existence directly from the Eternal Son as he interacts with the Universal Father on Paradise.

The Urantia Book calls the Creator Son who made our own Local Universe *Michael* because he belongs to the *Order of Michaels*, another name for the Paradise Corps of Creator Sons. (By a false etymology, the name Michael is said to be Hebrew for "Who is like God?" that is, no one.) The Michael of this Local Universe, registered under the name of Nebadon, was the very being who through an exceptional process incarnated as Jesus of Nazareth in Palestine in 7 BC, for complex reasons that we will later discuss. The Universal Father is present directly within the human mind as the Thought Adjuster, but otherwise the Universal Father has designated authority to the Creator Sons, through whom the invisible and unknowable Universal Father presents himself to the worlds of time and their inhabitants: that is, to us.

WHAT GOD IS LIKE

The Universal Father is infinite and perfect in every way, hence shut off from direct communication with us, because we dwell in a world defined by limitations. God is a self-conscious being, perfect in goodness, who shares in every experience of every creature in the universe, including the commission of evil: But in God himself there is no evil, which is a deficiency of the good.

Evil is a turning away from God, an absence of God and alignment with the unreal; hence it is self-destructive. Perfect alignment with evil is perfect alignment with nothingness and nothingness is death. Evil is inevitable in a universe filled with adventure and intelligent creatures of will, but because of God's perfect righteousness, the pursuit of evil brings the separation from God and the obliteration of personal consciousness. If that's what you want, it's there for you. Such formal obliterations take place periodically at the end of a planetary dispensation, of which there have been several on Urantia, the last at the time of the execution of Jesus of Nazareth in AD 30, the incarnated Michael of Nebadon.

Evil can exist because the qualities of divine mercy delay any immediate adjudication. An adjudication of nonexistence must come from the headquarters of the Superuniverse, in this case the planet Uversa, headquarters of Orvonton. There is no resurrection from such a fate, which is everlasting. If the individual was indwelt by a Thought Adjuster (Thought Adjusters do not indwell every human), any experience of spiritual value becomes part of the Thought Adjuster. In this way some Adjusters acquire experience, while others are without experience. As always with the Thought Adjusters, one must suspend logic, now to accept that a thing without qualities somehow may acquire them! And that things which are all identical, now are different! Such persisting memory within the experienced Adjuster is not, however, a basis for identity, because the survival of identity depends on the morontia immortal soul, born during the mortal life, in which the personality may reside. If the morontia soul does not survive because the personality has rejected survival, then identity does not survive.

We use the word love even for base desire, but God's love is rooted in God's goodness, which invokes the creature's love for God. God might not have been *good*, but only *mighty*, as an early Jewish conception portrayed him. Goodness, however, is part of the personality of God, knowable only through personal religious experience. Jesus of Nazareth lived an exemplary life, but as a teacher taught only two things: that the relationship between human and God is analogous to that between a child and its father; and that eternal life through faith is possible. Such teachings ran deeply contrary to traditional Jewish understanding, which saw God as a king and judge and worldly advantage to be the fruit of religious behavior.

In the teaching of Jesus, God rules not *like* a father, but *as* a father, the Universal Father of all personality. The divine father requires no mediator or sacrifice to establish good relations, as in Saint Paul's Christian doctrine of the atonement, according to which the blood of Jesus washed away the sin of Adam to permit good relations between man and God. Paul's unfortunate false doctrine stands between traditional Christianity and those trained in philosophy and is an assault on the unity and free-will of God. The doctrine is a bone in the throat of contemporary Christianity.

God is a creator. There was nothing inevitable about the universe, which came into being through the free will and choice of the Universal Father. To the limited vision of lower intelligence, the universe appears to be a chaos of misdirection, cruelty, and suffering. Such may be true of the part, but the whole is nonetheless governed according to a divine plan. Doing the "will of God" is living according to this plan; it is virtue. Because of the reality of free will many things that take place within the creation are not the doing of the Universal Father.

God is everywhere at once, *omnipresent*, because God is infinite, without boundaries. All that we know is finite, bounded.

God is all-powerful, *omnipotent*. There is absolutely nothing that God cannot do, except act contrary to God's own nature, which is truth, beauty, and goodness.

God knows all things, is *omniscient.* Only God knows how many stars there are and how many planets. The Infinite Spirit, the third person of Trinity, which is mind, is everywhere at once.

God is without limit, is *infinite.* Even if the creation were to continue on and on and even become infinite, nonetheless God would not be lessened one iota. Mortal humans cannot know God's infinity but can experience his love and they can reciprocate it. Love is the surest way to God.

CHAPTER 2. GOD'S RELATIONSHIP TO THE UNIVERSE AND THE INDIVIDUAL

THE CREATOR HAS A PURPOSE, but only the Paradise deities know what that purpose is, except that in some way the purpose is to fulfill the divine nature. The training of evolutionary mortals like ourselves to survive, reach Paradise and enter what is called the Paradise *Corps of Finality*, then receive further training for an unrevealed purpose, is an important concern throughout the Grand Universe.

God did not create the world, then stand back and allow it to run like a clock. He sends forth light, life, and energy to uphold the creation. What we think of as nature, and its hidden laws, is usually a Creator Son's working out of local modifications of universal laws; nature is therefore different in different Local Universes, constellations, systems, and planets. The laws of Paradise underlie all local modifications, but local modifications are liable to error, distortion, perversion, and mistakes, as has occurred in our own *Local System* of Satania. Through evolution the errors and disharmony of nature are eventually brought into line with the perfection of patterns on Paradise. God is not present personally in nature, which may include the imperfections of progressive evolution and, certainly on Urantia, the distortions of willful insurrection. Nature is not the expression of an all-wise and all-powerful God, nor a suitable subject for devotion.

In history, humans have thought of God as being like themselves, angry at their own imperfections and their constant "sin," but God's perfection hardly allows for such demeaning emotion. God regrets nothing, is angry at nothing. God is the only unchanging thing in the universe. For God there is no past, no present, no future. God is energy with a purpose. God is will that is self-existent. God *is* will. To deny God's power to limit voluntarily his infinity would deny the reality of his infinite will. Thus did the world come into being.

THE ADJUSTER AND THE INDIVIDUAL

Because the mysterious Thought Adjuster, which somehow shares the divinity of the Paradise Father, resides within every Urantian human of normal endowment, no other road to God is needed except through the evolving soul's contemplation of the pure spirit within. All churches, mosques, synagogues and the like may be useful organizations, but in no sense does anyone need them to find God.

As the bestower of personality on all universe creatures that possess personality, the Universal Father can also make contact in another way, through the *personality circuit*. As all material gravity is within the circuit of the Isle of Paradise, and all mind is within the circuit of the Conjoint Actor (the Infinite Spirit), and all spirit within the circuit of the Eternal Son, so is all personality within the circuit of the personal presence of the Universal Father, the source of all personality. This circuit unerringly transmits the worship of any individual personality directly to the original and eternal personality of the Universal Father.

Urantia mortals enjoy a wide diversity of intellectual and personal qualities, and the material environment can be grotesquely different. Yet because of the indwelling Adjuster, and because of the personality circuit, all humans are equal in their opportunities for spiritual discovery. By choosing to do the Father's will, by wanting to know God and become like God, no external conditions of mortal life can prevent assent in the journey toward Paradise. The enemy of progress is the misty world of intolerance, willful decision, and self-assertiveness into which humans readily sink, sometimes wandering so far as to lose the desire to do the Father's will.

The mortal mind consents to worship; the evolving immortal soul craves and initiates worship; the divine Adjuster conducts such worship on behalf of the mortal mind and the evolving immortal soul. True worship therefore takes place on four levels: intellectual, morontial, spiritual, and personal—the consciousness of mind, soul, spirit, and their unification in personality.

UNIVERSAL FATHER AND ETERNAL SON

The three aspects of godhead are present in some personal way on the Isle of Paradise, for as beings with personality, they are actually somewhere. It is hard to remember (and to understand) that the Isle of Paradise has location although it is not in space: You could not travel in a space ship to the Isle of Paradise. The persons of the Trinity are resident on Paradise.

The Universal Father is first a creator, then a controller; the world comes from him and he sustains it. The world is his thought, whereas the Eternal Son is the divine word that the thought makes possible. The Word is the outside, the thing that touches, and the thought is the inside, the thing shown in the word, but they are really the same thing. That's why Jesus said that "he who has seen the Son has seen the Father." According to *The Urantia Book*, on Paradise the Eternal Son enshrouds the personal presence of the Universal Father at the center of things.

In our world all things come in sequence, first this, then that, and it is natural for us to think that the thought of the Universal Father comes first, expressed later as the Eternal Son, but such a sequence never took place. The Universal Father and the Eternal Son dwell in a realm where there is no time, hence no sequence. The Eternal

Son had no beginning but always was and always will be. Although he derived from the Universal Father, nonetheless he is coeternal. You must accept paradox if you want to understand the nature of things.

The Eternal Son is first a cocreator, working in conjunction with the Father, then he is the administrator of all spiritual reality. "God is spirit," and the Son is a revelation of that spirit. The Universal Father is God as will, the Son is God as personality.

The Gospel of John begins with the famous words "In the beginning was the Word [= Greek *logos*, something ordered, with shape] and the Word was with God, and the Word was God. All things were made by him, and without him was not anything made that was made." That is true of the Eternal Son. The Universal Father never functions personally as a creator except in conjunction with the Eternal Son.

Constant confusion in Christian traditions has come from the mixing-up of the Eternal Son, one person of the Trinity, with his offspring, the Creator Son Michael of Nebadon, who lived on Urantia as the mortal Jesus of Nazareth. Michael represents the Eternal Son, is cocreated by him and by the Universal Father, but he is a different being.

As the Father is love, the Son is mercy. Mercy is love in action. In a crude comparison with the human family we might say that the love of the Universal Father is like a father's love, the love of the Son like a mother's love. They are the same, but expressed differently.

Because the Eternal Son is spirit (the invisible, vivifying breath of life), all spiritual realities and impulses go directly to him. He is the center of gravity of all spiritual reality. The Eternal Son is everywhere and all around you as spirit, but not within you and a part of you, as are the Mystery Monitors, which come from the Father.

Although omniscient, omnipresent, and omnipotent the Eternal Son does not function personally in the physical domains, nor in the realm of mind. The Eternal Son derives personality from the Father but does not bestow personality, the prerogative of the Universal Father. He may do so only in conjunction with the Father, as in their joint creation of the Creator Sons. The Creator Sons are themselves creators, but cannot pass on creative powers to their own creations. Nor can the Eternal Son fragmentize himself, as does the Father in the issuing of the Thought Adjusters.

As the center of spirit gravity in the Universe, the Eternal Son is a counterpoise to the Isle of Paradise, the center of material gravity. In this way the First Source and Center is revealed in both the exquisite beauty of the central Isle and in the spiritual omnipresence of the Eternal Son.

The Eternal Son is the upholder of all spiritual beings and values in the universe. Spiritual gravity, of which the Eternal Son is the center, operates outside of time and space, hence is undiminished in transmission. Spiritual gravity is real, and not a figure of speech; and it pulls harder in response to spiritual achievement. As a mortal becomes a spiritual being, it becomes possible by submitting to this gravity to know the Son.

The personal presence of the Eternal Son is strong on the Isle of Paradise and in the perfect universe of Havona that surrounds it. Outside the Central Universe, however, within the seven Superuniverses, the Eternal Son is not personally present; these are the domains of the Creator Sons. Because the administration of the Eternal Son in the Superuniverses (which surround the Central Universe of Havona) is exclusively spiritual, creatures present there do not discern the Eternal Son. Yet the spiritual-gravity pull of the Eternal Son is the motive force behind a career of universe ascension, the thing that pulls one along. The more one follows the spirit, the less real become the pull of physical gravity on the mind of the individual personality.

The spirit-gravity circuit is how genuine prayer from the human heart goes to God. The circuit is rather like the human nervous system, a network directing an impulse to the right office. There are no reception centers for selfish prayer for material profit, however, which is a waste of time. Nor are fancy words of any use in effective prayer—it's the thought that counts.

THE PLAN OF PROGRESS

The Universal Father in union with the Eternal Son created the overall *plan of progress*: the creation, evolution, ascension, and perfection of will creatures. It is useless to wonder why they did this because we are ourselves the direct products of this plan and are in no position to evaluate its reason.

In any event, the divine plan has three parts, independent but overlapping:

(1) Emanating from the Universal Father is the *plan of ascension, evolution*. Man becomes like God through experience and ascension. The Father's bestowal of the Thought Adjusters and the bestowal of personality on material beings are features of this plan.

(2) Emanating from the Eternal Son is the *bestowal plan* whereby representatives from the class of beings known as Paradise Sons of God (or just Sons of God, of which Creator Sons are one category) appear among the ascending races to reveal the love of the Father and the mercy of the Son.

(3) The Infinite Spirit put forth the a third plan, the *plan of mercy ministry*, which is not so easy to understand. As we will see, an immense variety of beings proceeds directly from the Third Source and Center, abroad in the universe to assist through their mercy in the Universal Father's arduous plan of ascension.

THE ADVANTAGE OF EVOLUTIONARY LIFE

The pain and uncertainty of existence does not diminish God's overcontrol, but makes possible the evolution of qualities with high spiritual values:

Courage is only possible in the face of challenge and adversity.

Altruism—selfless behavior—is only possible where there are social differences.

Hope, the grandeur of trust, is possible only in the middle of uncertainties.

Faith is only possible where one must believe more than one can know.

Love of truth requires a world where falsehood is possible.

Idealism, a longing for higher values, can only exist where there is room for improvement.

Loyalty requires the possibility of betrayal and deceit. Devotion to duty is a high quality only where default is possible.

To be *unselfish*, one must know the temptations of the self.

To understand *pleasure*, you must live in a world where its opposite is a clear alternative.

A billion worlds make up the Central Universe of Havona, inhabited by perfect beings who were created perfect; yet in their lack of the experience of the evolutionary plan they can never hold the unique possession open to successful mortals. We should be happy with our lot. God the Father loves humans; God the Son serves humans; God the Spirit inspires the children of the universe (human and otherwise) to the ever-ascending adventure of finding God the Father by ways ordained by God the Sons.

THE CREATOR SONS

Because the Eternal Son is wholly personal, he cannot be fragmented, but reaches man as an omnipresent spiritual force or in the person of the Sons of God, according to the bestowal plan. Humans cannot contact the Eternal Son as they do the Father through the Thought Adjuster, but the Son comes toward them through a series of downsteppings in these bestowals. In the unique case of Michael of Nebadon's incarnation as Jesus of Nazareth, a Son of God actually lived as a human among humans. The divine will comes from the Father within, but the Eternal Son comes as a personality, as did Jesus of Nazareth.

The Eternal Son comes from the first completed thought of the Father. Whenever in union the Son and Father project in turn a new, absolute, and personal thought there comes into being a Creator Son, a Paradise Son of God who shares perfect divinity and creative power with the Universal Father and with the Eternal Son. A Creator Son is not just the offspring of God, he is God.

The Creator Sons go out from the timeless Central Universe of Havona, with its billion perfect spheres, into the Superuniverses of time. There in conjunction with the Third Source and Center, the Infinite Spirit, they work to perfect these universes of progressive evolution. Creator Sons are not concerned with the centralized control of matter, mind or spirit, but administer only that which they themselves bring into existence.

A second category of the Paradise Sons of God is the *Magisterial Sons*, personalizations of the Father and the Infinite Spirit (as the Creator Sons are personalizations of the Father and the Son). A third category of the Sons of God is personalized through the joint action of the three persons of the Father, Son, and Infinite Spirit working together and are called *Trinitized Teacher Sons*, who roam the universe as teachers to divine and mortal personalities. There are still other categories of the Paradise Sons of God.

A special circuit centering in the Eternal Son on Paradise permits instantaneous communication between the Eternal Son, who lives in eternity, and the hosts of

Paradise Sons of God scattered throughout the universe, who live in time. They are in direct and constant communication.

THE INFINITE SPIRIT

It is only a manner of speaking to say that the Universal Father had one perfect thought, which was the Eternal Son, because they were always in perfect union and identical in divinity, power, and absoluteness. From their perfect union came the desire for an agent of combined action and unified expression, the Third Source and Center, or the Infinite Spirit. In reality all three persons have always existed in eternity and always will.

The Infinite Spirit is the God of action and sets the stage for the divine drama of creation. The Paradise Trinity now exists (and always has) and the Father's purpose may be fulfilled by the personality of the Son and the action of the Spirit.

Simultaneous with the divinization of the Third Source and Center as the Infinite Spirit, the one billion perfect spheres of Havona instantly came into existence. Before this hypothetical time, eternally long ago (which never existed), there was no material universe, but at the instant that the one billion perfect spheres appeared there was sufficient gravity from Paradise to hold them in place. Before this hypothetical moment, the energies of space were only potential.

There flashes through the now material creation, held by the gravity of Paradise, the spiritual circuit of the Eternal Son. Thus do two systems of gravity hold the material creation. Through the spiritual circuit of the Eternal Son the energy of eternity and divinity touch the material creation.

Once the Third Source and Center is eternalized and Havona comes into being, the Universal Father no longer participates directly in further creation. All postHavona creations come from the interactions of the Eternal Son and his derivatives (the Sons of God), and the Conjoint Actor and his derivatives (the *Creative Spirits*). Our own universe of Nebadon came into being through the interaction of just such a Son of God and a Creative Spirit, as did we ourselves. For every Creator Son that came from the Eternal Son, a Creative Spirit came into being through the union of the Eternal Son with the Infinite Spirit. A Creative Spirit is to the Infinite Spirit as a Creator Son is to the Eternal Son.

You can come to the Father only through the Son and to the Son only through the ministry of the Spirit. All the myriad spirit offspring of the Conjoint Actor work through mercy to minister and serve and assist the progression from living in time to living in eternity. According to the bestowal plan, the Sons of God, who embody the Eternal Son, sometimes appear in time and space as creatures of flesh and blood, as did Michael of Nebadon as Jesus of Nazareth, but the representatives of the Spirit are never so embodied. However, spirit beings of the Third Source and Center do downstep in gradations until they appear as *angels*, beings close to humans (but invisible to humans) who help the creatures of time in the Father's plan of ascension. Through this series of downsteppings the Conjoint Actor stands close as a person (an individual) to the creatures of will of animal origin, to human beings.

The Infinite Spirit, which is rather like a universal divine mind, is everywhere at once. So are the Eternal Son and the Universal Father, but omnipresence is the Spirit's outstanding attribute. The Father is present *with* the Son and *in* the Spirit. In the biblical tradition the phrase "spirit of God" confuses the Third Source and Center, one of the Trinity and a person on Paradise, with the Creative Spirit of the Local Universe of Nebadon, in conjunction with whom Michael of Nebadon created this universe. Such local mother spirits are projections of the Infinite Spirit, but their sphere is limited to the local creation while the Infinite Spirit is omnipresent. The multiplicity of these spirit influences upon Urantians, however, including the activity of the Thought Adjuster, is always perceived as a single influence by the individual: Their manifold sources are never apparent.

MIND AND THE CONJOINT ACTOR

As God of action, the Conjoint Actor (Infinite Spirit) manipulates energy (but does not originate it). The Third person of Trinity also controls antigravity, the suspension of the universal force emanating from Paradise that holds the physical universe together. Various creatures at all levels and with diverse functions control the physical forces within the physical universe, always as agents of the Third Source and Center.

The mind of the Third Source and Center is the absolute mind, present everywhere in the universe. Human mind, and all creature mind, derives directly from the Third Source and Center. The Thought Adjuster cannot indwell any individual until mind has prepared the way. The Father draws forth all personalities, the Son draws upward all spiritual reality, and the Conjoint Actor is the center of all mind and personally conscious of every act of mind in the universe.

As Paradise is to physical reality, and the Eternal Son to the realm of the spirit, so is the Conjoint Actor to mind. From the Conjoint Actor comes intelligence to the minds of material, morontial, and spiritual beings. The ability to coordinate things and thoughts, through intelligence functioning in mind, is not material. Mortal humans have a spirit nucleus, surrounded by a personal energy system, mind, living within a material environment. The body must one day go, but the mind may survive in the form of the morontial soul, unite with the spiritual center, and create an eternal personality.

The Infinite Spirit is not directly concerned with physical gravity (emanating from the Isle of Paradise), spiritual gravity (the circuit of the Son), or with the personality circuit (the circuit of the Father), but the Spirit participates in most other universe activities. There are many different orders of being that represent the Infinite Spirit throughout the universe; ordinarily human eyes cannot see them. Their mission is always service. They are a ladder to God. Once released from the physical form, and endowed with the vision of the morontia realm, we will see these beings easily.

COSMIC MIND

From the *Seven Master Spirits*, who represent the seven aspects of the Infinite Spirit and control the seven Superuniverses from foci on the periphery of Paradise (the Spirits themselves dwell on spheres that surround Paradise), comes cosmic mind, pervading all. We partake of cosmic mind and for this reason respond to three functions of the cosmic mind:

causation, the reality of the physical world where there are facts and nonfacts, the scientific realm of logical continuity.

duty, the response to morals, the realm of relative right and wrong.

worship, the spiritual domain of the reality of religious experience.

These three functions of cosmic mind make it possible for individuals to take account of the realities of science, philosophy, and religion. A strong character comes to personalities who unify these three understandings—causation, duty, worship—really self-revelations of the Infinite Spirit. Education refines these innate endowments and civilization expresses them.

The human intellect far transcends the animal, but nonetheless it is the innate human moral and religious nature (deriving from the cosmic mind) that distinguishes human from animal. Intellect alone cannot explain the moral nature, which is a basic human endowment, as is scientific curiosity (the desire to know) and spiritual longing. Animals are not moral beings; they cannot analyze the value of what they are doing. They learn only from experience and cannot predict outcomes based on insight.

The personality lives in mind, makes decisions, chooses God or rejects God, achieves eternal life or dies with the death of the body. The body is the product of material evolution, a working out of patterns first imparted by beings called the Life Carriers. The Adjuster is pure spirit, a gift of the Father. But in mind the individual personality is free, in control, responsible for results. Mind is lent to the individual for just this purpose. In all the universe only the individual will controls mind. In mind are made the fatal decisions that lead to life, or death.

Human awareness rides on the substrate, below, of the electrochemical processes in the body and, above, in the morontia-spirit realm, but the individual is not especially aware of either. The individual is conscious of mind alone. It is not what mind knows, but what mind wants that counts, that makes a difference in the game of survival that God, we might say, is playing with himself. The individual mind can be twisted into something ugly and evil, or exalted into something beautiful and good: Such is the power of the free and individual will. The Adjuster guides in subtle ways, but is always subject to that personal will.

The mind is like a ship. The Adjuster is the pilot, and the human will is captain. If the captain takes the pilot's good advice, the ship will sail into the harbor of eternal life. If not, the ship may strike against the rocks of deliberate defiance of divine will and so perish beneath the waves.

THE PARADISE TRINITY

The Paradise Trinity of God the Father, God the Son, and God the Infinite Spirit was the only inevitability within the eternal oneness of God and the eternal oneness of the three persons of Trinity. Only the Trinity could release the One God from the limitations of eternity: changelessness, universality, absoluteness, and infinity. Because the One God is illimitable, the Trinity was inevitable. The Trinity, although made of three persons, is the unity of Deity.

Although the three persons of Trinity act both independently and in conjunction with one another, the Trinity is One. There are seven possible independent or conjoint actions of the persons of Trinity:

1. Father
2. Son
3. Spirit
5. Father/Son
5. Father/Spirit
6. Son/Spirit
7. Father/Son/Spirit

For this reason the number seven is basic to the design of the universe, to its values and structure.

Law comes from the Father, but its administration as justice belongs to the Trinity and is carried out by an order of beings called the *Sons of the Trinity.* The Father is love, the Son is mercy, and the Spirit is ministry, but justice belongs to the Trinity. Justice is always a plural function.

There are various orders of the Sons of the Trinity (parallel to the Creator Sons and the Creative Spirits), but of especial importance are the *Ancients of Days*, who reside on the capitals of the Superuniverses, in this case on the planet Uversa, capital of Orvonton, our Superuniverse. To judge from various attributions attached to the Urantia papers, which are "signed," the Ancients of Days of Uversa were involved in organizing the Urantia revelation, and one must wonder why the commission came down from so very far on high.

The Trinity is responsible for the evolution of the *Supreme Being*, a difficult concept. Although absolute in nature, the Supreme Being somehow exists *within* the creation as the complex and universal synthesis of the emerging phases of all finite realities, the sum total of all finite growth. What Michael is to Nebadon, the Supreme Being is to the finite cosmos; his Deity is the avenue through which the Father's love flows outward to all creation, and he is the avenue through which finite creatures pass inward in search of the Father, who is love. In the Supreme Being the Father/I AM has achieved relative liberation from the limitations inherent in infinity of status, eternity of being, and absoluteness of nature, but in achieving liberation from eternity, Deity encounters the barriers of time and the incompleteness of nature.

All this must follow the Father's plan, which has predicated finite progress on effort, creature achievement on perseverance, and personality development on faith. The Supreme Being personifies the finite shadow cast across the Grand Universe by the infinite unity of the unsearchable Paradise Father, the First Source and Center. The Mystery Monitors are to human beings what the Paradise Trinity is to the Supreme Being. Anyone who attains the level of Godlikeness can find God the Father, but no *one* creature will personally discover the Supreme Being until that far-distant time when, through the universal attainment of perfection, *all* creatures will simultaneously find him.

CHAPTER 3.
THE DESIGN OF THE UNIVERSE

THE *GRAND UNIVERSE*—THAT IS, THE ORGANIZED UNIVERSE TAKEN AS A WHOLE—is vast beyond human comprehension. Nonetheless, its size is limited (although God is infinite). It is a work in progress, unfinished (it will never be finished). The Grand Universe is not an infinite plane, but has dimensions. Hence all cosmic energies travel in a circle and under the influence of gravity return upon themselves.

WHERE IS URANTIA?

On its horizontal plane extending from and around the Isle of Paradise at the center of the universe (paradoxically Paradise is not in time and not in space), space is divided into concentric elliptical bands, each band separated by a zone of quiescence (see Figure 1 opposite title page). Paradise is surrounded by the perfect universe of Havona. Havona is surrounded by the seven Superuniverses, each having around one million galaxies (that is, organized aggregations of stars).

Each Superuniverse has a center, a capital. As we have seen, our own Superuniverse is Orvonton and its capital is Uversa. Portions of *The Urantia Book* dealing with events beyond our Local Universe, and such theological matters as the nature of God, were composed by messengers from Uversa called *Divine Counselors* (who are Sons of the Trinity), under the supervision of the Ancients of Days (also Sons of the Trinity). Both groups—Divine Counselors and Ancients of Days—are immensely powerful beings who reside on the Superuniverse capital of Uversa. Three Ancients of Days rule each Superuniverse from an immense and glorious artificially constructed sphere, in the case of our Superuniverse of Orvonton from the headquarters sphere of Uversa. The Ancients of Days supervise the Superuniverse courts and play an important role in the story of *The Urantia Book.*

Each Superuniverse has ten *Major Sectors* (see Figure 2, next page).

PORTION OF UNIVERSE	CENTER/ HEADQUARTERS	TOTAL IN GRAND UNIVERSE
CENTRAL UNIVERSE, called Havona	Paradise	1
1 of 7 SUPERUNIVERSES, called Orvonton	Uversa	7
1 of 10 MAJOR SECTORS, called Splandon	U-Major the Fifth	70
1 of 100 MINOR SECTORS, called Ensa (= Milky Way)	U-Minor the Third	7,000
1 of 100 LOCAL UNIVERSES, called Nebadon	Salvington	700,000
1 of 100 CONSTELLATIONS, called Norlatiadek	Edentia	70,000,000
1 of 100 Local Systems, called Satania	Jerusem	7,000,000,000
1 of 619 inhabited planets, called, Urantia		around 7,000,000,000,000

FIGURE 2. Urantia's Place in the Universal Scheme

Our Major Sector is called *Splandon*, whose center is *U-Major the Fifth* (we do not hear much about Major Sectors in *The Urantia Book*). The Major Sectors are governed by high personalities of Trinity origin.

The Major Sector consists of one hundred *Minor Sectors*, ruled from an artificially constructed headquarters sphere—a so-called architectural sphere—by three beings of Trinity origin. Our Minor Sector is called *Ensa* and its center is *U-Minor the Third.* Our Minor Sector is virtually the Milky Way galaxy.

The Minor Sector consists of one hundred *Local Universes,* each of which has around ten million inhabitable planets. Our Local Universe of *Nebadon* has its capital on the architectural sphere called *Salvington*, home of the Michael who appeared as Jesus of Nazareth on our very unusual planet.

There are one hundred *Constellations* in a Local Universe (Sectors, Constellations, and Systems are not shown in Figure 1). Each constellation is ruled from an architectural headquarters sphere by three *Vorondadek Sons*, members of an order of sonship engendered by union of a Creator Son and the local Creative Spirit (Universe Mother Spirit). The Vorondadek Sons belong to the general order Sons of God (but not Paradise Sons of God because they are created within the Local Universes). The rulers of Constellations are called the *Most Highs*. The capital of our own Constellation of *Norlatiadek* is called *Edentia.*

Each Constellation has one hundred *Systems.* In all the Local Universe of Nebadon there are ten thousand Systems of inhabited worlds. Each System consists of around one thousand inhabited or inhabitable planets; young Systems may have many fewer. Of course many planets cannot support creatures with will because they are too close to their suns, or too far away, or for other reasons.

Each System is administered from an architectural headquarters sphere by a *System Sovereign.* Our own System is *Satania* (Lucifer was its Sovereign) and its headquarters sphere is *Jerusem,* surrounded by seven groups of satellites on which mortals who survive death are rematerialized in morontia forms (Jerusem is the heaven of popular lore).

Satania, our local and fairly typical System, is not a single astronomic unit or organization, but consists of 619 inhabited planets scattered over five hundred different physical conglomerations. Only five of these conglomerations have more than two inhabited worlds, only one has four, and forty-six have two inhabited worlds. Satania's headquarters of Jerusem, the System's astronomic center near an enormous dark island in space, are a far, far distant two hundred thousand light years from Uversa, the capital of Orvonton deep in the dense cloud of the Milky Way. The System of Satania is on the periphery of our Local Universe of Nebadon as Nebadon is on the periphery of the Superuniverse of Orvonton.

Inhabited planets within any System are ruled by a *Planetary Prince,* a member of the order of Lanonandek Sons (although in a different category from the Lanonandek Sons who rule the Systems). The original Planetary Prince of Urantia was *Caligastia.* Our planet Urantia is registered as number 606 (of 619 within this system). Two hundred more planets in Satania are nearly ready for life implantation. In the catalogue of inhabited worlds in the Grand Universe Urantia is registered on the Superuniverse capital Uversa and on Paradise as number 5,342,482,337,666.

FORCE-ENERGY

In Satania there are over seven thousand discrete physical systems, but few originated as did our solar system. The System of Satania includes over two thousand suns. The suns in all universes are similar. Our sun, of average size, is about six billion years old, has stabilized, and will shine at its present rate for another twenty-five billion years. Its interior temperatures reach thirty-five million degrees Fahrenheit and its pressures are so great that hydrogen atoms in the presence of carbon are transformed into helium atoms, releasing oceans of energy; but the sun's energies derive from other sources too.

The basis of the universe is material, as matter is a form of energy that proceeds from and is under the control of the Universal Father. Life, however, is spirit. The *ultimaton,* the first measurable form of energy, is not yet described by Urantia scientists (but presumably has something to do with particle physics). The ultimaton is indestructible, unlike the atom, and somehow has the timeless Isle of Paradise as its nucleus! The ultimaton is not responsive to gravity. It is the building block of the electron, made up of no fewer than 100 ultimatons.

Force-energy is derived from self-existent Deity, from the God that always was and always will be. Therefore force-energy exists forever, although manifestations of the First Source and Center may be transmuted endlessly. Still, in no sense, not even to the slightest degree, can force-energy suffer extinction. Force-energy, although it springs from the First Source and Center, is not found everywhere because there are outer limits to the present Grand Universe.

THE ISLE OF PARADISE

The eternal Paradise is the center of the universe (and of the ultimaton) and the dwelling place of the persons of the Universal Father, the Eternal Son, and the Infinite Spirit and their coordinates and associates. Paradise is the geographic center of infinity. It is both material and spiritual and by far the largest organized body in the universe. All of the Father's intelligent creations dwell on material spheres, hence must the model for such spheres be partially material too. Paradise is the model for all the material creation.

It is quite impossible for mortal intelligence to imagine the splendor and beauty of this dwelling place of God, where God is focalized, personalized, and geographically present. All circuits, force, and life lead back to God's presence on the Isle of Paradise. Yet Paradise does not exist in time and it does not exist in space. If you were on Paradise you would nonetheless be aware of distance between different points on the Isle, and of a "nontime" sequence of events. We really cannot understand any of these contradictions.

Paradise is elliptical in shape, being one sixth longer in its north-south diameter than in east to west. It is nearly flat: The distance from the upper to the lower surface is one tenth the distance from east to west. These dimensions, taken with its stationary condition and a greater emanation of force from the north end, make it possible to establish absolute direction in the universe.

Paradise has three spheres of activity: on its upper surface, on its lower surface, and on its periphery. The upper surface is the realm of personalities, the lower surface of impersonal forces, the periphery somewhere in between. The Trinity dominates the upper, personal plane, and a being known as the *Universal Absolute* dominates nether Paradise. We might think of the Universal Absolute as God without personality, functioning within the universe.

Paradise is composed of a unique substance (called *absolutum*) that is uniform in consistency, neither dead nor alive, and found nowhere else in the universe. Time appears to originate from upper Paradise, space from lower Paradise. Paradise is the original nonpersonal expression of the First Source and Center. Paradise is *Paradise.*

It is incongruous to think of God as resident in any place because, as ever repeated, God is everywhere at once. Nonetheless, in some strange primordial and incomprehensible reality God *is* resident on upper Paradise. At its center are the Deities (all the aspects of God we have discussed). Surrounding the center is an enormous area completely without structures of any kind, material or intellectual, wherein take place things we cannot imagine, wholly spiritual in nature; we are nearly altogether material. It is an intractable paradox that to the material mind the spiritual is nonexistent, while in reality spirit is the essence of existence. We do not need to say, "If you were thrust into eternity, you would not see a thing," because in fact we *are* thrust into eternity at this very moment.

Beyond the area on Upper Paradise that surrounds the eternal presence of the Deities lie seven concentric zones wherein dwell beings of all kinds. Within the second zone, itself with seven divisions, are creatures who have risen out of time and space from the seven Superuniverses, including beings who began like ourselves. Upper Paradise ends abruptly at the edge of the ellipse, but so enormous is the Isle that the incline is indiscernible at any one point. On the periphery of Paradise take place the transference of beings to destinations in space: Space travel never begins nor ends on upper or lower Paradise.

One area of peripheral Paradise is assigned to historic and prophetic exhibitions of some kind. Although seven trillion exhibitions are already in place, only four percent of the total area assigned is taken, helping us to imagine what lies ahead in the as-yet unorganized outer space levels beyond the seven Superuniverses (see Figure 1, opposite title page). Paradise is large enough to accommodate the activities of an *almost* infinite creation.

SPACE RESPIRATION

Nothing directly is known about lower Paradise because no personalities dwell there. Lower Paradise is the source of all physical energies, and material gravity. There may be some kind of circulatory system whereby all force-energy comes in at the south end of lower Paradise and goes out at the north, for all forms of force and energy appear to be in a single circuit—force-energy goes out to the edges of space, then returns. For around one billion years of Urantia time the space force-energy of lower Paradise is outgoing, extending through all pervaded space (the space that contains the central and surrounding Superuniverses), then for a similar length of time it is incoming.

All physical force, energy, and matter are one. They come from lower Paradise (although matter may be organized independently within space). Space comes from lower Paradise too, but is something different. Space is not responsive to gravity, the force-energy that holds the universe together. Space breathes, it respires, and its respiration corresponds with (but is not caused by) the two-billion year cycle of pulsations of force-energy. The mechanism of space respiration is unknown.

Pervaded space extends horizontally from Paradise, whereas unpervaded space extends above and below Paradise, something like the shape of an hour glass. When pervaded space expands in respiration, unpervaded space contracts and vice versa. The meaning of all this is completely unknown. We are presently near the middle point of an expanding cycle of space respiration.

THE SHAPE OF SPACE

A vertical cross-section of the totality of space would look something like a Maltese cross (Figure 3) with the center being Paradise and the horizontal arms describing the extension of pervaded space (the Grand Universe) and the vertical arms describing unpervaded space. The two realms are separated by midspace zones that increase in extent as you proceed outward from the central Isle of Paradise, until at the end of the arms the midspace zones expand to enclose both the horizontal realms of pervaded space and the vertical realms of unpervaded space.

FIGURE 3. The Shape of Space

Imagine a V-shaped plane extending outward from peripheral Paradise at right angles to upper and lower Paradise; then imagine that this plane circulated in an ellipse around Paradise, its point always touching peripheral Paradise: This would describe pretty well the volume of pervaded space. In theory you could reach the outer limits of space at any point within any Superuniverse by traveling at right angles to its plane of expansion.

MORE ON PARADISE

Having said all that, we need to remember that Paradise is not really part of the universe, even of the eternal Central Universe of Havona (in spite of our drawing in Figure 1). It is the source and goal of all spirit personalities, including ourselves. It is unique, without parallel, incomprehensible.

A way to think about this problem is to imagine that in the eternal past when God the Father, the giver of personality, engendered the Eternal Son, who is the perfection of personality, there simultaneously came into being a nonpersonal counterpart to the Son: Paradise. The tension between the personal and the nonpersonal, combined with the Father/Son's will to action, brought into being the Conjoint Actor (the Infinite Spirit) and with it the central and eternal universe of Havona and all its perfect spiritual beings. Of course there is no sequence of events in eternity so that such a description must remain a manner of speaking.

God is spirit, but Paradise is not. All spiritual beings live in a physical world with a material basis. Spiritual activities take place in a material world. Physical energy is the one reality subject to unchanging law. Only in the realm of creature volition is there deviance from universal law.

Paradise is not Deity, not a creator. Paradise is not conscious as we would think of it. Paradise is the center for the activity of all personality and the source of all forces in the universe. As far as we are concerned, Paradise is the goal toward which we strive, which we can reach, which untold millions have reached before, and will continue to reach in the unlimited progress of time.

THE THREE SACRED CIRCUITS OF PARADISE

Between the Isle of Paradise and the spheres of the seven-circuited Central Universe of Havona lie three intervening bands, located in space, that contain special worlds. These are the *three sacred circuits of Paradise*. In the first band or circuit are the seven secret spheres of the Universal Father. In the second circuit are the luminous seven spheres of the Son. In the third circuit are the immense seven spheres of the Conjoint Actor, which serve as the executive headquarters of the Seven Master Spirits, each one of which controls a Superuniverse from foci on the periphery of Paradise.

There are no records of these spheres having come into being and, like Paradise, they appear to have always existed. They are physical, but composed of materials of unknown composition. Every sphere is immense, splendid, and unparalleled, except for the seven spheres of the Son, which are all made in the same way.

The Urantia Book devotes a good deal of space to describing the glories and importance of these twenty-one spheres, on which reside a wide range of beings of unimaginable splendor and power. The spheres of the Father, in the circuit closest to Paradise, have names suited to their functions. Of special interest to us is the first sphere, called Divinington, where are contained the secrets of the Thought Adjusters, their origin and mission. Divinington is important to us because at the time of a mortal's death the indwelling Adjuster actually travels to this sphere by instantaneous transmission: Thought Adjusters exist outside of space and time. The second sphere of the innermost circuit, so-called Sonarington, holds the secret of incarnation by members of the orders of the Sons of God in the worlds of time (as Michael of Nebadon incarnated as Jesus of Nazareth). The sixth sphere in the circuit of the Father is called Seraphington and hides the secrets of *seraphic transport*, the ordinary means of interplanetary travel whereby *seraphim*, an order of angel, wrap their spirit forms around nonmaterial personalities to carry them immense distances through space. We will later have more to say about the seraphim and other orders of angels, which originate from the Third Source and Center. Near the end of the long career of ascension, when preparing to go to Paradise, you will spend time on the seventh sphere in the circuit of the Father, Ascendington. On Ascendington reside the secrets of raising a material and mortal mind to the level of a divine and spiritual entity, one of the most perplexing mysteries of the universe.

On the spheres of the middle circuit, assigned to the Eternal Son, there are no personal beings (which belong to the circuit of the Father), but only a spiritual power. Hence there is little to say about them. If a personal being were transported to these spheres, they would seem empty and deserted. But so far as is known, no personal being has ever been on these seven spheres.

Many types of beings visit or dwell on the spheres of the outermost circuit, belonging to the Conjoint Actor/Infinite Spirit and the Seven Master Spirits. This circuit separates Paradise from the billion Havona worlds. These executive worlds provide for all phases of Superuniverse administration.

THE CENTRAL UNIVERSE OF HAVONA

The Central Universe of Havona consists of one billion perfect spheres that surround and encircle the Isle of Paradise and its three sacred circuits of twenty-one satellites. Havona never began and it will never end; it is absonite. At its outer edge are enormous dark gravity bodies that make the mass of Havona greater than the mass of all the rest of creation combined. Paradise/Havona provides patterns for everything that happens in the master universe. All seven Superuniverses and the reaches of the outer space levels revolve around Havona. In the Superuniverses the Creator Sons attempt to recreate in time and space, through evolution, the perfection of Havona.

The spheres of Havona are organized into seven concentric circuits that surround the three concentric sacred circuits of Paradise. Beyond the outermost circuit of Havona lies a quiescent space zone, then a band of dark gravity bodies, then another space zone, then a second band of dark gravity bodies circulating in the opposite direction to the first, then the innermost circuits of the seven Superuniverses (see Figure 1, opposite title page).

The gigantic dark gravity bodies neither reflect nor absorb light and create a kind of veil that hides Havona from the view of the inhabitants of even nearby Local Universes. The dark gravity bodies are unique, being found only here in all the universe. The Havona circuits lie within a single plane and the millions of planets in each circuit are arranged in a linear progression. The three sacred Paradise circuits and the Havona circuits are part of a single system.

Time does not exist on Paradise, but the Havona worlds live in time. A standard Havona day is determined by a single complete circuit around Paradise of any sphere in the first Havona circuit (the one nearest the island of Paradise)—a single revolution requires about one thousand earth years. "A day is as a thousand years with God, as but a watch in the night." The Paradise/Havona standard day is the standard measurement of time in all the seven Superuniverses.

The physical organization of the Havona worlds is different from anything known in the evolutionary universes of time. If a Urantia mortal were somehow transported to a Havona sphere, he or she would see nothing and hear nothing, nor register any other sensation. Because of the perfection of its inhabitants, there are no courts or legislative assemblies on the Havona worlds. Beings called the *Eternals of Days*, personalities of Trinity origin, are the chief planetary administrators.

Ascenders from the worlds of time will land on the outer circuit of Havona, then proceed ever inwards until they reach Paradise itself and join the Corps of Finality. But most activities on these supernal worlds are far beyond mortal comprehension, as are its native inhabitants, who as physical beings conduct personal lives, advance in knowledge and wisdom, and serve as patterns for life forms within the Superuniverses. Native Havoners do not die and can scarcely be said ever to have come into being, nor do they reproduce, although they may change form.

To reach and be schooled in the absonite Central Universe of Havona is man's destiny. Graduating from the billion worlds of Havona and graduating to Paradise, beings of mortal origin become *finaliters* destined to pursue an undisclosed adventure in exploring the infinity of the Universal Father.

SEVEN SUPERUNIVERSES AND BEYOND

The Grand Universe, then, consists of the seven trillion inhabitable planets of the Superuniverses, plus the one billion planets of Havona. The Superuniverses, which occupy the first space level beyond Havona, are arranged like a wheel whose spokes are rays of the Seven Master Spirits, whose hub is the Isle of Paradise, and whose rim is the beginning of the outer space levels. The Superuniverses exist within time and space. Each occupies about the same amount of space and each has about the same number of Local Universes. Although each Superuniverse can potentially hold one trillion planets inhabited by creatures that possess will, the Superuniverses are by no means complete.

Our Local Universe of Nebadon is a fairly recent creation. Nebadon also stands in proximity to physical creations still being organized. Nearly all stars visible to the naked eye in the night sky belong to the Superuniverse of Orvonton. Urantia travels on its periphery, far, far from the central Isle of Paradise. The stars in Orvonton lie in an elliptical plane, far wider than deep and longer in one direction than in the other. When you view the Milky Way, you are looking toward the center of the plane so that the star density appears to be far greater than on either side.

It has been hard for Urantia astronomers to discover the true organization of space because: (1) the earth is rotating around the sun; (2) the sun and its planets are rotating around the nucleus of the Andronover nebula, where these bodies took origin; (3) the Andronover nebula is rotating around the gravity center of the star-cloud of Nebadon; (4) the star-cloud of Nebadon is rotating around the center of the Minor Sector within distant Sagittarius; (5) the Minor Sector is rotating around its Major Sector; (6) the Major Sector is rotating around the Superuniverse headquarters of Uversa; (7) Uversa is rotating around Paradise and Havona.

A zone of quiet, about four hundred thousand light years across, surrounds the seven Superuniverses. Beyond that, in the first outer space level, are taking place stupendous energy actions that intensify as the zone extends outward for twenty-five million light years, a continuous belt of cosmic energy wheeling around the known, organized, and inhabited Grand Universe.

Scientists on Uversa, the headquarters sphere of Orvonton, also detect stupendous activity in the second outer space level, no doubt in preparation for some unknown future development. Millions upon millions of new universes are in process of formation, their mass far exceeding the combined mass of the present seven Superuniverses. But no material or spiritual beings dwell there yet. Perhaps one day the Paradise Corps of the Finality, toward which humans aspire, may be active in a new kind of creation.

As far as the Universal Father is concerned, the universes do not exist; he is concerned with personalities. As far as the Eternal Son and the Conjoint Actor are concerned, who are cocreators, the universes are an assemblage of local organizations, each under the leadership of a Creator Son (such as Michael of Nebadon) and a Creative Spirit (Universe Mother Spirit). As far as the Trinity is concerned, there are just seven Superuniverses, controlled from the periphery of Paradise by the Seven Master Spirits.

The administration of the seven Superuniverses from specially created spheres has always been the same. The head administrator of a Superuniverse is therefore rightly called an Ancient of Days. The whole of a Superuniverse, and its myriads of millions of planets inhabited by creatures possessing will, swings in an elliptical arc in a counterclockwise direction around the First Great Source and Center, the central hub of Havona (contrary to descriptions by Urantian/Earth astronomers who see us as hurtling outward into uncharted space). Our universe has hurtled through this very space many times before.

SPACE BODIES

With the exception of the architectural spheres, built to order to serve as administrative centers, all space bodies are brought into being through similar evolutionary processes, in the sense that initial creative acts set in motion forces and energies in time-space that result in the appearance of suns, dark islands in space, planets, and such minor space bodies as meteors and planetesimals. Nebular formations, which are independent of the organization into Local Universes and Superuniverses, give birth to many space bodies. Various orders of intelligent beings, called *Force Organizers* (and various other things), are involved at many levels in the organization of these space energies. The behavior of physical energy is nearly universal, although local conditions apply. Forms of life are inexhaustibly diverse, but certain patterns will characterize individual systems; there are also universal patterns of intelligent life.

There are ten trillion suns in the Superuniverse of Orvonton, as many suns as glasses of water in the oceans of the earth. The unimaginably huge dark bodies of space, sometimes burned-out suns, help through their immense gravity to stabilize formations of stars and other bodies in space. In Orvonton not one planet in forty is suitable for supporting habitation by creatures such as ourselves. Only three planets in our own solar system are suitable for supporting life. Urantia, in size, density, and location is ideal for supporting human life.

Being special constructions, the headquarters spheres of the Superuniverses, and of its segments, are independently lighted and heated, having each its own sun that gives forth light but no heat. These spheres are warmed by the circulation of energies on the surface of the body. The satellites of these paradisiacal worlds are also architectural spheres. All kinds of material, morontial, and spiritual beings reside on them.

Jerusem, the architectural headquarters of the Local System of Satania, is encircled by seven architectural planets, which are each in turn encircled by seven spheres. The seven spheres encircling the innermost sphere are the first morontia residence of mortals, where you go if you survive death. These seven subsidiary spheres are the "seven heavens" of popular lore.

There are many architectural spheres in the Grand Universe. When all the projected universes are developed, there will be around five hundred billion of them in the seven Superuniverses. The architectural headquarters spheres channel force and energy to their domains. They balance and control the physical energies circulating throughout the universe. Additional control is provided by Power Controllers, difficult beings (like the Force Organizers) to understand: They are intelligent but they have no will, they cannot make decisions.

CHAPTER 4. SOME CREATURES OF THE GRAND UNIVERSE

LIFE FORMS MAY BE ORGANIZED, sometimes loosely, in accordance whether they take origin in one, two, or three of the Deities, either directly, or down the line of generations. *The Urantia Book* spends a good deal of time on these personalities, of whom there are very many. Let us consider those beings most important to us and try to understand how some of these personalities take their origin in one level of the Grand Universe, but are effective on another, and in what capacity they are effective (see Figure 4).

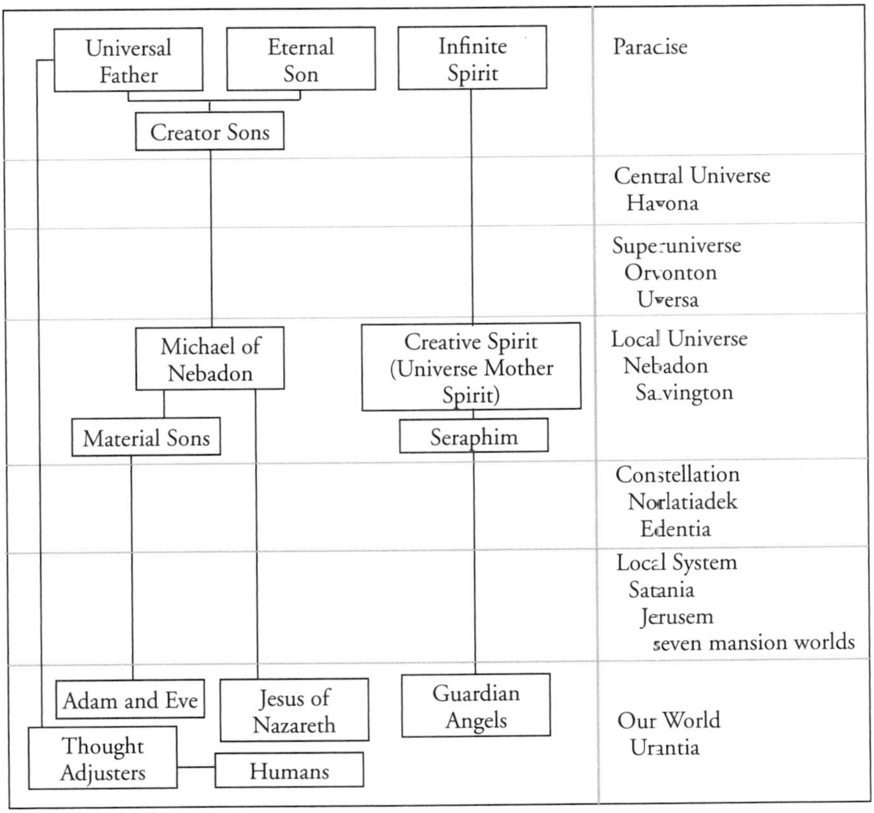

FIGURE 4. Some Beings of the Grand Universe and Where They Function

BEINGS OF TRINITY ORIGIN

These beings either spring directly from the Trinity or have been "trinitized," embraced by the Trinity. We can think of these creatures as "higher," close to the Trinity itself, or "lower," further down the line of descent.

The higher Trinity-origin beings include the Seven Master Spirits, whom we have already encountered dwelling on the seven planets of the third sacred circuit that surrounds Paradise, whose energies control the central eternal universe of Havona and the seven Superuniverses from foci on the periphery of Paradise. Also of Trinity origin are the seven *Orders of Reflective Spirits*, creatures of great power who dwell on the headquarters planets of the seven Superuniverses and through a reflective technique enable instant communication between the Superuniverse capital and Havona and Paradise. Through these spirits Michael sends communications from Salvington, capital of our Local Universe of Nebadon, to Paradise.

The "lower" Trinity beings include:

- the Eternals of Days, who rule each of the one billion planets of Havona
- the Ancients of Days, who in groups of three rule each Superuniverse
- the *Perfectors of Wisdom, Divine Counselors*, and *Universal Censors*, hard to distinguish, but who all serve in some advisory capacity to the Ancients of Days on the Superuniverse capital; representatives from each of these groups dictated numerous early papers of *The Urantia Book*
- *Havona Natives*, who have always lived on the planets of Havona, and who always will
- *Paradise citizens*, who have always lived on Paradise.

BEINGS OF DUAL ORIGIN

These beings can trace their origin to the functioning of any two of the three Deities. These beings are best grouped according to whether their journey is a descent from Paradise into the worlds of time and space, or an ascent from those same worlds. We will later have more to say of the ascending orders, but the descending orders of dual-origin beings include Creator Sons, like Michael, creator of Nebadon. Also in this category are the following orders of descending dual-origin beings, all belonging to the general category Sons of God (some of Paradise origin—the Paradise Sons of God—some of origin in the Local Universes):

- *Magisterial Sons,* Paradise personalizations of the Universal Father and the Infinite Spirit, who appear at critical junctures on inhabited planets as teachers and demarcators of planetary ages; there are about one billion in the Grand Universe.
- *Melchizedeks,* offspring in the Local Universe of a Creator Son and a Universe Mother Spirit (in collaboration with a unique being, the Father Melchizedek, himself the product of the union of the Creator Son and the Local Universe Mother Spirit). These versatile beings fulfill many tasks and are available for

emergency functions, as when a Melchizedek appeared to Abraham around 2000 BC. The same Melchizedek is today the vicegerent Planetary Prince of Urantia (in substitution for the titular Prince, Michael of Nebadon, presently on Salvington).

- *Vorondadek Sons,* offspring of the local Creator Son and the Local Universe Mother Spirit, three of whom govern each Constellation as the Most Highs; there are one million in Nebadon.
- *Lanonandek Sons,* also offspring of the local Creator Son and the Local Universe Mother Spirit, who serve as System Sovereigns (like Lucifer, now imprisoned, in the system of Satania) and, in a different class, as Planetary Princes (like Caligastia on Urantia, now deposed); there are twelve million Lanonandek Sons in Nebadon.
- *Archangels,* offspring of the local Creator Son and the Local Universe Mother Spirit (other angels are of single origin from the Third Source and Center); there are about eight hundred thousand in Nebadon.
- *Life Carriers* (more later), offspring of the Creator Son and the Universe Mother Spirit in collaboration with one of the three presiding Ancients of Days; there are one hundred million Life Carriers in Nebadon; they are critical to the evolution of life on an inhabited planet.

SOME BEINGS OF SINGLE ORIGIN

A large number of beings of single origin come from the Infinite Spirit:

- *Creative Spirits,* who proceed from the Infinite Spirit on Paradise; called Universe Mother Spirits when in union with the Creator Sons they create the Local Universes (and many other beings too).
- *Solitary Messengers,* a corps of spirits with special powers of interplanetary travel.
- *various orders of angels,* some belonging to the ascending order.

Of course the Thought Adjusters come from the Father as single-origin beings; when personalized, they belong to the dual-origin ascending orders. The angels, single-origin beings from the Third Source and Center, are a large category in their own right, discussed later in this chapter.

FUSION IN THE ASCENSION CAREER

Such ascending creatures as ourselves (potentially Adjuster-fused mortals) are working from the bottom to the top. We will be rewarded handsomely for our suffering and our experience, if we survive, but it is easy to lose sight of the end when bemired and confused in the lowest level of reality. In fact only some do survive, so the stakes are very high.

Adjusters may indwell primitive humans, but the relationship is usually transitory and ends at death. Few humans reach the status of Father/Adjuster-fusion in the days before the arrival of the planetary Adams, the *Material Sons* who come to step-up progress in biological evolution. The Material Sons and Daughters are semimaterial sex beings who live on the System capital as mostly permanent residents. The original pair in each system is brought into being by the Creator Son alone without assistance from the Universe Mother Spirit; they then reproduce themselves sexually. A Material Son and Daughter came to this planet about thirty-eight thousand years ago, remembered as Adam and Eve (the first humans go back about one million years). "Adam" is the name of the original Material Son, who never leaves Satania, and his male offspring, when dispatched to the evolutionary worlds, ordinarily bear his name. In Hebrew Adam is said to mean "man" and Eve to mean "life," but the names are far older than the Hebrew language.

All Adjuster-fused creatures are of animal origin. In the Local System of Satania, fusion ordinarily takes place on the satellites that surround the System capital of Jerusem, although fusion may take place as late as on the educational worlds of Salvington, capital of the Local Universe of Nebadon and far beyond Jerusem, or sometimes never. If you were to witness such a fusion on Urantia, you would see a bright flash of light, and then the person would be gone "in chariots of fire."

Adjuster-fusion is a kind of Son-seizure whereby the individual is removed from among the living, glorified, and appears immediately on Salvington in the presence of the Creator Son. The Hebrew patriarch Enoch was the first human on this planet to fuse with his Adjuster while in the flesh. Such Adjuster-fused mortals are afterward clothed in morontia bodies in the resurrection halls of the higher morontia mansion worlds, thence to pursue their careers toward Paradise. Natural, physical death is therefore not inevitable, and on planets of advanced evolutionary status most mortals do not die, but are fused with their indwelling Adjuster while still alive.

After Adjuster-fusion, the new creature is a unified being in which all the human aspects have been divinized and the divine aspects humanized; yet the Adjuster is never known in completeness until attaining knowledge of the Father on Paradise. The Adjuster *is* the Father. Before Adjuster-fusion, even after the creation of a morontia soul, the personality is free to reject survival; after fusion, the new creature is an immortal being, an ascending child of God.

On other planets other categories of evolutionary ascending sons of God may progress through fusion with elements of the Spirit or of the Son when, for unclear reasons, fusion with the Father/Adjuster cannot be attained. Such fusion-creatures do not progress to Paradise but remain in the Local or Superuniverse as its permanent residents. Different destinies for different categories of the survivors of mortal death might appear unfair, but the universe has no interest in fairness as such. Only in this way is it possible to build, in a community of permanent residents, the experience necessary to oversee the progression of the whole universe into a settlement in light and life. The Father-fused mortals have a high destiny on Paradise and in the creations to come, yes, but can never savor the progress of local achievement.

MORE ON THE MATERIAL SONS

The dual-natured Material Sons (male and female) is a category of central importance to the unhappy history of our planet. On Jerusem, where they are permanent residents, each family lives on its own estate prior to departing to an evolving world. In most matters Material Sons and Daughters govern themselves on Jerusem through universal suffrage and representative government. They are material and easily visible to mortals when present on the evolving worlds, an intermediary between material creatures and the invisible Planetary Prince. The semiphysical Material Sons and Daughters are so similar to mortal ascenders that ascenders reaching Jerusem from the mansion worlds easily socialize with them and enjoy being taught by them.

The special mission of the Material Sons is to incarnate on evolving planets when natural evolution has run its course in order through admixture of their own DNA with human DNA to step-up the process of natural evolution. They are biologic uplifters and serve on every planet that has an evolving human life. Their offspring constitute usually the eighth race on a planet, the *Adamic* or *violet race*.

Material Sons and Daughters, who are descending beings, are not indwelt by Thought Adjusters. When in ordinary conditions a planet is settled in light and life, they may, however, register as ascending sons of God, if they choose, and like humans aspire to become members of the Corps of Finality on Paradise. The fate of the Adam and Eve of this planet took a rather special course, as we will see.

MIDWAY CREATURES

A puzzling category are the mysterious *Midway Creatures*, who for various reasons might, like humans, also be accepted as children of God and traverse the same universe career as ascendant mortals. They play an important role in *The Urantia Book*. They come into being in two ways: (1) through union between members of the material corporeal staff that always accompanies a Planetary Prince, called *primary midwayers*; or (2) as a sort of byproduct of union between the progeny of the Material Sons and Daughters, called *secondary midwayers*. Thus they are "midway" between the Material Sons and the evolutionary races.

Midway Creatures, or midwayers, are semimaterial and not ordinarily visible to the human eye. On Urantia, most of the primary midwayers joined Prince Caligastia in the Lucifer revolt and are today interned, with Caligastia, on the prison world of the System capital, as are some of the secondary midwayers. In ordinary conditions, both orders of midwayers are of exceptional importance in the advancement of culture and civilization. The apostate midwayers on Urantia are responsible for many stories about ghosts and spirits. As so-called unclean spirits, they did sometimes take possession of weak or feeble individuals, but not after the incarnation of Michael as Jesus of Nazareth. There are still about one hundred thousand midwayers on this planet of both kinds, primary and secondary. They are the permanent residents of Urantia, intelligent beings who have been here for millennia and, except when translated as ascenders, do not leave.

OTHER BEINGS

On Paradise many orders of beings do not fit into these categories. They are neither creators, nor creatures, not infinite nor finite, but absonite. They are not created, but "eventuated." They proceed somehow directly from Deity. *The Urantia Book* calls them Transcendentaliters. There are also hosts of prepersonal or superpersonal beings who perform myriad functions in the universe, but lie beyond the understanding of personalities such as ourselves.

MAGISTERIAL SONS AND THEIR MISSION

Although they are Paradise Sons of God, like the Creator Sons, the dual-origin Magisterial Sons (from the Universal Father and Eternal Son) are not creators, but ministers who work closely with the Michaels. They are called Magisterial Sons because they often perform judicial duties, including judgment of the sleeping survivors at the end of a dispensation. There are about one billion Magisterial Sons in the Grand Universe. A principal form of service is to materialize as a fully grown material being in the likeness of a planetary race. In this capacity they perform a variety of functions. They may also appear in invisible form.

The materialization of a Magisterial Son on an evolutionary world ordinarily precedes the appearance of a *Bestowal Son*, when another Magisterial Son is born from an earthly mother, as Jesus was born from Mary. Every Magisterial Son undergoes this experience at least once. Such an incarnation always marks the end of a dispensation—a large unit in the evolutionary plan for creatures of will—and ordinarily takes place but a single time on any one planet.

The technique whereby such incarnations take place is completely unknown, a secret held on the sphere of Divinington, home of Adjusters in the first circuit of spheres that surround Paradise. Such a bestowal child, always male, grows up and experiences life as a mortal of the realm. They do not, however, father children, a restriction placed on all incarnated Sons of God. Such beings become teachers to the mortal races about the realities of the spirit.

When incarnated on a bestowal mission, a Magisterial Son is always indwelt by an Adjuster, but such an Adjuster cannot hope for personality fusion. Sometimes, however, the Universal Father will personalize such an Adjuster by fiat, as in fact happened with the Adjuster that indwelt Jesus of Nazareth. A Magisterial Son has never visited Urantia on a magisterial mission because of the confused conditions here and because of the unique bestowal of the Creator Son Michael of Nebadon as Jesus of Nazareth. After entering the portals of death, such an incarnated Son of God always reappears on the third day in morontia form, as Jesus did on Urantia.

Bestowal is a necessity to a developing planet, and a necessity to the individual careers of the Magisterial Sons and of the Creator Sons. Because death is part of the experience of the ordinary mortal, an incarnated Son of God must also go through the portals of death. By no means, however, is the cruel and savage death inflicted on Jesus of Nazareth either required or common.

Before a Paradise Son of God, a Magisterial Son, has been born into the evolutionary and mortal races, Thought Adjusters do not indwell every normal human (but only special humans). After a mortal incarnation, the Magisterial Son journeys to Paradise, is accepted by the Universal Father, then returns to the Local Universe of assignment, in this case Orvonton. At this time the local Creator Son, in conjunction with the Magisterial Son, sends the Spirit of Truth over the human races of that planet, as happened at the festival of Pentecost after the execution of Jesus of Nazareth. Henceforth all normal humans are Adjuster-indwelt. Such is ordinary procedure.

CHAPTER 5. MAKING A UNIVERSE AND MAKING IT WORK

THE LOCAL UNIVERSES ARE EVOLUTIONARY AND PROGRESSIVE, whereas the Central Universe of Havona and its central Isle of Paradise are perfect and unchanging. The universes of time and space wish to be like the perfect Central Universe, and after long evolution to become "settled in life and light." Whereas the Father is personally present on Paradise, he is not present within the minds of the Havona natives; in the evolutionary creations, by contrast, he is present within the minds of the mortal ascenders as the Thought Adjuster. Otherwise he is represented in the Superuniverses by the Creator Sons. The Universal Father is not present personally outside of Paradise.

THE LOCAL UNIVERSE

A Local Universe is defined by the specific presence of the Creative Spirit. The Creative Spirit pervades the space of the Local Universe of Nebadon (the so-called Holy Ghost of popular lore). Where the presence ends, that universe ends. Beyond the space pervaded by the Creative Spirit of Nebadon are the extra-Nebadon space regions, where other Local Universes may reside.

The bodies in space—suns, planets, dark bodies, other things—ordinarily come from nebulae, immense spiral-shaped systems of energy fashioned and controlled by beings who dwell on Paradise called the *Master Architects of the Universe.* A local creation, such as Nebadon, may descend from several or many nebulae, as the space bodies in Nebadon spun off from the Andronover and other nebulae.

When the space bodies acquire a certain level of gravity presence, control of the physical creation is turned over to Force Organizers, who work with the Creator Son assigned to this space sector to design the lines of communication, circuits of energy, and lanes of power that will bind the creation together into a single administrative unit. Such power directors proceed from the Infinite Spirit.

First the Son, while still resident on Paradise, must solve problems of materialization, whose patterns are preset by the Infinite Spirit, and he must establish an equilibrium in the materialized realm. After construction of architectural spheres to house the universe headquarters, the Creator Son and Creative Spirit go there, in this case to Salvington where Michael and his staff reside today.

Next comes the organization of the architectural headquarters of the one hundred Constellations and ten thousand Systems that each Local Universe contains. The Creator Son comes together with the Creative Spirit to generate the hosts of Local Universe children. At this point the Creative Spirit becomes the Universe Mother Spirit to the local creation, changes from a principle and power to a person. Because each Creator Son is unique, and to a Local Universe virtually God, the patterns of life within his creation differ from those in any other Local Universe.

The power organizers, who first organized the energy of the space sector, continue to function from the Local Universe capital. The Creator Son is not able to leave the headquarters planet until gravity is stabilized. The Creator Son may then range through his universe. However, until he has experienced at first hand the terms of his own creation, he cannot possess full authority of it, but must govern as vicegerent of the Universal Father. Those are the rules. The Creator Son could as a freewill agent claim full sovereignty at any time over the universe of his own creation without undergoing experience in a series of seven creature bestowals, but if a Creator Son were to do so, Paradise personalities on the headquarters of that Superuniverse would withdraw to Paradise. In any event, it has never happened.

The seventh bestowal of a Creator Son within the Local Universe of his own creation must, like that of a Magisterial Son, be as a mortal of the realm, the lowest order of universe freewill creature. Following this final bestowal the Creator Son receives from the Universal Father absolute authority over his universe.

MINISTERING SPIRITS

The Creator Son could not govern the Local Universe without the assistance of groups of single-origin beings that originate in the Third Source and Center and whose mission in governance is to minister and to serve. These are the "hosts of angels" (the Midway Creatures are for convenience grouped with the angels) who work throughout the Grand Universe, from Paradise to the outermost regions, fostering the careers of ascendant beings such as ourselves. Hence they are called *Ministering Spirits* and comprise:

- omniaphim ("allers," serve on the seven spheres in the Paradise sacred circuit of the Spirit)
- supernaphim ("top ones," serve only in the Central Universe of Havona and on Paradise)
- seconaphim ("seconders," serve in the Superuniverses)
- tertiaphim ("thirders," serve the Creator Sons and the Ancients of Days in the Local Universes)
- seraphim (from the Hebrew "fiery ones," serve in the Local Universes)
- cherubim (from the Hebrew *cherubim*, of unclear meaning, serve the seraphim in the Local Universes)
- the Midway Creatures (native to individual planets, not really angels but grouped with them)

Most important to humans are the *supernaphim*, *seconaphim*, and *seraphim* who function respectively in the Central Universe, in the Superuniverses, and in the Local Universes, an unbroken chain of spirit ministers from the Infinite Spirit to the individual human, assisting those who desire to find the Universal Father through the Eternal Son. The angels are not ordinarily ascendant creatures like humans, but to some extent their status may change and progress. The angels are highly individualized, with distinct personalities. They are a separate order of created beings, entirely different from the material order of mortal creatures, a distinct group of universe intelligences.

MORE ON ANGELS

Angels are not the glorified spirits of mortal men, and they do not reproduce themselves. The angelic hosts, slightly more spiritual than humans, have only a spiritual kinship with the human race. Angels possess surprising powers (they somehow know how many hairs are on your head!), but they do not judge humans. As humans progress in the journey to the Father in Paradise, they do traverse a state of being analogous to the state of angels, but mortal humans are a different and ascending order and never become angels.

The angels are immortal unless, perhaps, they become involved in sin, as did those who followed the deceptions of Lucifer in the system of Satania. The angels are the spirit servants in heaven, neither all wise nor all powerful. They are not objects of worship. By the ministry of the angels even a quarantined world, such as our own because of the Lucifer rebellion (much more later), may be kept in touch with other worlds. Angels are not the spies of the spirit world who watch and tell the Father the thoughts of your heart and the deeds of your flesh; the Father does not need such service because his own spirit lives within you.

Angels are much concerned with the means whereby the human personality is released from the flesh and the soul escorted to the resurrection halls. Angels are the sure and heavenly guides of the soul during the uncharted and indefinite period that comes between the death of the flesh and the new life in the morontia and spirit realms.

Whereas Urantia mortals must derive their energy from light through vegetable and animal intermediaries, these brilliant creatures of light are sustained through direct intake of spiritual energy from the primary circuits of the universe. Although discrete beings, angels do not have material bodies. Although invisible to mortals, they see mortals without assistance and they understand mortal ways of thought.

They share the same emotions as humans, whom they love, except they do not experience feelings attached to the senses, and they enjoy music, art, and humor. Angels experience some emotions unknown to humans. Their emotional lives are similar to those that surviving humans will experience in the morontial world where we "neither marry nor are given in marriage but will be as the angels of heaven" (Matthew 22:30). They do not share or understand the afflicting emotions of dread and fear that humans inherit from their animal ancestors.

Although angels have no sexual identity as such, as offspring of the Universe Mother Spirit they are referred to by the feminine pronoun on planets inhabited by sex creatures. They do appear with either a positive or a negative charge, and for this reason most angels ordinarily work in pairs. They are present in large numbers on this planet and assist constantly in the mortal enterprise. In the morontia world to come, you will easily see them, where they will be your visible friends and companions.

Angels receive extensive training on the planets surrounding Salvington, capital of the Superuniverse of Orvonton. They are rigorously organized into a hierarchy of groups whose numbers increase exponentially. Jesus alluded to such a group of angels at night in the garden of Gethsemane when he said, "I can even now ask my Father, and he will straightaway give me more than twelve legions of angels" (Matthew 16:53). There are 248,832 pairs or 497,664 individual seraphim in a legion, so that twelve legions of angels would number 2,985,984 pairs or 5,971,968 individuals. Twelve such hosts—35,831,808 pairs or 71,663,616 individuals—make up the largest operating organization of seraphim, an angelic army, commanded by an archangel or other high being.

Archangels are dual-origin beings, asexual offspring of the Creator Son and the Universe Mother Spirit, the highest type of spirit produced in large numbers in a Local Universe. They are not seraphim, which have a single origin in the Universe Mother Spirit. At the last registry there were nearly eight hundred thousand archangels in Nebadon. The archangelic commanders take their orders from Gabriel, the local Bright and Morning Star (more later) and Michael's chief executive on the Local Universe capital of Salvington, or from Gabriel's agents.

SERAPHIM AND THE CLASSES OF HUMANKIND

Seraphim show little variation throughout the seven Superuniverses; they are the most uniform of all personalized spirit beings. They are still being created as the Local Universe grows and matures. Unlike humans, they have no infancy or childhood; nonetheless they are experiential creatures, like humans, learning and progressing through direct experience with the world and through education. The seraphim have ministered to the spiritual needs of humans from the beginning of humans on this planet one million years ago. They are affectionate and come to love the individuals and groups over which they stand guard.

Certain categories of seraphim are ascenders and can attain Paradise. Mostly these are the *guardian seraphim* who accompany the career of an individual human. Such service is the surest way for a seraphim to reach Paradise as an ascendant being. Humans of a certain stature and achievement receive such individual guardian seraphim. Humankind falls into three classes: (1) the subnormal, who cannot make average decisions and who cannot know God: Over all these a group of seraphim presides; (2) the average human, divided into seven subclasses according to which circle of human attainment they have reached (more later on the seven circles of attainment):

Seraphim are distributed over all seven classes; (3) the supernormal, those who have more or less regular direct contact with their indwelling Adjusters Each has a personal guardian seraphim.

In the seven subclasses of average humans, those in the outermost ring receive a pair of seraphim (they work in pairs) and a group of subordinate cherubim per one thousand individuals. Those in the sixth circle fall into groups of five hundred, and so forth until the third circle, after which everyone receives a personal guardian. Many of such seraphic destiny guardians have long served on other worlds, often of the nonAdjuster fusion series, and are of wide experience.

An important task of the guardian seraphim is to unify and focus the diverse and usually impersonal spirit forces that surround the evolving human being. They are the personal agency of the Infinite Spirit. The seraphim, both group and personal, are the custodians of the memories, mind patterns, and identity of the mortal ascender between the time of physical death and resurrection on the morontia worlds. Ordinarily they function on the spiritual and the intellectual levels, but are capable of effects on the physical level through the cooperation of the *Physical Controllers* (and the Midway Creatures); the Physical Controllers belong to the same class of intelligent nonwill creatures as the Force Controllers and are involved in the organization of matter at the planetary level.

The Adjuster is essence, working from the inside out; the seraphim are teachers, working from the outside in, educating the evolving soul and encouraging the individual through mind to make decisions that raise that individual into a higher circle of attainment. On the mansion worlds the now visible seraphim will continue to guide the evolving morontial soul. On Earth, the seraphim guide the individual into paths of progressive experience, not necessarily pleasant. They do not communicate with the Adjuster, but somehow their efforts seem coordinated. They inspire the urge to pray, but do not themselves heed prayer. They are under instructions from superiors and act on your behalf in spite of your own inclinations.

The seraphim never interfere with personal choice. They are guardians. You make the choices, they do their best to protect you from the consequences. They rarely act in the material world or communicate directly to you; but sometimes they do both. Group guardians who serve repeatedly may gather many thousands of sleeping souls under their care. All guardians, group and personal, who participated in the Lucifer rebellion have now been imprisoned and their charges turned over to another order of angels.

Before leaving the mansion worlds all ascenders have permanent individual guardians assigned to them who will accompany them through the Constellation, the Local Universe headquarters, the Minor and Major Sectors, to the Superuniverse headquarters, Havona, and even to Paradise, where such a guardian of destiny may find her own exaltation as an evolutionary seraphim.

SERAPHIM AND PLANETARY GOVERNMENT

There are around one billion seraphim and as many cherubim on Urantia. Not only do they serve as personal and group guardians, but they are integral to the supermaterial government of the planet. They work closely with the Midway Creatures.

But the government of Urantia is unique in the entire universe of Nebadon because of the following factors: (1) Urantia is a life-experiment world isolated (2) by the Lucifer rebellion and (3) suffering from the default of the Material Sons in providing racial uplift. Urantia alone (4) is the site of the seventh bestowal of the Creator of the Local Universe of Nebadon, so that Michael of Nebadon is the titular Planetary Prince of this planet. The twenty-four planetary directors (5) of Urantia on the System capital of Jerusem, who look after the welfare of the planet, is an unusual institution, as is (6) the location on Urantia of an archangels' circuit. Finally, (7) Michael has made a certain Machiventa Melchizedek planetary vicegerent (much more about him later), while turning over the actual governance of Urantia to the twenty-four planetary directors on Jerusem. One director is always present on Urantia as resident-governor. The planetary vicegerent Machiventa Melchizedek remains on Jerusem, but some think that he may soon arrive on Urantia to take up his duties in person for now and the distant future.

The commission of twenty-four on Jerusem is made up of former Urantians who have interrupted their Paradise ascent to serve in this special capacity. The council include members from each of the eight Urantia human races (several of them extinct), men who in the flesh knew the truth and brought it to their people, including Enoch (the first mortal to fuse with his Thought Adjuster while in the flesh), Moses, John the Baptist (presently chair of the council), and the planetary Material Sons called Adam and Eve (reduced to human ascender status after their default). Perhaps the commission will be disbanded at the end of a dispensation, or at the final adjudication of the Lucifer rebellion (itself now settled), probably imminent, or by Machiventa Melchizedek's assumption of full authority over Urantia, or at the reappearance of Michael on Urantia, according to his promise. All is speculation. For the time being the commission of twenty-four also supervises commissions responsible for the other thirty-six planets in the Local System of Satania quarantined because of the Lucifer rebellion.

The resident-governor of Urantia, a member of the commission of twenty-four, is more advisor than executive and serves a term of one hundred years. The resident-governor was last changed around AD 1930 at the time of the dictation of *The Urantia Book*. The resident-governor represents Salvington, capital of the Local Universe of Nebadon, as much as Jerusem, capital of the System of Satania, because the commission of twenty-four directly represents Michael and Gabriel on Salvington.

Planetary authority is made still murkier by the presence on Urantia of a Vorondadek Son dispatched as observer from Edentia, the capital of the Constellation of Norlatiadek, at the time of the Lucifer rebellion. This observer remains a "trustee of planetary sovereignty" and the Most Highs of Norlatiadek retain direct control over certain planetary affairs. In a crisis the Vorondadek Son would take temporary control of this planet, and in fact has done so thirty-three times in the last two hundred thousand years.

THE RESERVE CORPS OF DESTINY

An important responsibility of the chiefs of the resident seraphim is to watch over the *Reserve Corps of Destiny*, consisting of men and women specially trained, usually unconsciously, to preserve the planetary evolutionary civilization against disaster. About one thousand individuals today serve in this corps, divided into twelve groups as are the seraphim themselves. The largest group has 172 members, the smallest 41. Members of the Reserve Corps are rarely aware that they have been chosen for this role, and they are hard to find in the pages of history. Yet their role is critical to insuring the continuance of the growth of civilization on Urantia.

SPACE TRAVEL

Certain angelic orders are dedicated to transporting personalities unable of themselves to journey from one sphere to another. One group of seraphim headquartered on Salvington, capital of the Local Universe of Nebadon, functions as space traversers to and from Salvington. Like other subdivisions of the superior seraphim, some were created as such while others have risen from the lower or less endowed groups.

The "energy range" of seraphim is wholly adequate for Local Universe and even for Superuniverse requirements, but they could never withstand the energy demands entailed by such a long journey as that from Uversa, capital of the Superuniverse of Orvonton, to the Central Universe of Havona. Such an exhausting journey requires the exceptional powers of a special seconaphim of transport endowments. Transporters take on energy for flight while in transit and recuperate personal power at the end of the journey. Even on Salvington ascending mortals do not possess personal transit forms. Ascenders must depend upon seraphic transport in advancing from world to world until after the last rest of sleep on the inner circle of Havona and the eternal awakening on Paradise. Subsequently you will not be dependent on angels for transport from universe to universe.

Being enseraphimed is similar to death or sleep except you are "consciously unconscious." The Thought Adjuster is wholly and fully conscious and efficient because you are not in your usual position of opposing or resisting its initiatives. When enseraphimed, you sleep for a specified time and awake at a designated moment. You are not directly aware of the passing of time, as if you went to sleep on a transport vehicle in one city and, after resting in peaceful slumber, awakened in another distant metropolis. The transit sleep is induced by a liaison between the Adjuster and the seraphic transporter.

The angels cannot transport combustion bodies, flesh and blood such as ours, but they can transport all other forms from the lowest morontia to the higher spirit forms. When you finish your earthly career, your body remains on this planet, like a pupa that has served its purpose. On the first mansion world you receive a new body in morontia form, one that can enseraphim. You "sow a mortal body" in the grave to "reap a morontia form" on the mansion worlds.

The limit of velocity for most nonenseraphimed beings is 186,280 miles per second, although the Midway Creatures and certain others can attain double velocity —372,560 miles per second. An enseraphimed being cannot exceed the velocity of 558,840 Urantia miles in one second of our time. Massive stars, crosscurrents, and detours, as well as attraction tangents, will all tend to retard such speed so that on a long journey the velocity will average about 550,000 miles per second. When it develops that it will require hundreds of years for a native ambassador to reach a far-distant Local Universe, a being from the order of *Solitary Messengers*, direct offspring of the Infinite Spirit, is often asked to proceed there immediately to act as ambassador. Solitary Messengers can go in very short order, nearly independent of time and space. The universe is well supplied with spirits who utilize gravity for purposes of transit, for example the very special *Gravity Messengers,* but they are not persons, except for the Solitary Messengers.

From time to time humans have been allowed to observe seraphim being prepared for transport service, which to some extent has determined the Urantian concept of angels. When a transport seraphim is made ready to receive a passenger, energy insulators, friction shields resembling a double sets of wings extend from the head to the foot of the angel. Beings about to be enseraphimed are brought to the headquarters of the sphere and, after registration, are inducted into the transit sleep. The transport seraphim moves into a horizontal position immediately above the energy pole of the planet. When the energy shields are opened wide, officiating seraphic assistants deposit the sleeping personality on top of the transport angel. The upper and lower pairs of shields are closed and adjusted. The seraphim prepares to swing into the energy currents of the universe. The seraphim becomes elongated, pointed at either end, and bathed in an amber light. The transport dispatcher makes final inspections, encircuits the seraphim, and it departs with lightning speed, leaving a trail of luminescence.

CHAPTER 6. AT HOME IN NEBADON: OUR LOCAL UNIVERSE

THE MICHAEL WHO CREATED NEBADON dwells in a "threefold house of light" on Salvington, architectural headquarters sphere of Nebadon, threefold because in his career he experienced directly the realities of the spiritual, morontial, and material worlds. He was the 611,121st simultaneous personification of the identity of the Universal Father and the Eternal Son and is sometimes known as Christ Michael because of his experience on Urantia, where he incarnated as Jesus of Nazareth in the year 7 BC.

MICHAEL OF NEBADON

Michael of Nebadon is not the Eternal Son (a common confusion in Urantian theologies) who dwells on Paradise, but he possesses all the powers and attributes that the Eternal Son himself would have were he resident in Nebadon. A Creator Son, who represents the Father within the creation itself, is the most powerful, versatile, and influential of all divine beings capable of direct administration of an evolutionary universe and of contact with the personalities of immature beings.

Michael's headquarters are officially on Salvington, capital of Nebadon, but he visits often the Constellations and System headquarters and even individual planets, and periodically journeys to Paradise and often to Uversa, capital of Superuniverse Orvonton, to counsel with the Ancients of Days. A Creator Son is not bound by time, but he is conditioned by space; he cannot personally be in two places at the same time. The Creative Spirit, by contrast, is not bound by space and is present everywhere in a Local Universe at once, but she is limited by time. Working together, they are independent of both time and space. When Michael is away from Salvington, his place is assumed by the being known as *Gabriel*, who governs as regent over the universe of Nebadon.

Gabriel is unique in this Local Universe. Belonging to the class of *Bright and Morning Stars*, he is the first product of the creative union of the Creator Son and the Creative Spirit. No more than one exists in any Local Universe. This brilliant and versatile being possesses qualities of both parents. He shares their divinity, but does not have their creative powers. He is the head executive administrator of Nebadon and may decide on any matter, if it does not pertain to personality. He supervises an enormous cadre of other administrative beings. Gabriel has had intimate contacts with the remote planet of Urantia since the time of Michael's bestowal here, when he was in direct communication with Jesus of Nazareth and a sponsor of his mission. He has been personally present on Urantia numerous times.

In the creation of a Local Universe the local Creative Spirit or Universe Mother Spirit (after the engendering of the Bright and Morning Star) materializes patterns generated by the Son. To the Mother Spirit of Nebadon, Michael grants loving equality of authority in every administrative decision, while from the time of his seventh bestowal on Urantia she acknowledges his sovereignty. She is the source of the myriad spirits and angels that populate the Local Universe, who represent, like the Universe Mother Spirit herself, aspects of the Third Source and Center. In human terms, the Son acts as the father and the Spirit acts as a mother in governing the world as an enormous family. The successful pattern of voluntary human marriage between man and woman distantly echoes this ideal.

Michael has three concerns: creation, sustenance, and ministry. He does not participate in the adjudication of his creatures—which creators never do—a responsibility given to highly trained judges with creature experience, under the supervision of Gabriel. The Nebadon courts hear cases of general importance and appellate cases coming up from the Systems. They are not qualified to pass judgment on eternal life and death, however, which are handled by courts on Uversa at the level of the Superuniverse, where judgments of extinction are carried out. Readmission of a System isolated through rebellion (such as our own through the Lucifer rebellion) must also be approved by agents of Orvonton on Uversa.

There are no legislative bodies on Salvington. Local legislation takes places on the Constellation headquarters. It is the responsibility of the System Sovereigns to carry out legislation passed at the Constellation level and juridical decrees made at the Superuniverse level on Uversa. One must remember, however, that juridical decrees are something very different, in the realm of spiritual beings, from what is familiar to us in our own experience of courts and assemblies.

CIRCUITS OF THE LOCAL UNIVERSE

The Creator Son may come and go, but the Mother Spirit never leaves the universe headquarters. Through her proceed three spiritual circuits that permeate the Local Universe. The Spirit of Truth comes from the Son and is like a magnet, drawing truth-seekers towards knowledge of the Universal Father through the person of the Son. The Mother Spirit's own pervasive, ministering spirit is familiar in popular religion as the Holy Ghost. Finally, the seven *Adjutant Mind Spirits*, which impart intelligence to the evolving races, also come from and function through the Universe Mother Spirit. The seven Adjutant Mind Spirits are the spirits of *intuition, understanding, courage, knowledge, counsel, worship,* and *wisdom*. Where you find these qualities, you find the direct influence of the Adjutant Spirits.

Sometimes called the Divine Minister in this capacity, the Universe Mother Spirit fosters evolutionary progression on the planets, beginning with lifeless material, progressing to vegetable life, animal life, and the first will creatures, human beings. The transition from animal to human is made through the direct working of her agents, the seven Adjutant Mind Spirits, who lead humans to higher and spiritual ideals. The Adjutant Mind Spirits that elevate the animal mind into the human are the sixth and seventh, the spirits of *worship* and *wisdom*. Through the effects of the sixth and seventh Adjutants on the evolving material mind, the first humans are included in the circuit of the Universe Mother Spirit and are able to choose the will of the Universal Father—the Thought Adjuster that may now enter the individual being. Not, however, until a Bestowal Son has liberated the Spirit of Truth do all normal human beings receive Thought Adjusters, as we have seen. The Universe Mother Spirit is the water of life, refreshing those who receive it, fulfilling them independently of their material circumstance.

Because of the disorder in the governance of Urantia the native races suffer far more from the conflict between the animal nature and the spiritual impulse than do mortals on ordinary planets. In fact the natural and physical urges and demands of the body are not in conflict with even the highest spiritual achievement, except in the minds of the ignorant, foolish, and intolerant; those who in faith live in the spirit know this. There is no need to fear or be confused, for the Spirit of Truth always makes clear that "this is the way."

ORIGINS OF LIFE

Life does not originate spontaneously, but ordinarily is implanted on a planet by the Life Carriers, themselves an order of the local Sons of God. They design life forms, carry them to planets, and then remain on those planets to foster the development of their designs. Life Carriers have a triple origin in the union of a Creator Son, the Local Universe Mother Spirit, and one of the three Ancients of Days who govern the Superuniverse, the only beings with the power to extinguish intelligent life. In Nebadon there are one hundred million Life Carriers.

Although the Creator Son lays down the general plans for life forms, the Life Carriers refine these into special designs, always attempting improvement. Evolution is neither arbitrary nor magical. Often it is not predictable, but its course is never accidental. The biologic basis of material life is the protoplasmic cell, whose combinations of chemical, electrical, and other energies differ in different systems and in different Local Universes. All such patterns of life development reflect initial decisions and designs of the Life Carriers.

There are twelve distinct concepts of transmissible life, and the number twelve, with its multiples, stands behind all life forms in all seven Superuniverses. Hence on Urantia, there are forty-eight chromosomes that encode and transmit life patterns (from the possible 12, 24, 48, 96, 192, 384, and 768). One inhabited planet in ten is designated a life-experiment world, where fairly large deviations from the general pattern are allowed. Urantia is one of these worlds.

The initial life plasm on an evolutionary world must contain within it the full potential for all the evolutionary variations that will one day grow from it. Many apparently pointless forms of animal and vegetable life are inevitable, but always through the stream of physical evolution runs the intelligent formulations of the original design for any planet's life. As the initial patterns of life unfold into the higher forms, intense conflict emerges with lower and often antagonistic life forms.

The Life Carriers travel in a large corps, with 100 senior members, to a planet ripe for implantation. They may carry actual life plasm with them, or in some cases they organize on the spot physical patterns in accordance with formulas previously approved. Such was the case on the life-experimental planet of Urantia: There are only sixty-one such planets in all the Local System of Satania.

Such organized physical patterns, inert and lifeless, are then infused with the flash of life from the Life Carriers themselves. This spark, which enlivens the bodies and presages the mind, comes not from the Life Carriers but, through them, from the Local Universe Mother Spirit. The Life Carriers are not creators; they are catalysts.

The Life Carriers ordinarily spend a half million of the planet's years in implantation, then cease their efforts. During the long period between implantation and the appearance of human-will creatures of moral status, the Life Carriers manipulate the environment to encourage the progress of evolution. Once will creatures have appeared, the work of the Life Carriers is ended. The Life Carriers may not influence or interfere with will creatures. From this point, evolution is restricted to the working out of the patterns described in the design of the original plasm.

When the Planetary Prince arrives, the Life Carriers prepare to leave. Sometimes a handful remains. Today on Urantia are present two senior Life Carriers and twelve associates, who remained behind to advise on the conservation of the local life plasm.

MORE ON THE SEVEN ADJUTANT MIND SPIRITS

The seven Adjutant Mind Spirits accompany the Life Carriers in a journey to any planet. For this reason evolution is purposeful and not accidental. The Adjutant Spirits represent the ministry of mind of the Infinite Spirit as directed toward lower material life forms through the intermediary of the Local Universe Mother Spirit; she brought them into being. They are not so much entities as circuits. They are really a level of consciousness of the Universe Mother Spirit.

Humans share the first five Adjutant Mind Spirits with the animal realm: *intuition* (self-preservative response), *understanding* (coordination of facts), *courage* (basis of fidelity), *knowledge* (the discovery of facts), and *counsel* (harmonization with others). When the sixth spirit, *worship* (the religious impulse), comes to bear, the animal becomes human. The seventh Adjutant Mind Spirit *wisdom* (the inherent tendency of moral creatures to evolve) coordinates the other Adjutants and is the goal of mind at this level of life. The Adjutants are related to intelligent life in a local creation much as the Physical Controllers are to the nonliving forces of the universe and to the preadjutant levels of mind, which is mechanical and nonteachable, before evolution of the capacity to learn from experience.

The Creator Sons create the patterns of life, which the Life Carriers execute, but the "breath of life" itself comes directly from the Spirit. More precisely, we might say that life flows from the Father through the Son and *by* the Spirit. When life leaves the dying plant, animal, or human, it has neither identity nor personality. It survives only as a part of cosmic force. The survival of mortal creatures is based on the evolution of an immortal soul within the mortal mind, through interaction of the personality, a gift of the Father with a longing for God, and the divine fragment of the Father within.

THE LOCAL CONSTELLATION OF NORLATIADEK

We can scarcely describe conditions on Uversa, capital of Superuniverse Orvonton, or on Salvington, capital of Local Universe Nebadon, but can say some things about life at the constellation headquarters of Norlatiadek, on Edentia. Norlatiadek is one of one hundred constellations in Nebadon. Constellations are the administrative center linking the Local Systems with the Local Universe capital of Salvington. Salvington is itself linked by means of the Reflective Spirits to Uversa, capital of Superuniverse Orvonton, and to the Ancients of Days who govern Uversa.

Government of the Constellation Norlatiadek takes place on 771 architectural spheres. On the central sphere of Edentia, about one hundred times the size of Urantia, reside the three Constellation Fathers, the Most Highs of Norlatiadek. They are of the order of the local Sons of God called the Vorondadek Sons, as we have seen, and are appointed to this service by Gabriel of Salvington, chief executive of Nebadon. Seventy planets, each ten times the size of Urantia, surround Edentia, and ten Urantia-sized planets surround each of these. All are morontia training worlds, supervised by an order of seraphim.

Scarcely imaginable forms of gorgeous life inhabit these spheres. On Urantia there is plant and animal life, but on Edentia there are ten categories of living material forms. The animals are of an exquisite intelligence and gentleness; there are no carnivores. The material plants are green, but the morontia varieties are a violet color. When eaten, the morontia plants, being an energy growth, leave no residue. About half of Edentia is given to gardens, whose supernal beauty is unsurpassed in the entire Local Universe. No human eye has seen such glory. For this reason the extraordinarily beautiful places on an inhabited world in Norlatiadek are often called "the garden of Eden."

On Edentia there are charming highlands, but no mountains and no oceans. Tens of thousands of sparkling lakes are joined by streams. The water there is the same as water here. It is possible to circumnavigate Edentia by means of these water-courses; material and morontia beings do just that, but the spiritual inhabitants travel through the air, made up of three gases, two the same as on Urantia and the third the basis for morontia respiration. There are no storms or seasons, allowing for the embellishment of all the outdoors. On one highland are built the resurrection halls wherein ascendant mortals awake after transit from the headquarters of the Local System.

The Sea of Glass, a kind of receiving area for interplanetary travel found on many spheres, is located near the administrative center. Its floor is an enormous circular crystal one hundred miles in circumference and thirty miles thick. The crystal greatly enables seraphic transport and also plays a role in the reception and distribution of various cosmic energies. Most architectural spheres have such crystals. An amphitheater surrounds it. Seventy centers for administration of the Constellation are grouped nearby; most of the rest of the planet is one vast garden of God.

Constellations are the legislative units in a Local Universe, while the Local Systems are the executive or enforcement units. Universe government on Salvington, capital of Local Universe Nebadon, is the supreme judicial authority. Several chambers participate in drafting legislation, which becomes the law of the Constellation. The Constellation government has little to do with the individual, whose interests focus on the Local System, but because of the Lucifer rebellion the Most Highs of Norlatiadek have intervened directly in some planetary affairs and continue to do so. When Michael of Nebadon incarnated as Jesus of Nazareth, the Most Highs of Norlatiadek supervised the mission.

Ordinarily the three Vorondadek Sons who make up the Most Highs rotate out of office after around fifty thousand Urantia years. However, because of the Lucifer rebellion about two hundred thousand years ago, these three Sons have remained in place. Because of this very serious disturbance in the governance of the Constellation, an additional body of Vorondadek Sons is now present on Edentia to advise in the settlement of the disturbance. Every quarantined or isolated world, including Urantia, now also has on the planet a Vorondadek Son as observer who, if ordered to intervene, "rules in the kingdoms of men." When one of the Melchizedek order appeared in material form on Urantia about 2000 BC—this was Machiventa Melchizedek—he paid homage to the Vorondadek observer, remembered in the biblical report that "Melchizedek, king of Salem, was priest of the Most High" (Genesis 14.18).

THE LOCAL SYSTEM OF SATANIA

The administrative center of the Local System of Satania consists of fifty-seven architectural spheres: the headquarters planet of Jerusem surrounded by seven satellites, each of which has its own seven subsatellites. Jerusem is about one hundred times the size of Urantia; the seven satellites, the transition worlds, are about ten times larger than Urantia; the forty-nine subsatellites are about the size of Urantia. The subsatellites of the first planet of the seven transition worlds, on the innermost circuit around Jerusem, are the "seven mansion worlds." The whole system is lighted, heated, watered, and energized by the same standard technique used for all architectural worlds. During full light the temperature is about seventy degrees Fahrenheit, dropping to fifty degrees during the recession of the light.

The System Sovereign on Jerusem is always a Lanonandek Son, offspring of the local Creator Son and the Universe Mother Spirit. Such sovereigns have unusual powers of personal authority and freedom to act, and three times they have revolted against the authority of Michael: By far the most widespread was the revolt of Lucifer on Jerusem. Today a Lanonandek son called *Lanaforge* is sovereign over Satania, an experienced administrator tried by rebellion in another system. He will apparently continue as Sovereign until the residue of the Lucifer rebellion has been cleared away. Lanaforge is a frequent visitor to Urantia.

Lanaforge is assisted in his project of reconstruction by the unusual number of six other Lanonandek Sons, an emergency council to deal with the repercussions of the Lucifer rebellion. We have already mentioned the twenty-four member advisory council on Urantia affairs, who work with Lanaforge. John the Revelator, taken to Jerusem in a vision, saw the members of this council: "And round about the throne were twenty-four seats, and upon the seats I saw twenty-four elders sitting, clothed in white garments" (Revelation 4.4).

RECORDS AND COMMUNICATION

Everything significant that happens in the organized and inhabited universe, in the Grand Universe and in Local Universes, is a matter of record. If the matter is of local interest only, then it is recorded locally. Events of importance on the planets, Systems, and Constellations of Nebadon are all posted on the universe capital of Salvington. From there important events are passed on to the Sector and Superuniverse headquarters, and from there to Paradise, where archives are under the care of a special kind of angel (a type of supernaphim).

On Local Universes most records are material, but some are in the form of living beings; as you ascend higher, and inward, the recorders are less material and they are more usually living beings. Another specialized kind of creature called Celestial Recorders makes duplicate copies of all records, one spiritual and one semimaterial. In this way it is possible for ascenders, once they have entered the morontial realm, to obtain information about the history of their own planet and about an individual they may have known there. Much of the information in *The Urantia Book* comes from such records.

On the Superuniverse capital of Uversa, there are about eight million Celestial Recorders, able to report on every event of cosmic significance in all Orvonton since the days when the Ancients of Days arrived on Uversa from Paradise. On Paradise similar recorders preserve the records of all that has happened on Paradise and in Havona since the time of the personalization of the Infinite Spirit (before then was eternity).

Each Superuniverse has its own language, in our Superuniverse known as "the tongue of Uversa." Because each Local Universe also has its own language, the higher orders of Nebadon are all bilingual, speaking both the language of Nebadon and that of Uversa. When two personalities from different Local Universes meet, they communicate in the tongue of Uversa; if, however, one hails from another Superuniverse, they must use a translator. In the Central Universe of Havona there is little need of a language because communication takes place in some direct way.

For purposes of recording, Uversa uses a system of concepts consisting of more than a billion characters, used side by side with a basic phonetic system of only seventy symbols tied to the tongue of Uversa. The much simpler writing system of Nebadon, tied to the language of the Local Universe, has forty-eight symbols.

As mysterious as is the notion that there are living "books," actual beings who preserve all knowledge in the universe, are the beings called Reflective Spirits, whom we have already encountered: living mirrors that enable the Ancients of Days (but only they) on the Superuniverse headquarters of Uversa to see and hear responses from other beings a thousand or a hundred thousand light-years away, instantly and without error. Thus can the Ancients of Days, positioned midway between the inhabited worlds and Paradise—between man and God—instantly look both ways, hear both ways, and know both ways. The Reflective Spirits do not, however, allow communication with other Superuniverses, but only with Havona and Paradise, the center of creation.

From the capital of a Local Universe, in our case from Salvington, broadcasts go forth simultaneously to the Constellation headquarters, to System headquarters, and to individual planets. All higher orders of beings can use this service to communicate with their fellows throughout the universe. The universe broadcasts go forth to all inhabited worlds, but are denied to worlds such as Urantia that are under quarantine. Constellation broadcasts are also sent out periodically from the headquarters of the Constellation. In the vision of John the Revelator he saw that "out of the throne proceeded lightnings and thunderings and voices" (Revelation 4:5)—the universe broadcasts to the Local Systems.

CHAPTER 7.
WHAT HAPPENS WHEN YOU DIE: THE ASCENDANT CAREER

By no means are all normal humans the same. Their destinies are different. As we saw earlier, a human's spiritual status is determined by how far the individual has penetrated the *seven circles of attainment,* associated with qualities of intellect, socialization, and spiritual and cosmic insight.

THE SEVEN CIRCLES OF ATTAINMENT

The mortal's relationship to the Adjuster measures that person's progress toward God. In the seventh and outermost circle the creature enters the realm of personality growth as a human being. In the first circle the person reaches the attainment of mortal possibilities, precedent to actual fusion with the Adjuster.

Attainment of the various circles depends on achievement by the whole personality, not just one part of it. An emphasis on the intellectual side can leave the spiritual impoverished, but a glimpse of the spiritual, even sent by the Adjuster, can lead to fanaticism and delusion, the origin of many religious sects. To progress inward through the seven circles, you need poise, all-around development, generosity, and a body not poisoned or abused. It is hard to describe each of the seven levels because they vary in the individual according to moral decisions, stalwart faith, and behavior in a crisis, but there are three critical stages:

The seventh circle. The human being has reached the level of moral choice and received an Adjuster. Ordinarily, in our contemporary world, the individual receives an Adjuster just before the sixth birthday. The person is now under the direct influence of the seven Adjutant Mind Spirits, encircuited in the Holy Spirit, and on Urantia in the present day a recipient of the Spirit of Truth that Christ Michael poured out on humankind shortly after the execution of Jesus of Nazareth. The person is a potential candidate for immortality and citizenship in the Local Universe.

The third circle. The personality has evolved far enough toward the Adjuster to receive a personal seraphic guardian, enhancing the candidate's progress in the spirit.

The first circle. Ordinarily the Adjuster cannot speak directly to the person indwelled until reaching the first circle, when direct communication becomes possible. The first circle represents the highest possible coordination between the mortal personality and the divine Adjuster, until the clothing of the flesh is left behind.

There are many on Urantia who have reached the first circle but have not fused with their Adjusters. Adjuster-fusion requires more refinement, a perfect attuning of the mortal will with the will of the Father as embodied in the Adjuster. Once the two wills become as one, an order is issued from Uversa, capital of the Superuniverse, for Adjuster-fusion, usually in the morontia world after mortal death, but sometimes in this world. In planets advanced in light and life, such fusions while in the flesh become ever more common. On Urantia, Adjuster-fusion while in the flesh occurs generally when the Adjuster is highly experienced through having indwelt, on other planets, personalities who failed to survive.

DEATH AND THE THIRD CIRCLE OF ATTAINMENT

Those who have penetrated from the seventh to the third circle of attainment and received a personal guardian seraphim may be repersonalized on the third day after death on the first of the seven morontia mansion worlds, the innermost subsatellite of the innermost of the seven spheres that go around Jerusem. In the early days of an evolutionary planet, few individuals reach the third circle and receive personal guardians, but as the planet grows older ever more individuals achieve this status. Those Adjuster-indwelt individuals who have chosen survival but have only group guardians must await millennial and dispensational roll calls that take place at the recommendation of the Planetary Prince or the System Sovereign (*The Urantia Book* is vague about how many millennial roll calls have taken place on Urantia). Such individuals are the *sleeping survivors.* Mass resurrections also take place at the end of an epoch when a Son of God appears, as after the execution of Jesus when Christ Michael "led a great multitude of captives" (Ephesians 4.8), that is, released the sleeping survivors.

Advanced ascenders with personal seraphic guardians may at death skip the first mansion world altogether and land on whatever satellite is appropriate to their development, even going directly to the seventh mansion world, or to the System capital of Jerusem, or in exceptional cases even to the Constellation capital of Edentia.

TERMS OF RESURRECTION

The sleeping survivor is utterly unconscious. On repersonalization one who has slept five thousand years will respond no differently from one who has slept five days. Except for the lag in time, the careers of the sleeping survivors, who have group guardian seraphim, are the same as those with personal guardian seraphim, who pass directly to the mansion worlds of Satania. Together with the arrival of the first Thought Adjuster on any planet come these guardian seraphim, without whom terrestrial escape is impossible. The guardian seraphim preserve the lasting values and realities embodied in the evolving soul.

After death the Adjuster travels instantaneously to Divinington, the planet of the Father in the first of the three sacred circuits that surrounds Paradise, then returns outward to the System capital. There the identity and personality of the ascender is recreated in association with the guardian seraphim. When the Adjuster does not return, because the associated personality cannot be saved, that person ceases to exist: The extinction of human personality therefore takes place on the first mansion world and not on the planet of origin.

Whereas the guardian seraphim keeps the soul, which preserves the mind and personality of the exmortal, the Thought Adjusters preserve the life memories of the ex-mortal, translated from the material to the spiritual level. Such memories are now a permanent part of the Thought Adjuster. The uniting of the Thought Adjuster, detached at death, with the morontia soul constitutes the resurrection of the individual. If the individual does not survive and the Adjuster does not return from Divinington, any experience of spiritual value of the mortal remains a part of the Thought Adjuster, now an "experienced Adjuster." Many great men and women in human history have had such experienced Adjusters.

MORONTIA AND SPIRITUAL TRANSFORMATIONS

The plan of progressive education on seven mansion worlds is nearly universal in the Superuniverse of Orvonton. Each System capital, which controls around one thousand inhabited planets (but Jerusem has 619), has its seven educational spheres. The journey of the ascender after mortal death through the Local Universe is the morontia adventure, preparing the ascending being for the spirit life in the Superuniverse. The existence of this level of reality has been known on Urantia since the days of the Planetary Prince, who came to Urantia about five hundred thousand years ago, and knowledge of it exists in distorted form in some religions. Paul knew about the morontia realm and wrote of it, "They have in heaven a better and more enduring substance" (Hebrews 34).

All fifty-six satellites surrounding Jerusem (seven main satellites—called the "transition worlds"—each with its own seven subsatellites), as well as the seventy primary satellites surrounding the Constellation headquarters of Edentia (770 spheres when you count the subsatellites), and the 490 primary and secondary satellites surrounding the Local Universe headquarters of Salvington, belong to the "morontia world." All these spheres are architectural made-to-order worlds and contain twice as many elements as found on the evolutionary planets. Morontia material is entirely real but not the kind of matter with which you are familiar.

The first transition world circulating around Jerusem (the sphere closest in), surrounded by the seven mansion worlds, is the headquarters of the local corps of finaliters, glorified mortals who have returned from Paradise to assist in the mortal progression. More than one hundred thousand companies of finaliters reside on this sphere with one thousand finaliters in each group. At least once a year (based on the Local Universe time) the ascender, resident on one of the seven subsatellites of the first transition world, will enseraphim to visit the first transition world itself; although morontia vision does not ordinarily permit perception of the spiritual beings who live there, the finaliters, or their dwellings, the ascender will nonetheless gain a glimpse of what achievement is possible, of what lies ahead.

The resurrection halls are located on this first subsatellite of the first transition world. The physical body is gone, but under the care of guardian seraphim a new morontia form is fashioned there as vehicle for the eternal soul (see Figure 5).

LOCAL SYSTEM

- seven mansion worlds surrounding innermost satellite surrounding Jerusem
- corrective training and cultural education; eight morontia transformations

First mansion world

- awaken in resurrection hall; pick up where you left off in the flesh
- Thought Adjuster returns from its headquarters in Divinington and unites with morontia soul
- new morontia form
- exploration of environs
- plan for future training
- correction of deficiencies resulting from family life, sexual life, and environment suffered on the evolutionary world

Second Mansion World

- resolution of mental disharmonies and conflicts of the mind

Third Mansion World

- learning of morontia mota, a system for organizing knowledge that is parallel to but higher than logic
- understanding of cosmic meanings and universal relationships

Fourth Mansion World

- training in morontia social organization based on sympathy with the other being
- learning of standard language of the Local Universe of Nebadon

Fifth Mansion World

- learning language of the Superuniverse of Uversa
- becoming aware of the real scheme of things and your place in it (cosmic consciousness)

Sixth Mansion World

- first lessons in the spirit career
- instruction in universe administration
- usually, fusion of the morontia soul with the Thought Adjuster
- entry to eternal career of universe service

Seventh Mansion World

- preparation for life on Jerusem

CONSTELLATION

- visit the worlds surrounding Edentia
- training in ethics, administration, and socialization
- fusion with Thought Adjuster if not previously attained

LOCAL UNIVERSE
- visit the worlds surrounding Salvington, the school of the Melchizedeks
- review of earlier universe career
- first-stage spirit level on Salvington

SUPERUNIVERSE
- progress through Minor and Major Sectors in spiritual form, to Uversa
- schools of philosophy, divinity, and pure spirituality
- perfection of purpose

CENTRAL UNIVERSE
- progress through the billion worlds of Havona
- travel inward, each of the seven circuits specializing in understanding and achieving a certain level or aspect of divinity
- understand and perceive the Infinite Spirit, Eternal Son, and Universal Father

ISLE OF PARADISE
- further training, experience, adventure
- admission to the Corps of the Finality and endless service

FIGURE 5. Some Highlights in the Ascendant Career

The Thought Adjuster, which separated at death, returns from Divinington to join the soul in the new morontial body and to reconstitute the individual identity, a resurrection. From this point on the pilgrims of time are *morontia progressors*, passing upward through the spheres of the System, Constellation, and Local Universe as morontia beings, gaining rich experience and knowledge of unimaginable kinds. As morontia progressors, they are wards of the Local Universe. Only when the spirit career begins can they enter the schools of the Superuniverse.

The cosmic mind, emanating from the Third Source and Center through the Seven Master Spirits, working through the Adjutant Mind Spirits, evolves by means of biological process a suitable physical vehicle for the mortal human, our bodies. Just so does morontia mind generate a morontia form for the mortal survivor. As every mortal body is personal and different, so will the morontia form reflect the creative mind that controls it so that every morontia form is different. After the morontia life, spirit forms are equally diverse, personal, and characteristic of the personalities that indwell them. Most beings in the universe, including spiritual beings, have form, although the Thought Adjuster, who separates at death, does not.

In the mortal life the divine spirit lives within as a thing apart, almost as a theory more than reality, but in the progressive refinement of the morontia nature the spirit becomes always more a part of you. The morontial material becomes less physical and more spiritual as you advance up through the spheres, and for this reason you must constantly renew your body with morontia material of a more rarefied nature.

The transition from the highest level of training of the Local Universe to the spirit level of the Superuniverse is slight. The mind, personality, and character of the ascender remain unchanged, but the form is again changed, yet as real as was the morontia form. The ascender becomes a ward of the Ancients of Days on the Superuniverse capital of Uversa from the moment the pilgrim reaches the receiving worlds of the Minor Sector, through the study worlds of the ten Major Sectors, at last to the training spheres that surround Uversa. Different pilgrims receive different forms of training, but all gain experience in the administration of local and Superuniverse affairs, perhaps to prepare for an unknown role in the future destiny of the Corps of Finality.

In traversing the realms of the Superuniverse the ascender is an evolving spirit, but as a Havona Pilgrim the ascender's education in the spirit begins in earnest. Seconaphim transport the ascender from the Superuniverse headquarters to the receiving world of the outer Havona circuit. After completing various formalities, the ascender will be free to look up old acquaintances on the long journey from the worlds of time and space.

At last the pilgrim will reach Paradise as a perfected spirit. The ascender has found God and will be mustered into the Corps of the Finality, restricted to Father-fused beings. As finaliters they are assigned tasks in the far-flung creation (such as tending schools on the first transitional world surrounding Jerusem).

Most members of the Corps of the Finality are of mortal origin, or they are glorified seraphim, but the corps also includes a small percentage of glorified Material Sons, glorified Midway Creatures, Havona natives, and members of a unique messenger order able to function independently of time and space (the Gravity Messengers). All members of the corps are fused with a fragment from the Father. It is a completely self-governing order, answerable to the Trinity alone. More than one corps of mortal finaliters are always present on Urantia itself, and they perform every kind of universe service, awaiting their unknown ultimate destiny in the levels of outer space.

MORE ON THE MANSION WORLDS

On the mansion worlds a mortal survivor is given a chance to make up for deficiencies resulting from heredity, bad environment, and early death. Further experience of sexual behavior, however, and the knowledge that such experience can bring is no longer possible; sexual life is over. If in mortal life the ascender for whatever reason did not explore the deep and entangling world of sexual relationships, compensation for that loss can now only be gained through eventual close association with the Material Sons and Daughters resident on Jerusem, who are sex creatures.

When you awake on the first mansion world, if you survive death, you simply pick up where you left off in the life of the flesh. Mortal death changes nothing as such. Dying does not make you more spiritual. You will notice that you have a different body, but otherwise everything will seem much the same in this morontia world as it did in your former world.

The center of activity on the first mansion world is the immense resurrection hall, where personalities from the evolutionary worlds are reassembled through the activity of the seraphic guardians of destiny and individual Thought Adjusters returned from Divinington. Life Carriers and other beings help in the resurrections.

The immense temple of resurrection on the first subsatellite of the first transition world has seven radial wings. There are one hundred thousand chambers in each wing and large halls at the end of each wing for the mass resurrections that take place at the end of a dispensation. For ten days after regaining consciousness, you will be free to explore your environs and will first learn of the plan for your future training. You will be able to consult the registry in case you wish to find those you knew during the life in the flesh, if they have survived.

Mostly experience on the first mansion world is dedicated to correcting deficiencies of experience suffered on the evolutionary worlds. No surviving mortal, midwayer, or seraphim may ascend to Paradise, attain the Father, and enter the Corps of the Finality without having experienced a parental relationship to an evolving child of the worlds, or had some equivalent experience. The ascender cannot understand God the Father without having been a parent, for God's relationship to his creatures is like that of a father or mother to their children. The nonreproducing midwayers and seraphim gain this experience through helping to raise the children of the Material Sons and Daughters on Jerusem, and so must mortal ascenders who were never parents in the flesh.

Such individuals must make up part of their deficiencies by working in the System nursery located on the first transitional world, where one half of the planet is devoted to child rearing and one half occupied by finaliters. One thousand couples of Material Sons and Daughters supervise this nursery. If a child dies before achieving spiritual status as an individual, and one of its parents survives, then that child is reassembled on the first transitional world and allowed to make free will choices that could lead to its own survival. There is, however, no sexual differentiation between such children.

If you are not to be detained on the first mansion world, after ten days you will enter the translation sleep and progress to the next sphere. Every ten days thereafter you will advance to the succeeding mansion world until you reach your world of assignment. On each mansion world you will receive a new morontia body suited to the local conditions. The procedure is much as after mortal death except that the Thought Adjuster does not leave as you enseraphim for transport; your personality remains intact.

During your transit through the seven mansion worlds, you eat, drink, and rest. Morontia food is a kind of living energy unknown on the material worlds. There is no residue; you do not excrete. The first mansion world is close to the material world with which we are familiar. Each subsequent mansion world is less material. Progress through the seven worlds is intellectual and to some extent spiritual. You will be allowed periodically to visit Jerusem itself, whose nature and organization you will come more and more to understand.

To summarize: On the *first mansion world*, deficiencies resulting from family, sexual life, and environment are corrected. While resident on this world, you are allowed to visit the first transitional world (around which the mansion worlds revolve), which itself goes around Jerusem. As well as being the headquarters of the finaliters, this sphere holds the System probationary nursery.

On the *second mansion world* are resolved conflicts in the mind, mental disharmony. From here you can visit the corresponding second transitional world, headquarters of the morontia supervisors for all Satania.

The *third mansion world* is headquarters of the corps of morontia teachers that accompany ascenders through the mansion worlds, the Constellation centers, even all the way to the Superuniverse capital. From the third mansion world you can visit the third transitional world, headquarters of the local corps of angels and site of their training schools.

Training on the first two mansion worlds is of a remedial nature, to correct deficiencies inherited from the life in the flesh. On the third mansion world new morontia training begins and the first exposure to *mota,* a system for organizing knowledge that is parallel to but higher than logic. Learning morontia mota, the ascender on this sphere comes first to understand cosmic meanings and universal interrelationships, a kind of metaphysics.

On the *fourth mansion world* you learn a form of morontia social organization based not on the aggrandizement of the individual, but on sympathy with the other being. You learn the standard language of the Local Universe of Nebadon. Visiting the fourth transitional sphere from here, you can witness activities of a class of superangels whose origin lies in the Third Source and Center, but they are higher than angels.

The *fifth mansion world* provides a foretaste of Jerusem. Here you learn the language of Uversa, of the Superuniverse of Orvonton. All ascenders on Jerusem with residential status are bilingual. The languages of Havona and Paradise are expansions of the Superuniverse tongue of Uversa. On the fifth mansion world you first meet instructors from the training worlds of the Constellation and for the first time enter a kind of cosmic consciousness, aware of the real scheme of things and your place in it and of your unimaginable destiny. From this sphere you are allowed to visit the fifth transitional world, dedicated to orders of the Sons of God.

On the *sixth mansion world* you receive your first lessons in the spirit career that immediately follows graduation from the morontia training of the Local Universe. Instruction in universe administration begins. At this time commonly takes places the fusion of morontia soul and divine Adjuster. Even if such fusion has taken place earlier, often not until the sixth mansion world does the perfect working identity of the new being emerge. The ascending mortal has now entered the eternal career of universe service. One may visit the sixth transition sphere, where the ascender learns about the high spirits of the universe resident there, although they are not ordinarily visible.

Instruction on the *seventh mansion world* prepares the ascending pilgrim for life on Jerusem. Differences between candidates that depend on the status of their evolutionary planets of origin are here obliterated. There is a temple on the seventh transitional world, which can be visited, dedicated to the unseen Father. When training on the seventh mansion world is complete, the ascenders are advanced in groups to Jerusem, enseraphimed on the local Sea of Glass.

The seven mansion worlds are really dematerializing or demortalizing spheres. You are mostly human on the first world, a mortal being without a material body, a human mind housed in a morontia body. You pass from the mortal state to immortal status at the time of Adjuster fusion. A principal purpose of the morontia career is to eradicate permanently from mortal survivors such animal vestigial traits as procrastination,

equivocation, insincerity, the avoidance of problems, unfairness, and the seeking of ease. By the time you finish your career on Jerusem, you will be mortals no more, but full-fledged morontians.

The seven subsatellites that surround each transition world, including the mansion spheres surrounding the first transition sphere, are mostly devoted to education of various kinds, except the seven prison worlds that surround the seventh transition world of the Father. Here now are detained the leaders of the Lucifer rebellion, awaiting execution. No such prison worlds exist at the Constellation headquarters. The ascender may visit these various satellites, including the prison worlds.

JERUSEM

John the Revelator (c. AD 3–89) saw in his vision the advance of a class of graduates from the seventh mansion world to the Sea of Glass on Jerusem: "And I saw as it were a sea of glass mingled with fire; and those who had gained the victory over the beast that was originally in them and over the image that persisted through the mansion worlds and finally over the last mark and trace, standing on the sea of glass, having the harps of God, and singing the song of deliverance from mortal fear and death" (Revelation 15.2). What he saw as "harps" were evidently a communication device for those not yet able to receive space communications directly.

After residence on Jerusem, there are no more resurrections. The ascender keeps the same morontia form to the end of the Local Universe experience (although its substance becomes ever more refined). On Salvington you bid that form farewell to take on a first-stage spirit form for your adventures in the Superuniverse. Mortal death is a technique of escape from the material life in the flesh. Progressive life through the seven mansion worlds of corrective training and cultural education has introduced the mortal survivor to the morontia career, the transition life intervening between the evolutionary material existence and the higher spirit attainment of the ascenders of time destined to pass through the gates of eternity.

On Jerusem there are no days and nights as such and no seasons. Most structures are roofless for this reason. The planet is lighted by a reflective technique whereby energy projected upwards is returned down in an even haze with the intensity of an earth day in summer around 10:00 AM. For this reason such architectural spheres are not luminous in space. If Jerusem were near us, we could not see it. The lighting remains full for about seventy-five percent of the Jerusem day (the Satania day is three days of Urantia time less one hour, four minutes, fifteen seconds, the time of one axial revolution), then recedes to the illumination of a full moon.

The ascender may on Jerusem miss the drama of climate, earthquake, and violent storm so characteristic of Urantia. Instead, much of the planet is given to natural highlands of exorbitant, one might say heavenly beauty. The sphere is more material than spiritual; like all System capitals it exhibits perfectly the three phases of universe existence: material, morontial, and spiritual. There are many thousands of lakes, but no oceans or raging rivers or rainfall.

During the period of light recession a refreshing dew covers the ground. A subsoil system provides water for the material and morontial life, sometimes rising to the surface as canals that connect the many lakes. The atmosphere is similar to that of Urantia, except that there is a third gas respired by morontia beings. Various forms of high speed mechanical transport are available to the material and morontia beings; resident spiritual beings make use of spiritual currents.

The highest mountain, at 15,000 feet, is Mount Seraph, from which depart the transport seraphim. One departs every three seconds, taking off at about twenty-five miles per second of Urantia time and not attaining standard velocity until over two thousand miles away from Jerusem. Transports arrive on a crystal field located on one of the poles, the "sea of glass" in the vision of John, such as are found on all architectural worlds. There is an observatory there and an immense relief map of the planet.

A gigantic central broadcast amphitheater made of materials unknown on Urantia seats five billion material and morontia beings, functioning in connection with the great crystal Sea of Glass. Communications are received from the local planets, the Constellation headquarters of Edentia, and the Local Universe headquarters of Salvington. These communications are displayed for all to view, and watching them is a favorite pastime of the many kinds of beings who live on Jerusem.

Watching these broadcasts is the only activity not slowed down during the periods of the recession of light. The Jerusem sending station is located on the opposite pole of the sphere. All messages sent to individual planets come through this station, except for exceptional communications coming from Michael himself, which travel over the archangel circuit.

Portions of Jerusem are devoted to the residential housing of the immense numbers of celestial beings who live there. These beings belong to all categories and most stay in their own enclaves. Jerusem has seven major capitals and seventy minor administrative centers. Of especial interest is the quarter of the Material Sons and Daughters, the highest type of sex-reproducing beings found on the training spheres of the evolving universes, with whom the mortal ascenders of course feel a natural sympathy.

EDENTIA, HAVONA, PARADISE, AND BEYOND

On the mansion worlds surrounding the first satellite of the System capital of Jerusem the ascender unifies the evolving mortal personality and either there or on the capital of Jerusem ordinarily unites with the Thought Adjuster. On the training worlds of the Constellation one achieves socialization of the evolving personality through progressively complex social interactions with other morontia ascenders and with the local inhabitants (called univitatia), who except for intelligence are in every other way dissimilar beings.

Seventy worlds surround Edentia, and ten planets surround each of the seventy. You will visit every one of these 770 worlds, and in the end achieve citizenship on Edentia. From there the ascender visits the worlds surrounding Salvington, called the school of the Melchizedeks, offspring of the Creator Son and the Universe Mother Spirit and a Father Melchizedek (there is only one, offspring of the Creator Son and the Universe Mother Spirit; the Father Melchizedek himself presides over the first primary sphere in the first cluster and the ascender's experience there). When the course is complete, the ascender takes up residence on Salvington itself. The entire morontia adventure is a relaxed period in the ascendant career, midway between the material world of Urantia and the almost spiritual world of Uversa, the Superuniverse capital.

The 490 satellites surrounding Uversa, capital of Orvonton and the ascender's next goal, are organized in ten clusters of seven primary planets and seven subsidiary spheres to each primary planet. These spheres are the finishing schools of the universes of time, a preparatory course before reaching the seven circuits of the university of Havona. Each cluster of seven primary planets, which circulates around Uversa as a unit, is dedicated to a certain category of being, and on these 490 spheres the ascendant mortal will gain knowledge and experience as the spirit progresses inwards.

Further training takes place in the many schools for advanced study on the headquarters world of Uversa itself; the universe is a gigantic school. When the course of education on Uversa is complete, an order of transport seconaphim carry the ascendant pilgrim through the zone of dark bodies to the pilot world of the outermost circuit of the seven circuits of Havona. The candidate has achieved perfection of purpose. From this sphere the ascendant will-creature begins an immensely long journey through all one billion worlds of the Central Universe of Havona, traveling ever inward from circuit to circuit.

Each circuit specializes in understanding and achieving a certain level or aspect of deity. There is no given time limit to successful ascension, but failure at any point may send one back to the universes of time for more training. We are not in a position to know anything about what really happens on these planets, except that the individual will-creature comes to understand and perceive the Infinite Spirit, the Eternal Son, and the Universal Father.

When the billion planets have been successfully traversed, including the planets of the three sacred inner circuits surrounding Paradise, on the last station the ascender falls into a sleep and in this condition is transported to the timeless Isle of Paradise. After further training, experience, and adventure there, the ascender may be admitted to the Corps of the Finality. The ascender is a finaliter now, a new kind of divine being, rich in the record of experience in the worlds of space and time with a presumed destiny in the vast areas of outer space now being organized.

The ascendant career can be thought of as a series of seven jubilees. At the *first jubilee* the mortal makes an agreement with the Thought Adjuster that the individual intends to survive. The *second jubilee* is the awakening in the morontia realms. The *third jubilee* is fusion with the Thought Adjuster. The *fourth jubilee* is falling asleep on Uversa and awakening in Havona. The *fifth jubilee* celebrates the finding of the Universal Father, which takes place on the third circuit of the seven Havona circuits (counting outward from Paradise). In the *sixth jubilee* the ascender awakens on the shores of Paradise from a metamorphic sleep; the ascender is no longer in time and space. The *seventh jubilee* marks entrance into the Corps of the Finality. The endless service of the eternal Trinity is about to begin.

So great, so unfathomable, so vast, so unimaginable is the destiny of the mortal human who survives death!

CHAPTER 8. PATTERNS OF HUMAN DEVELOPMENT

All worlds inhabited by mortals are evolutionary in origin and design. For administrative purposes, all inhabited planets are grouped in the Local Systems, which when fully developed control about one thousand planets, each supporting mortal humans of survival status and evolutionary origin. That does not count planets where evolutionary life is still primitive or on planets "settled in light and life," the goal of all evolutionary material worlds.

THE ORIGIN OF HUMANS

With 619 inhabited worlds, Satania remains an unfinished system. Thirty-six planets even now are reaching the stage where they are ready for life implantations, and a further two hundred are in preparation for development within the next few million years. Planets need to be of a certain size in order to foster evolutionary life, not too small with a swift axial rotation, nor too large, where the gravity overwhelms. Satellites of very large planets can make ideal locations.

On atmospheric worlds implantation usually takes place in the seas, but not always. It may take place in a single place, or in three, as on Urantia. The Life Carriers have great latitude in how they behave. In human development, evolution is the absolute rule, but it works out quite differently on different worlds. Vegetable life, however, always precedes the animal, and animal life always comes from and retains the patterns of the earlier vegetable life; it is never a separate development.

Common views about the accidental origin of mortal humans are inaccurate. A universal law governs the unfolding of a planetary life plan, which is always orderly and controlled. Decisive factors are not the numbers of offspring and the vast sweep of time. For example, there are more mice than elephants, but elephants evolve more rapidly than mice. Sometimes desirable strains in the evolving life plasm are destroyed, requiring immense ages to recover, but recovery always takes place.

HUMAN TYPES

There are seven varieties of physical form that a mortal human might inhabit. The planet's atmosphere is decisive. Most mortals are midbreathers (over ninety-one percent), like ourselves, but some are superbreathers and some are subbreathers. If there were mortal life on Venus, such beings would be superbreathers; on Mars, subbreathers. Humans can also inhabit spheres without atmospheres. They are the nonbreathers, constituting about one and a half percent of all inhabited planets in the system of Satania. Everything about the life of nonbreathers is alien to our understanding (except sexual reproduction). Neither do they eat or drink. They can achieve Spirit fusion, but not Adjuster fusion. Yet in mind and character they are the same as all mortal humans. Such a race of beings inhabits a sphere close to Urantia.

Different strains of humans dwell in different elements. We are of course land dwellers, but for long eras our forebears lived in the trees, as did the apes. Still today young children delight to swing from the monkey bars on playgrounds. But on other planets humans also dwell in a watery environment, in the air, or on air and land. Like mammalian bats, some can fly. In Satania, seven percent of the human races live in water, ten percent in the air, seventy percent on the land, and thirteen percent both on land and in the air. All are erect bepedal animals. All humans, regardless of physical and chemical differences, share a common mental life and a common fate after death.

Although the average height of land-dwelling mortals in the System of Satania is around seven feet, on large planets they can be as small as two and a half feet and as tall as ten feet on small planets with small gravity fields. About twenty percent of inhabited planets in Satania exhibit such variations. Mortals may also live within different ranges of temperature, either much colder than ourselves (about eighteen percent) or warmer (about twelve percent). There are six different ways of obtaining nutrition (one is familiar to us).

There are six basic evolutionary racial types based on the spectrum, the three primary red, blue, and yellow races, and the secondary orange, green, and indigo. Some planets have all six, but on the planets of three-brained humans sometimes only the primary races appear. Urantians are of the two-brained type, being more imaginative, adventurous, and philosophical than the one-brained mortals but less spiritual, ethical, and worshipful than the three-brained orders. Such differences in the construction of the human brain characterize even the prehuman animal existences. On planets with three-brained humans you can find astounding achievements in material and philosophical progress, but two-brained humans can equal such achievements, and all three orders stand on an equal footing in the ascension career and conduct their lives in similar ways.

Single births are the rule, and the family life is much the same among all human groups. Sex equality characterizes all the advanced planets; males and females are intellectual and spiritual equals.

ADVANCEMENT IN HUMAN LIFE

All planets lighted by suns have seasons and variations in temperature. Humans on atmospheric planets always become agriculturalists. Disease in its many forms afflicts all humans at early levels, although disease has been especially severe on Urantia. The span of human life varies from twenty-five years on primitive planets to five hundred on advanced and developed spheres. Humans are all gregarious by nature, identifying with tribe or race. Social, economic, and governmental problems change with time and through the influence of the divine Sons of God, who come to live among human groups.

On an ordinary planet occur seven epochs, or dispensations, in the development of human life:

1. prePlanetary Prince man
2. postPlanetary Prince man
3. postAdamic man
4. postMagisterial Son man
5. postBestowal Son man
6. postTeacher Son man
7. the era of light and life.

Let us survey characteristics of each.

PREPLANETARY PRINCE MAN: PRIMITIVE HUMAN LIFE

From the moment when creatures of animal origin become human, capable of recognizing their creator, until the arrival of the Planetary Prince is the era of *primitive humans.* Such primitive humans live in caves or in cliff shelters or in crude huts built in large trees. At this time the planet may be overrun with ferocious animals, but the early mastery of fire and tools and weapons places primitive humans in an advantageous position.

Huge birds that live at this time are able to carry one or two passengers for as far as five hundred miles. On some planets such birds, obedient and affectionate, are of so high an intelligence that they can speak a few words of the native languages. Such birds have long been extinct on Urantia, but once they did exist here.

With the development of language comes the moral will, and some individuals are temporarily indwelt by Adjusters. After death the Adjusters depart, but archangels select some humans for survival by means of Spirit fusion; their identities are kept under seal until the appearance of the Prince and the first planetary adjudication. Although the Adjusters' indwelling makes worship possible, few individuals rise above a religion made wholly from animal fear and tribal superstition.

Sometime during this period, which may last between one hundred thousand to one million Urantia years, appear successively the six races of the spectrum, beginning with the red race. The first, third, and fifth are superior strains, the *red, yellow*, and *blue* races. The second, fourth, and sixth races—*orange, green*, and *indigo*—are of somewhat lesser endowment and do not appear at all on some worlds, or on others are soon exterminated during interracial warfare. Thus were the orange and green races exterminated on Urantia.

The system of six colored races provides certain desirable variations in mortal types and offers an otherwise unattainable expression of diverse human potential. The primitive races of primary colors will ordinarily subdue the races of secondary colors and use them as laborers and slaves, regarding them as disposable commodities much as if they were cattle or sheep, or they will destroy them utterly. On a normal world such servitude ends with the arrival of the Material Son and Daughter (introducing "postAdamic man").

These colored races, which walk erect, are mighty fighters and fine hunters and live in tribal organizations. They engage in constant vicious and unrelenting racial and tribal warfare. Only the fittest peoples survive. On Urantia the green and orange races were utterly destroyed, along with superior strains of the blue race (whose genes the so-called white race has largely inherited). From a later perspective, this is a long, dark, and blood-soaked era, an unending nightmare and realm of terror whose purpose we are ill-placed to judge (but there is a purpose).

POSTPLANETARY PRINCE MAN: THE BEGINNINGS OF CIVILIZATION

The Planetary Prince is dispatched by the System Sovereign on report by the Life Carriers that creatures of will have appeared on the planet. When he appears with his visible and tangible corporeal staff and many invisible helpers—the Prince himself is invisible—the native peoples ordinarily welcome the arrival with enthusiasm. In many cases the corporeal staff are unfortunately worshiped as gods. With the arrival of the Planetary Prince comes the first adjudication of the living and the dead. The Prince's mission is to inaugurate human civilization. In the case of Urantia, however, the Planetary Prince threw in his lot with Lucifer, setting the stage for this planet's troubled history.

The arrival of the Planetary Prince inaugurates a new dispensation. A system of schools is established at the headquarters of the Prince to train representatives from the outlying native peoples. On an undisturbed planet civilization begins to appear, and real government and rapid progress ensues (but not on Urantia). The dispensation of the Planetary Prince may last for around five hundred thousand years. Thought Adjusters arrive in increasing numbers and the seraphim enhance their regime of mortal supervision. The Prince first introduces higher religious concepts, above the primitive animal religion of fear. In many respects Urantians have not risen above this early stage in religious development. Subsequently religion advances by gradual revelation that goes in hand with evolutionary development.

During this period the different races develop their own languages and traditions of religion and continue to war against one another. By no means is the Planetary Prince a bringer of peace. Each expanding group seeks isolation from the others, or their total destruction. Towards the end of this dispensation national organization may replace tribal organization, but the great social achievement is the creation of the human family. On many worlds emerges the recognition of full sex equality, the basis for a strong life of the home. The home, uncomfortable to the wandering hunter, makes possible agriculture and a settled life and promotes the domestication of animals. Devices of a mechanical nature, however, are little developed.

POSTADAMIC MAN: CIVILIZATION BLOSSOMS

Once the native races have reached the pinnacle of biologic evolution, the System Sovereign dispatches the Material Son and Daughter, the Adamic pair, biologic uplifters whose mission is to create a race to interbreed with the evolutionary mortals and elevate their biologic development. They do not interbreed immediately with the native races (ordinarily), but for long ages labor to create a separate race, called the *violet race*, that will interbreed with the natives.

The Material Son and Daughter remain on the planet throughout its evolutionary course. Under usual conditions such a mission is not a hazard, but when the Planetary Prince is apostate and communications with the System capital are broken, as on Urantia, the mission is dangerous. Although Adam and Eve on Urantia failed in their mission, victims of the stratagems of the apostate Planetary Prince Caligastia, nonetheless they did bring great benefit to the human races, whose effects we feel today.

The Material Sons are from eight to ten feet in height and glow with a violet light. Although material blood flows in their veins, they enjoy a dual nutrition, receiving in addition to ordinary food a constant charge of celestial energy. Intake of such energy depends on an unbroken intellectual synchrony with the circuit of the Spirit. They are visible to mortal eyes, unlike the other orders of Sons, and can even procreate with the evolutionary races, although ordinarily this function is left to their progeny. The Material Sons and Daughters, a direct link between the evolutionary races and the incomprehensibly distant divine world, can see the lower orders of spirit beings, including the Planetary Prince and his staff.

Material Sons and Daughters always serve in pairs. They are immortal unless they fail in their mission, or deliberately rebel, when the currents of celestial energy are so diminished that they become mortal beings destined to die and to share the fate of other mortal beings. Such was the case with the Adam and Eve of Urantia. Since the creation of the system of Satania, thirteen Planetary Adams have been lost to rebellion and default, most at the time of the Lucifer rebellion. The children born to the Material Sons on the evolutionary worlds are not, however, immortal, but subject to natural death in accordance with the patterns that the Life Carriers have established for that planet. The longevity of such descendants decreases with each succeeding generation.

Because they are semimaterial creatures, the Material Sons and Daughters cannot be enseraphimed for interplanetary transport to their planets of assignment. First they must be dematerialized, then rematerialized on the planet of destination, with the assistance of the resident Life Carriers. Rematerialization takes from ten to twenty-eight days of Urantia time. There is, however, no comparable technique for returning them from the planet of assignment to the System capital, and for this reason they must remain on the planet of assignment throughout its evolutionary ages as the visible heads of planetary affairs.

When the Adam and Eve awake after rematerialization, they work to build a garden home, whose location has usually been selected in advance by the Planetary Prince. His corporeal staff help in the preparation. Later, the planetary offspring of the Material Son and Daughter expand the garden home, assisted by selected members of the native races. The garden home, called the Garden of Eden after the botanic glories of the Constellation's capital, is ordinarily in a secluded location near a tropical zone.

A corps of about one hundred thousand angels of a special order accompany the Adams and Eves. A member of just such a corps remonstrated with the Adam and Eve of Urantia when they violated the rules of their mission, remembered in the biblical tradition as the angelic "voice in the Garden."

PostAdamic humans increasingly become candidates for Adjuster fusion. The prime economic purpose of this dispensation is to oversee the completion of the transition from hunter-gatherer and pastoralist to agricultural and horticultural pursuits. Twenty-five thousand years of such supervision ordinarily makes the planet ready for the reception of a Magisterial Son (inaugurating "postMagisterial Son man").

During the postAdamic era rapid progress is made in the mechanical arts and the control of natural forces. It is the golden age of exploration, when the entire planet is subdued and populated. Our own planet, more than a full dispensation behind the usual plan of development, even now witnesses such a rapid development in mechanical devices. The evolutionary races, freed of regressive strains, are blended with the violet race into a single race, whose color is somewhat of an olive shade with a violet hue.

On an ordinary planet the offspring of the Material Son and Daughter, the *Adamites* or violet race, do not mate with the native stocks until their numbers have reached around one million; the planetary Adam and Eve do not themselves mate with the native people. The Adamites or violet race do not go forth into the world, but superior individuals are selected from among the native peoples and brought to the garden for voluntary mating, a high and much coveted honor. On Urantia, however, the entire scheme for human biological improvement was wrecked almost from the beginning.

The violet race are monogamous. Their children by the evolutionary races are educated in the schools of the Planetary Prince and of the Garden, then sent forth to the tribes of their evolutionary parent where they marry, produce children and inculcate monogamous union and other social practices as an ideal. Such unions lead to rapid social progress. In this way more is accomplished in one hundred thousand years than in the preceding one million years of evolutionary development. Even on Urantia the small amount of violet blood has led to marked human advancement.

On most inhabited worlds the Gardens of Eden continue as cultural centers, radiating outward patterns of advanced behavior and enhancing receptivity to spiritual influences. Along with the city of the Planetary Prince, the Garden serves as a model for the development of civilization and as an educational center. The schools of the Planetary Prince are primarily concerned with philosophy, religion, morals, and the higher intellectual and artistic achievements, whereas the Garden schools of Adam and Eve are devoted to practical arts, basic intellectual training, social culture, economic development, trade relations, and civil government. Eventually these world centers are amalgamated, but usually not until the times of the first Magisterial Son (who has never appeared on Urantia). Subordinate training centers are then established throughout the planet.

During the postAdamic dispensation internationalism replaces nationalism and representative government replaces the earlier monarchical and paternal systems. The languages of the races give way to a single tongue as ideals of the brotherhood of man become real. Systems of education become worldwide and great advances are made in art, music, and literature. When worldwide peace comes to the planet, the time has come for the appearance of the Magisterial Son.

POSTMAGISTERIAL SON MAN: THE LOCAL UNIVERSE REVEALED

On an ordinary world, the human races achieve the pinnacle of their biological evolution during the days of the Planetary Prince; then a Material Son and Daughter arrive from Jerusem to serve as biologic uplifters. When the pinnacle of ethical and philosophical evolution is reached, a Magisterial Son from Paradise appears as a teacher of higher truth; later comes the second Magisterial Son, a Bestowal Son, who establishes the planet's settled status and sheds the Spirit of Truth on all flesh (inaugurating "postBestowal Son man").

The Planetary Prince and the Material Sons and Daughters are of Local Universe origin but the Magisterial Sons come from the central Isle of Paradise. Although the first Magisterial Sons incarnate in the flesh of the realm, they are not born of woman, nor do they experience death. When their missions are complete, they give up their planetary lives and return to their previous status.

The first Magisterial Sons increase the revelation of truth to include affairs of the Local Universe. The advanced inhabitants of the planet now work about two and one half hours daily, being sufficiently developed to make creative use of their free time. Restrictions on reproduction placed on the less fit members further enhance the biological fitness of humans. Forms of representative government are further refined. Most humans, but not all, are now Adjuster indwelt. The physical sciences have reached the height of their development. On average this age lasts from twenty-five to fifty thousand years of Urantia time.

At the termination of the postMagisterial Son age comes a great religious awakening and spiritual enhancement of the peoples. The planet prepares for the arrival of a Bestowal Son, really a second Magisterial Son who belongs to the same order of Paradise Sons of God. He will inaugurate the fifth planetary epoch of "postBestowal Son man." On some worlds, however, a succession of Magisterial Sons is required before the arrival of the Bestowal Son. On the corrupt and isolated planet of Urantia no Magisterial Son has ever appeared, but instead at the end of the Adamic age Michael himself incarnated as a Bestowal Son, the terminal incarnation in his own career, a situation unique to a single inhabited planet from the ten million of our Local Universe. How lucky we are to live on Urantia!

POSTBESTOWAL SON MAN: THE SUPERUNIVERSE REVEALED

During the dispensation of a Bestowal Son the whole planet is given to the pursuit of moral culture and spiritual truth. The revelations of truth are extended to the realities of the Superuniverse. New and highly refined systems of education and government emerge.

The Bestowal Son is born of woman, lives a life in the flesh, and dies. He establishes a new and living way and after death sheds over the planet the Spirit of Truth, in whose knowledge all humans can be free. On Urantia Michael acting as Bestowal Son altered the destinies of all Urantians. Before Michael, because of the Lucifer rebellion, no human at death, even the great Moses, traveled immediately to the mansion worlds, but was compelled to sleep until the millennial or special roll calls at the end of a dispensation. Since the day of Pentecost (fifty days after the Jewish Passover) it is again possible for ascenders who have penetrated the inner circles of achievement to go straight to the mansion worlds.

After death, the Bestowal Son is resurrected on the third day, travels to the side of the Universal Father on Paradise where his mission is approved, then returns to the headquarters of the Local Universe and the company of the Creator Son. This Magisterial Son together with the Creator Son now send the Spirit of Truth over the bestowal planet. Henceforth all normal will-creatures of that world receive Adjusters when they reach the age of moral responsibility, of spiritual choice.

The postBestowal Son age may last from ten to one hundred thousand years, during which great advances are made in culture and spiritual receptivity. Disease and delinquency are eliminated and human life span extends to more than three hundred years. There is no more war and human character is so transformed that it is possible actually to live by the golden rule. Truly, this a wondrous age.

POSTTEACHER SON MAN: THE CENTRAL UNIVERSE REVEALED

On an ordinary world, the final stage in a planet's evolution comes with the arrival of a special teacher corps created by union of the Father, Son, and Spirit, the *Trinity Teacher Sons*, who have an important role to play in preparing an ordinary planet for entry into its evolutionary culmination and destination, a worldly utopia, the age of light and life. Again Urantia is out of step with ordinary progress, because Michael, who incarnated as Jesus, has promised to return to Urantia, but whether after visits by Magisterial Sons or Trinity Teacher Sons, no one knows.

The Trinity Teacher Son comes accompanied by a corps of eighty-five divine beings, some themselves of Son status. They remain on a planet for not less than one thousand years and often longer, to complete the journey of the inhabited planet from its savage origins to the era of light and life. They may depart, and then return again, even many times.

The Trinity Teacher Sons reveal the workings of Havona and Paradise. At this time something like the kingdom of heaven appears on the planet. Life is interesting and pleasant. Many mortals fuse with their Adjusters while living and are transported directly to the mansion worlds. Human life extends to around five hundred years. Rates of reproduction are intelligently controlled. Government remains for economic issues but is vanishing otherwise; individuals govern themselves. The planet is in close touch with the System broadcasts and closely monitors universe affairs. A thousand objects of interest unknown to us occupy their attention.

The golden age comes on, the era of light and life, when at the completion of the visit of a Trinity Teacher Son the Planetary Prince is elevated to *Planetary Sovereign* and the planet is registered on Satania as settled in light and life, the termination of its long, long evolutionary growth. "I saw a new heaven and a new earth and the new Jerusalem coming down from God out of heaven, prepared as a princess adorned for the prince" (Revelation 21.1).

The mortals of such an age are "a chosen generation, a royal priesthood, a holy nation, an exalted people; and you shall show forth the praises of him who has called you out of darkness into this marvelous light" (1 Peter 2.9). Even a planet like Urantia, torn by rebellion and evil, can look forward one day to the terminal visitation by a Trinity Teacher Son, the departure with him of the Trinity Teacher corps, and the era of light and life that will follow, when creatures of space and time dwell in the vestibule of eternity.

THE MORONTIA TEMPLE

Planets settled in light and life are no longer liable to physical dissolution but, evidently, will continue forever, included in the main circuits of the Superuniverse. As his last act on a terminal bestowal, a Trinity Teacher Son supervises the construction of a morontia temple at the headquarters of the inhabited planet. In this temple of immense beauty appears the Magisterial Son who earlier had lived as a Bestowal Son to elevate the Planetary Prince to Planetary Sovereign, with expanded powers and responsibilities. The System Sovereign is also present and confirms the proceedings.

At the center of the morontia temple are three shrines. The central shrine is dedicated to the Bestowal Son. The right-hand shrine is dedicated to the Planetary Sovereign, who can now be visualized when present in the shrine. The left-hand shrine is given to the chief of the resident finaliters, who assume an increasing role in the supervision of the ever-advancing stages of the era of light and life.

The morontia temple, which many mortals of this time can see, is dedicated to various planetary celebrations, including the witnessing of Adjuster fusion. Because the temple is composed of morontia material, it is not destroyed by the blazing and consuming fire that obliterates utterly the bodies of mortals who experience final fusion with their divine Adjusters. On a large world such departure flares are almost continuous. On worlds not settled in light and life, where there are no morontia temples, such fusion flashes often occur in the atmosphere, where the material body of a translation candidate is carried by Midway Creatures.

Where there is a morontia temple, the family and friends of a fusion candidate will gather around a central stage where the fusion candidates rest, or converse with their assembled friends. There are many of these stages in the large temple because there are likely to be many candidates for ascension at any time. A network of celestial personalities protects the material mortals from the energies, the "life flash" that will deliver the ascension candidate from the flesh. On such worlds, funerals are occasion for joy and hope.

After the flash, the candidate proceeds to the morontia resurrection hall on the appropriate mansion world, in accordance with the ascender's achievement.

PART II

HISTORICAL ISSUES

CHAPTER 9. THE LUCIFER INTERPLANETARY REBELLION AND ITS CONSEQUENCES

On Urantia nothing has happened in the normal order or according to plan. Only revelation can bring peace to this planet, one of the reasons that the Urantia papers were assembled and delivered to Urantians, and that this compendium was prepared in support of them. While Jesus showed the way to *spiritual* brotherhood, *social* brotherhood requires the extensive international exchange of teachers, scholars, industrialists, and philosophers. It requires a common language (such as English is becoming today), or very many who know several languages. Races and nations must learn the thoughts and traditions of other races and nations, for ignorance destroys sympathy and love. Only through a heightened moral awareness—mutual understanding and fraternal love—can intolerance and war be left behind.

The normal course of human evolution has not been followed on Urantia because of disruptions caused by insurrection of the System Sovereign of Satania, the Lucifer rebellion, and Urantia's participation in this revolt, to which we have already often referred.

BACKGROUND TO THE REVOLT

In the universe, life moves in two directions, downward from Paradise into the realms of time and space; and upwards, from the realms of time and space toward Paradise. One order of beings, as we have seen, is the descending Sons of God, of which there are two categories: the Paradise Sons of God, which include such Creator Sons as Michael of Nebadon, and the Local Universe Sons of God, which include the Vorondadek and Lanonandek Sons, brought into being by a Creator Son and his Creative Spirit associate. In general, Lanonandek Sons serve as the heads of Local Systems and as Planetary Princes. There are twelve million of these nonreproducing beings in Nebadon.

To this descending order belongs the being known on Urantia as Lucifer (whose Latin name means "light-bearer"), one of the most brilliant of his kind, once voted among the top one hundred from a corps of over seven hundred thousand Lanonandek Sons. Many times Lucifer sat in counsel with the three Most Highs of

Edentia, headquarters of the local Constellation of Norlatiadek. Lucifer is one of three System Sovereigns in all Nebadon ever found in contempt of the rule of Michael of Nebadon. Lucifer rejected his obligations and exalted a theory of personal liberty and the brilliance of self. At present Lucifer is the fallen and deposed sovereign of the System of Satania awaiting execution on a prison world near Jerusem, a highly unusual situation that has produced much human misery on this planet.

Little was known of Lucifer on Urantia because he assigned his lieutenant, Satan, also a Lanonandek Son, to represent his case on this planet. The "devil" of popular lore is, however, Caligastia, once the Planetary Prince of Urantia, charged to oversee the orderly development of intelligent life here. Also a Lanonandek Son, Caligastia threw in his lot with the Lucifer secession.

Lucifer with his first assistant Satan had ruled on Jerusem for half a million earth years when Lucifer began gradually to advocate a radical philosophy of "self-assertion and liberty." He disavowed the authority of the local Creator Son, Michael of Nebadon, vicegerent of the Universal Father (vicegerent because he had not yet assumed full independent authority). Lucifer increasingly criticized the system of universe governance and appears to have sincerely believed that a new philosophy was for the good of the system, or of the whole universe. Although repeatedly offered an opportunity to return to allegiance, Lucifer came actually to embrace evil, the conscious choice of what is known to be contrary to the will of the Universal Father.

LUCIFER'S DECLARATION OF LIBERTY

At last this brilliant being issued a Declaration of Liberty. He maintained that the Universal Father, whom no creature had ever seen, did not really exist, but was a myth generated by the Creator Sons to maintain their own authority and power. He denied that personality was a gift of the Universal Father and claimed that reverence was a form of ignorance. He declared that the Local Systems (he himself ruled Satania) should be independent and that the Creator Son Michael of Nebadon had no authority from any so-called Paradise Father or any right to expect anyone to worship such a fanciful being. True, Michael had created Lucifer himself (in conjunction with the Local Universe Mother Spirit), but was Michael therefore to rule Lucifer forever?

Furthermore, there was no power that could prevent Lucifer from carrying out his designs, Lucifer said. The Ancients of Days on Uversa, capital of the Superuniverse, may preside over personality extinction at any level, but they are certainly not entitled to interfere directly in the governance of Local Systems or of Local Universes. Lucifer claimed that the Ancients of Days had no real power over local affairs at all, and that he and his followers could do what they wanted without fear of repercussion, if only his followers would summon the courage to defy conventional behavior.

As for the executioners of the Ancients of Days, the beings who dissolve the elements of personality of those condemned to extinction, they could have no power to act within the Local Systems, Lucifer argued, if only the native beings would assert their freedom and independence. Immortality was a given, not something that could be taken away, and resurrection was automatic, part of the order of things. All beings would certainly live forever were it not for the malign injustice of the executioners of the Ancients of Days who sometimes interfered in local affairs.

Finally, he complained that the plan for training mortal ascenders (for example humans) in universe administration was a complete waste of time and energy, a pointless pursuit. Allegedly such training was to serve some unknown destiny in the faraway future of the universe, but Lucifer pointed to the many finaliters resident on Jerusem and the first mansion world as proof that no such destiny existed. The entire plan of mortal ascension as designed by the Creator Sons and implemented by the Infinite Spirit was wrong and misguided. Lucifer mocked that all the long preparation and suffering of the finaliters was for nothing and that they had become agents of a tyranny that wished to enslave the universe in an obscene denial of personal freedom. "And it was with such a Declaration of Liberty that Lucifer launched his orgy of darkness and death" (*Urantia* paper 53).

THE REBELLION BREAKS OUT

Lucifer made his declaration on the last day of the year at a conclave on the Sea of Glass on the planet Jerusem, headquarters of Satania, about two hundred thousand years ago Urantia time. Satan announced that henceforth all allegiance would be toward Lucifer as the Prince of the Local System of Satania. Lucifer was "the friend of men and angels" and the "God of liberty." From now on Local Systems must rule themselves as a collection of independent planets over which the Planetary Princes had absolute authority.

The rebellion was system-wide. Many Planetary Princes joined the rebellion, including Caligastia, who did all in his power to persuade his administrative staff. The philosophy of unbridled personal liberty and self-assertion appealed to the peoples of the still undeveloped planet, two hundred thousand years ago. But a crisis of communication swiftly followed when the Most Highs of Edentia suspended all broadcast circuits between Satania and the Constellation headquarters, and with other universe offices, a suspension that exists today. All interplanetary communication was restricted to the personal agency of two orders of seraphim and Solitary Messengers (the unique order of beings, first begotten of the Infinite Spirit in time and space, that can nearly travel independent of time and space). When the circuits to the fallen worlds were cut, Lucifer could no longer communicate directly with his adherents on these planets.

Of the many orders resident in Satania, some followed the rebellion, others remained loyal to the Local Universe administration. Losses among one order of angels were severe. John the author of Revelation referred to the loss of these angels when he wrote symbolically of the great red dragon, saying, "And his tail drew a third part of the stars of heaven and cast them down in darkness" (Revelation 12.14). On Urantia the dragon (Greek for "snake") symbolized the leaders Lucifer, Satan, and Caligastia.

The Lucifer rebellion was the most widespread and disastrous of all political disturbances in the Local Universe of Nebadon. After two years of turmoil, a new System Sovereign suddenly appeared on Jerusem on the Sea of Glass, sent from Edentia, a Lanonandek Son named *Lanaforge*. The rebels were stripped of administrative powers, although allowed to range freely on Jerusem, and even to travel to inhabited spheres in the System, promulgating their philosophy.

Lucifer himself, although stripped of power, could not, as a Lanonandek Son, be detained or destroyed, for Michael was not yet sovereign of the universe he had himself created; Lucifer's prediction of untouchability was for the moment accurate. The Ancients of Days on the Superuniverse capital Uversa sustained the actions of the Constellation Fathers in replacing the administration of Satania, but did not respond to requests to act in the matter of Lucifer, Satan, and their associates, who were left to roam unmolested. No new worlds, however, fell to their persuasion, including those peopled since the revolt began.

In the lifetime of Jesus of Nazareth, both Lucifer and Satan were present on Urantia and worked to frustrate his mission. After the execution of Jesus, Michael assumed full sovereignty of Nebadon, and even in the flesh had ended the Lucifer rebellion in all Satania, stripped the rebels of all power whatsoever and issued orders for the apprehension of Lucifer and Satan. Shortly before his death, Jesus referred to this when he said, "And I beheld Satan fall as lightning from heaven" (Luke 10.18). The thirty-seven apostate planets of Satania, however, remain under quarantine.

The Uversa tribunals have not yet issued a decree of execution on Lucifer and his cohorts, in spite of appeals that they do so, but the decree is expected at any time. Nearly up to his death, Jesus recognized Caligastia's continued formal rulership of this planet, until he said, "Now is the judgment of this world. Now shall the prince of this world be cast down" (John 12.31). But Caligastia never had the power to enter men's minds, unless they earnestly supplicate his presence. Such has always been true: The faith children of God are immune to the corrupting influence of fallen spirits because the armor of faith drives away evil.

PRESENT STATUS OF THE REVOLT

When assuming Local Universe sovereignty two thousand years ago, Michael offered reinstatement to all beings who wished to turn away from Lucifer, whom agents of the Ancients of Days took into custody at this time. Many of the lower orders accepted reconciliation, but leaders remained at large.

Those beings who accepted reinstatement are nonetheless now held on the prison world of the first satellite going around the seventh transition world that goes around Jerusem, a technical requirement until the legal case of Gabriel vs. Lucifer is decided in the Uversa courts. The seventh transition world is the sphere of the Father whom Lucifer denied. Presumably when the order of annihilation is given, the reinstated beings will be spared. Temporary emergency Princes now rule on the apostate worlds; we have seen that the present Prince vicegerent of this planet is Machiventa Melchizedek.

The broadcast circuits interrupted two hundred thousand years ago cannot be restored as long as the Lucifer insurrection remains in the courts. Once Lucifer and his cohorts are gone, the apostate planets should be restored to the larger universe community. "The wages of sin is death" (Romans 6:23).

SOME MORAL ISSUES

God does not create evil, but when freewill beings make choices in an imperfect world, then *sin*—the conscious deliberate doing of evil—becomes possible. It could be no other way.

The Lucifer rebellion exemplifies the difficulty in choosing between true and false liberty. Everyone wants true liberty, but true liberty is based on justice, fairness to others, and fraternity. Liberty is self-destructive when it is thoughtless and uncontrolled. True liberty follows cosmic reality. The growth of true liberty follows the conquest of the self, while false liberty exalts the self and prefers self-admiration to self-respect.

There is no greater moral error than when intelligent beings desire to obtain power over others to deprive them of their natural liberties. Lucifer's error was to attempt to do the undoable, to short-circuit time, to go against the experiential, evolutionary plan of the Universal Father. By breaking from the central administration, and isolating his System, Lucifer deprived the will creatures within his System of the liberty to make freewill decisions to follow the divine plan, to create themselves and in a collective effort to create the destiny of the Local System of inhabited worlds.

People like to wonder why an all-wise creator permits evil and sin, but a moment's reflection shows how evil and sin are inevitable if the creature is to be truly free. The choice between good and evil is a real choice.

Why were Lucifer and his cohorts allowed so long a reign of terror? Michael, a creator-father, will be reluctant to destroy one of his own sons, who may even return to the universe order. Such was the moral of Jesus' own tale of the Prodigal Son. Delay in the apprehension and execution of malefactors is in any event a function of the Son's mercy to the inhabitants of space and time. The restraint of justice by mercy proves that God is love and that such a God of love dominates the universe and in mercy controls the fate and judgment of all creatures.

Justice in a universe dominated by mercy is slow, but certain. The time lag in the disposition of the Lucifer rebellion has allowed every creature to declare himself within or without. By allowing the rebellion to go on unchecked, its bankrupt vision was made clear to all. A summary execution of justice would have encouraged many inclined to believe Lucifer's promulgations. Of course the perception of time, too, is utterly relative; one hundred thousand years on Urantia are but seconds on the Superuniverse headquarters of Uversa.

CHAPTER 10. EARLY HISTORY OF URANTIA

The Urantia Book's descriptions of the history of our planet depends on records kept on Jerusem. The planet originates in the sun, which belongs to the Andronover nebula, one of the primary nebulae of Orvonton. The Andronover nebula has undergone a typical if spectacular evolution from the moment that the Force Organizers instigated the energy whirl that grew into a vast cyclone of space. After initiating such nebular revolutions, the living Force Organizers withdraw at right angles to the plane of the revolutionary disk. The intrinsic qualities of energy insure the orderly evolution of a new physical system.

SOME IMPORTANT DATES IN THE HISTORY OF NEBADON

The following rounded dates are translated into Urantia time:

- *eight hundred seventy five billion* years ago: beginning of Andronover nebula.
- *five hundred billion* years ago: the first Andronover sun is born.
- *four hundred billion years* ago: Michael of Nebadon selects a space sector in the contracting and stabilizing nebula to construct a Local Universe. At once are placed under construction the architectural spheres of Salvington and the one hundred Constellation capitals, requiring about one billion years. Headquarters planets of Local Systems were not completed until five billion years ago.
- *three hundred billion* years ago: Michael and his staff arrive on Salvington and the Local Universe of Nebadon is entered in the records on the Superuniverse capital of Uversa.
- *two hundred billion* years ago: the first inhabited planets appear in Nebadon, now registered as a universe of progressive mortal ascension.
- *eight billion* years ago: beginning of terminal eruption of Andronover nebula.
- *six billion* years ago: break up of nebula and *birth of our sun*, one of 1,013,628 produced by the Andronover nebula.

ORIGIN OF OUR SOLAR SYSTEM

The origin of our solar system, registered on Jerusem under the name of Monmatia, was unusual; only one percent of solar systems in all Orvonton have originated in this way.

Four billion five hundred thousand years ago the enormous Angona system approached near our solitary sun. The center of this great system was a dark giant of space, solid, highly charged, possessing tremendous gravity pull. During periods of maximum expansion, the pulsating sun threw off long gaseous tongues. At first these solar tongues fell back into the sun, but as Angona approached, its immense gravitational pull so elongated the tongues of fiery gas that they broke away, resolved into meteorites, and began to revolve around the sun in elliptical orbits. When Angona came closest, the sun in coordination with one of its regular pulsations experienced a partial disruption: Immense quantities of matter extruded from opposite sides at the same time, but the matter torn on the side toward Angona formed into a great column, pointed at both ends and thick in the middle, and now free of the immediate gravity of the sun.

This column of gas evolved into the planets of this solar system. The gas that escaped by repercussion from the other side of the sun became meteorites and space dust, although much was recaptured by the sun once the Angona system receded. At first formed five smaller planets at either of the tapered ends of the column. Jupiter and Saturn formed in the middle, but contained so much solar material that they shone with a brilliant light and emitted enormous heat. Even today these planets are largely gaseous, not having condensed into solid matter. Some material was captured from the Angona system, reflected in the retrograde motion of some of the moons of Jupiter and Saturn; retrograde motion never occurs in any solar system except as the result of collision.

Once formed, the nuclei of the planets attracted through gravitation vast quantities of space material and so grew in size. The worlds of our solar system therefore had a dual origin: the condensation of a gaseous nucleus later augmented by the capture of enormous quantities of meteors. Planets still continue to capture meteors, although in greatly reduced numbers.

The planets do not rotate around the sun in the sun's equatorial plane, which they would if they had been thrown off by solar revolution. Instead, they travel in the plane of the solar extrusion caused by the passing of the Angona system, which approached the sun at a considerable angle to its equator.

By *three billion years ago*, the solar system had much the configuration it has today, although not all the planets survived; the orbit of the fifth planet was close to Jupiter whose gravity broke it into the present asteroid belt. Urantia was a tenth its present size, but grew constantly through meteoric accretion. The heat in the interior steadily increased from the sinking inward of the heavier and radioactive metals captured from space.

EVOLUTION ON URANTIA

On Urantia a primitive atmosphere appeared, now held by the sphere's increasing gravity. *One and a half billion* years ago the planet was two thirds its present size, a volcanic inferno. A crust, mostly granite, formed on the molten surface. As the surface cooled and the atmosphere stabilized, it began to rain. For thousands of years the planet was enveloped in mist as the rains fell on the hot rock. The sun could not be seen. Still, there was no free oxygen, and never would be until vegetable life began to release it.

By *one billion* years ago the planet had reached its present size. It was registered on Salvington, capital of the Local Universe, as Urantia. The constant rains cooled the earth's surface so that a crust was formed over the entire planet, covered with fresh water everywhere to a depth of over one mile.

After five hundred thousand years, the thickness of the surface had so increased that a great land mass emerged from the waters as a compensatory adjustment. The planet was assigned to the Local System of Satania for supervision and registered on the life registry of the Constellation of Norlatiadek. Thus was prepared this speck in the middle of nowhere that was later to become the site of Michael of Nebadon's human incarnation and of many extraordinary events to be discussed throughout the entire Local Universe.

Nine hundred million years ago a scouting party arrived on Urantia from Jerusem to examine local conditions. The commission made a careful survey, then returned to Jerusem to recommend to the System Sovereign Lucifer that the planet be placed on the registry of life-experiment spheres. The Life Carriers were notified of permission to fashion new patterns of mechanical, chemical, and electrical mobilization for implantation on Urantia, their sixtieth experiment in Satania. On Urantia the Life Carriers hoped to improve the Satania type of the Nebadon life patterns.

Eight hundred million years ago about ten percent of the surface was land. As the oceans became deep, the land was upthrust until one third of the planet was a gigantic single continent.

Around *seven hundred fifty million* years ago the single continent split north-south and east-west, preparing the formation of the present continents separated by water. During the next one hundred million years long fingers of water extended over the separating masses of land, preparing shallow seas, becoming ever more salty.

Six hundred million years ago a commission of Life Carriers arrived on the planet from Jerusem to inspect conditions. The Satania Life Carriers had decided on a sodium-chloride based form of life, so it was necessary that the seas on Urantia reach a sufficient degree of salinity. The corps returned to Jerusem to await the further breaking-up of the land masses and the development of more shallow inland seas.

Five hundred fifty million years ago the Life Carriers returned. Although they can and do carry life to the planets, in this case they formulated life here on the planet, a life-experiment world, so that there is no other world in Satania, in fact in Nebadon, where life is just like that on Urantia. The Carriers formulated the patterns of life for this planet and made three identical marine implantations in three widely separated shallow seas.

By *five hundred million* years ago marine plant life was well established, and by *four hundred fifty million* years ago, as the continents drifted ever further away from each other, developed the first animal life in the sheltered tropic bays and inlets (amoeba are typical survivors of this early stage of animal life). Although we can trace the evolution of vegetable life into animal life through a graduated series of plants and animals that progressively leads from simple to complex, we cannot find such connecting links between the major divisions of the animal kingdom or between the highest prehuman animals and the dawn men of the human races because such links never existed. Radically new species of animal life do not evolve from the gradual accumulation of small variations (as commonly believed), but appear *suddenly* as full-fledged new orders of life. The *sudden* appearance of new species is wholly biologic and natural, a procedure encoded in the life plasm initially implanted by the Life Carriers. This feature of evolutionary life will be important in the much later formation of the human races.

By *four hundred million* years ago marine life, animal and vegetable, was distributed widely and profusely over the planet. The first land plants appeared. *Suddenly*, without transition, appeared the first multicellular animals. The trilobite, predecessor of modern crustaceans, ruled the seas.

The lands had been rising and subsiding beneath the sea for millions of years, but around *three hundred fifty million* years ago virtually the whole planet was inundated except for the central Asian land mass. Periodic inundations, followed by reemergence, went on for over one hundred million years. Around *two hundred fifty million* years ago fish appeared, the first vertebrates. They appeared *suddenly* from their arthropod ancestors.

Soon fish ruled the seas. Modern sharks are close to these often huge primitive fishes. There are still no land animals and too much carbon in the atmosphere to make land life possible. At the same time appeared *suddenly* the fern family, which quickly spread over the earth, growing far inland from the water's edge. The prolific and often giant ferns delivered vast quantities of oxygen into the atmosphere.

About *two hundred ten million* years ago, when warm seas covered most of North and South America and Europe, there *suddenly* appeared the first land animals, air-breathing amphibians developed from arthropods whose swim bladders had evolved into lungs. This was the age of frogs and the giant ferns whose luxuriant growth was transformed into immense deposits of coal. Shortly thereafter insects appeared and together with spiders, scorpions, cockroaches, crickets, and locusts overspread the whole world.

Around *two hundred million* years ago were laid down the great beds of coal, in some places eighteen thousand feet thick in North America. These beds are the remains of once gigantic forests that grew along the bogs of ancient seas, transformed by pressure.

Around *one hundred seventy million* years ago profound climatological changes came from the rising of high mountain ranges, the subsidence of once warm seas, and the formation of glaciers on the mountains. From around one hundred thousand species in the seas, only five hundred survived. Still, only frogs and insects lived on the land.

The appearance of seed plants made possible a greater variety of land animals. Spreading vegetation increased the level of oxygen. A prereptilian form appeared and spread over all the continents that were at this time still attached to one another.

The first oceanic stage of evolution is complete, having lasted about two hundred fifty million years, one quarter of the history of the planet. Now begins the second stage of evolution on land.

Around *one hundred forty million* years ago *suddenly* appeared the first full-fledged reptiles, rapidly developing into crocodiles, sea-serpents, winged reptiles, and scaled land-dwellers, then into the fabled dinosaurs, sometimes gigantic, but always with tiny brains. Several million years later appeared the first mammals in an experimental program that did not use placentas; the experiment failed and the animals disappeared.

By *one hundred twenty million* years ago the dinosaurs had reached a pinnacle, having overrun the earth, the seas, and the air, with the largest species ever to exist on this planet, some land species seventy-five feet long and so huge that they required vast quantities of food. With tiny brains unable to cope with crisis, and because there were very large numbers of them, they starved to death and became extinct.

By *seventy five million* years ago continental drift ceased, leaving the continents much as they are today. *Suddenly* there appeared the flowering plants and many modern trees and seed-bearing grasses spread over the land.

As the land reptiles declined, around *fifty-five million* years ago appeared the first true birds. The lengthy marine invasions of the land masses are coming to an end. Around *fifty million* years ago *suddenly* appeared placental mammals, the single most important evolutionary development on Urantia up to this time. The first placental mammals sprang from a small, highly active, carnivorous, springing type of dinosaur. The early mammals held an easy advantage over other life forms because their offspring were born fairly developed, partners were affectionate and protective, they were agile, and their high intelligence enabled a creative response to their environment.

By *thirty-five million* years ago the placental mammals, through their high intelligence and agility, began to destroy the remnants of the dinosaur age, although turtles, crocodiles, and snakes, with the prereptilian frog, survived. By *thirty million* years ago modern mammals appear. Earlier mammals, clawed eaters of flesh, lived in the mountains, but *suddenly* appeared the hoofed animals of the plains, including an early form of the horse, who ate grasses.

Around *twenty million* years ago was a golden age for animals, species of all kinds roaming in great numbers over the entire planet. Huge elephants, with enormous brains, were triumphant, the rulers of the animal kingdom. The elephant is surpassed only by man himself. Horses come behind elephants for intelligence and adaptability, but could never overcome their instinct to flee when startled. Yet their service to humankind as food, transport, traction, and in war was to be incalculable: The horse has played a vital role in the evolution of civilization and our history would not be the same without them.

Around *ten million* years ago all the continents, except Australia, were connected by land bridges, and animals of all kinds passed from one continent to the other. Big cats roamed the earth and in their own sphere were lord of all. Around *five million* years ago the Americas were separated from Asia, and since then separate developments have characterized the two great world regions.

The raising of land elevations, the cooling of the climate, and constant precipitation caused enormous ice sheets to form, whose pressure was so great that they spread outward and southward, covering large portions of the northern landmasses. There were six advances of the ice, the first around *two million* years ago. During the ice ages was laid down the preparation for the appearance of the first humans.

BIOLOGICAL BACKGROUND OF THE HUMAN SPECIES

The origin of humans on Urantia took place in a then nearly paradisiacal peninsula near where Mesopotamia is today, bounded by the sea to the west and south, high mountains to the east, and ice to the north. There a new kind of mammal *suddenly* appeared, sprung from the North American lemur, which had migrated across the Bering Strait into Eurasia. This mammal was characterized by its lively, affectionate, and curious behavior.

Motivated by extreme fear of dangers in the forest environment, an emotion that remains powerful in the human species today, this easy prey to larger animals used its superior intelligence to construct shelters in the high trees. They were gregarious, nearly tribal, and extremely pugnacious. They organized and destroyed less fit bands of the same species, so enhancing the quality of the species, and they dominated or drove out all other small animals where they lived.

These were the *dawn mammals*. Among the dawn mammals were born one day brother/sister twins who were superior in every way, with thumbs similar to our own, who walked nearly upright and had little hair. They mated and produced many offspring, who grew into a different race that in a ferocious war exterminated their dawn mammal ancestors, killing every last one.

These were the *mid-mammals*, superior in every way to the dawn mammals. They utterly subdued the enclosed Mesopotamian paradise in which they lived, which the big cats had not penetrated and where the terrifying carnivores of earlier times had perished. The mid-mammals lived about twenty-five years. They were capable of showing disgust. They liked to hoard things. They competed in building treetop shelters, but also dug underground places to hide. They lived on the ground in the day, but slept in the trees at night. Groups warred against one another in campaigns of extermination over sex and ever scarcer supplies of food.

After narrow escapes from lightning and other dangers, a superior set of mid-mammal parents gave birth to the *primate twins*, the ancestors of human beings. At about the same time a mentally retarded mid-mammal mother and father produced another set of twins, who were themselves mentally impaired. These twins became ancestors to the families of monkeys and apes, some of which still survive; otherwise, the "apes" do not figure in the human family tree.

The primate twins stood erect from the start. They reached a height of five feet. They learned to communicate with one another by signs and sounds, a kind of protospeech, but could never make their mid-mammalian brethren understand their meaning. They ran away to the west coast of what was then the Mesopotamian peninsula projecting into the southern sea, while the mid-mammals lived around the southern point and along the east coast of the peninsula. The skeletal systems of these first primates were similar to those of primitive humans. Hands and feet were recognizably human. They walked and ran, but their minds were not human.

After twenty-nine thousand years from the appearance of the dawn mammals, around 900 generations, *suddenly* appeared among the tribe descended from the primate twins the first human beings, the brother and sister called *Andon* and *Fonta*. In five thousand years all the descendants of the primate twins would be gone, the last link between humans and their animal ancestry.

CHAPTER 11. THE FIRST HUMANS

FROM AD 1934, WHEN *THE URANTIA BOOK* CAME INTO BEING, back to the birth of the first two human beings, Andon and Fonta, is *993,419* years. These early humans, sprung from the descendants of the primate twins, lived on the ground and resorted to the trees only in times of danger (see Figure 6).

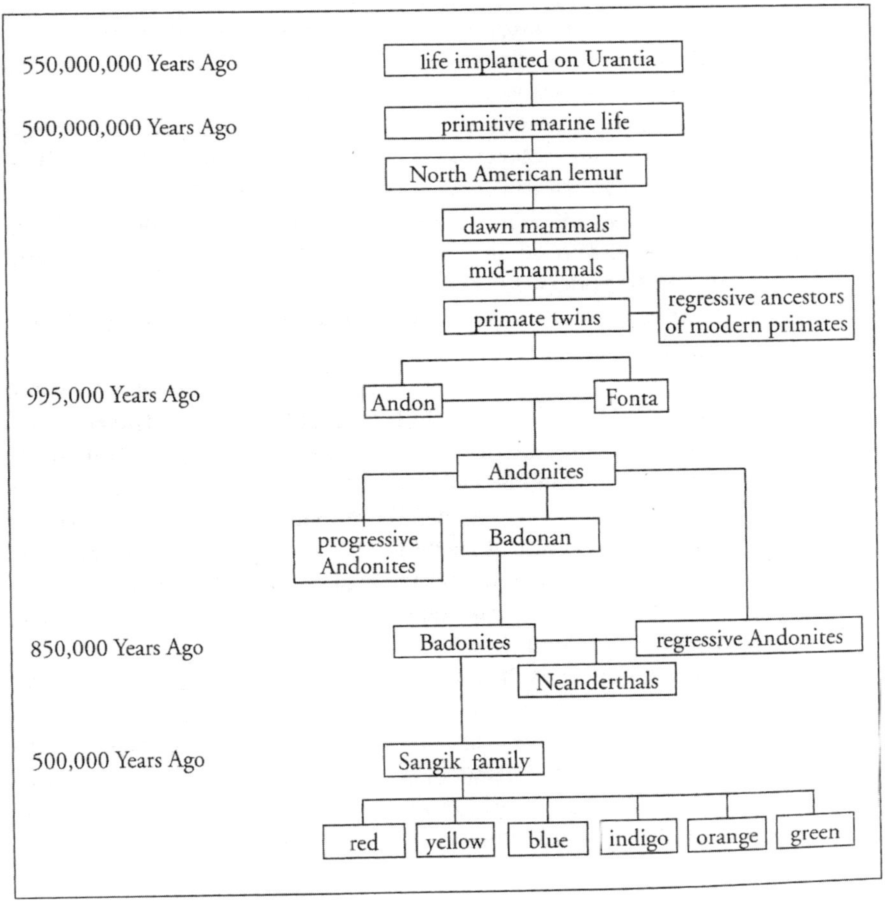

FIGURE 6. The Origin of Early Human Groups

They had completely lost the prehensile power of the big toe and climbed a tree just as we do. They reached maturity at age twelve and had a potential life-span of seventy-five years.

ANDON AND FONTA

In Andon and Fonta appeared a range of worshipful feelings, including awe, reverence, humility, and even gratitude. They also had feelings of pity, shame, and reproach and felt love, hate, and the desire for revenge. They were jealous. Their highly evolved brains allowed communication with the sixth and seventh Adjutant Mind Spirits, the spirit of *worship* and the spirit of *wisdom*. They were the first truly intelligent creatures of will on this planet. Their deep-seated fear, joined with their ignorance of natural phenomena, would soon give birth to primitive religion.

The curiosity and adventuresome personalities of the primordial twins got them into constant scrapes and many times they nearly lost their lives. They created a system of verbal communication and communication with signs amounting to around five hundred items, but were unable to communicate these symbols to their mid-mammal parents and others of the primate tribes.

At age nine the twins conspired to devote their lives to one another and to escape from the primate tribes. In communication with the Adjutant Spirit of wisdom, they laid elaborate plans for escape, then journeyed northwards, guided invisibly by resident Life Carriers, and settled in a secluded valley where their descendants could not degrade their heritage through interbreeding with the genetically close hairy and still partially tree-dwelling primates.

On the third day after the twins fled north, there arrived on Urantia the Nebadon archangel of planetary circuit establishment. Over the newly established mind circuit, the first message arrived at the planetary pole of space communication on Urantia, where the Life Carriers gathered: "To the Life Carriers on Urantia—Greetings! We transmit assurance of great pleasure on Salvington, Edentia, and Jerusem in honor of the registration on Salvington, the headquarters of Nebadon, of the signal of the existence on Urantia of mind of will dignity. The purposeful decision of the twins to flee northward and segregate their offspring from their inferior ancestors has been noted. This is the first decision of mind—the human type of mind—on Urantia and automatically establishes the circuit of communication over which this initial message of acknowledgment is transmitting."

Next came the beautiful voice of Lucifer, sovereign of the Life Carriers on Urantia, who were now given permission to return to Jerusem: The patterns of life they had established billions of years earlier had come to fruition! Biologic evolution had achieved the human levels of will dignity; humans had arrived on planet 606 of the star system of Satania. The Life Carriers were freed of all future criticism for their work to improve the life patterns of Nebadon as established for the system of Satania on the experimental planet of Urantia.

And so did they depart, except for two senior Life Carriers and twelve assistants, who still today reside on the planet. Life Carriers dictated several of the Urantia papers.

LATER LIFE OF ANDON AND FONTA

The twins were eleven years old when the planet was registered on Salvington. The report of an archangel on Salvington ended: "Human mind has appeared on 606 of Satania, and these parents of the new race shall be called Andon and Fonta. And all archangels pray that these creatures may speedily be endowed with the personal indwelling of the gift of the spirit of the Universal Father."

In the language of the Local Universe of Nebadon, Andon, the male twin, is said to mean something like "the first Fatherlike creature to show human perfection hunger." Fonta, the female, signifies "the first Sonlike creature to show human perfection hunger." These names were unknown to Andon and Fonta until the time of fusion with their Thought Adjusters. They called each other Sonta-an, "loved by mother," and Sonta-en, "loved by father," significant of their mutual regard and affection.

On the flight north, begun in the dark of night from fear of the hostile primates, Andon noticed that sparks came from clashing together two pieces of flint. The primate tribes had often fed fires started by lightning and kept them burning, but could never make a fire. As the weather cooled, Andon experimented for two months but could not create a fire until he climbed a tree to retrieve a bird's nest. Striking the flint near the inner lining, immediately it burst into flame. The first humans had discovered how to make a fire, the greatest single invention in the history of the species. Now they could live independently of their animal cousins.

Two years later was born their first child, whom they called Sontad, the first creature in the history of Urantia to be wrapped in coverings at birth. Henceforth, as the brain continued to evolve, human infants became ever more helpless.

Andon and Fonta had nineteen children, almost fifty grandchildren and half a dozen great-grandchildren before they died. They lived in adjoining rock shelters, three of which were connected by hallways excavated in the limestone with flint tools devised by Andon's children. At age forty-two Andon and Fonta were both killed at the same time when an earthquake dislodged a large overhanging rock. Five of their children and eleven grandchildren were killed at the same time.

The descendants of Andon and Fonta are the *Andonites*, the first humans. The Andonites were clannish and always did things in groups. They stayed close to their home territory. The Andonites lived together as a large family group until the twentieth generation, when food scarcity and personality conflicts began their dispersion.

The Andonites were small, had black eyes and swarthy complexions; the present-day Eskimo are mixed descendants of one branch of Andonites. Their bodies were mostly hairless and they were first to wear animal skins for protection and warmth. They had little sense of play and no sense of humor. The Andonite smiled but never laughed. Childbirth was not painful. The males fought fearlessly to protect their mates and their children, and females were deeply affectionate to their offspring. Before the dispersion, the Andonites had evolved a complex language, the tongue of the planet until the appearance five hundred thousand years later of the colored races.

Ever intensifying warfare between the various clans over often trivial matters led to the loss of some of the best strains of the Andonite people, some of it irreparable, and even the survival of this first human group was open to question. After all, humans are descended directly from fighting animals, and when forced to live in close quarters, they easily offend one another and violence erupts.

THE ANDONITES DISPERSE

When the Andonite clans began to move away from their home in what was then the Mesopotamian peninsula, they did not at first go east into Asia, blocked by the elevated Tibetan land mass, then thirty thousand feet high, or south or west, blocked by the Mediterranean Sea that at that time joined with the Indian Ocean. They traveled north until they reached the edge of the third ice sheet. From there they followed along the edge of the sheet west through northern Europe before the third ice sheet buried the British isles and northern France. They lived along the river Somme, whose course was not changed by the advance of the ice; evidence of their occupation is found today on the river's banks where they lived under overhanging cliffs or in hillside grottoes with a good view of who was coming.

The Andonites were mighty hunters and ate mostly raw flesh, but also berries and fruits when they could find them. Andon himself invented the stone ax, and his descendants invented the harpoon and the throwing stick. They coveted flint and traded it widely.

Andon and Fonta had no clear notion of spiritual truth. Their descendants, dependent on wild animals to survive, made gods of now this animal, now that, and inscribed their images in caves and manufactured images of them. A common practice was to avoid eating the animal chosen for worship, around whose sacrificed body various rituals were held, the origin of sacrifice in human religious cult. One great teacher did acknowledge a greater power and offered the prayer, "O Breath of Life, give us this day our daily food, deliver us from the curse of the ice, save us from our forest enemies, and with mercy receive us into the Great Beyond."

Andon and Fonta, founders of the human race, received Thought Adjusters during their lives and after death received recognition at the adjudication of Urantia on the arrival of the Planetary Prince about five hundred thousand years later. In due time they emerged from the mansion worlds and obtained citizenship status on Jerusem, from where they have followed developments on Urantia. On Jerusem they were fused with their Adjusters, as were some of their children. Their later descendants, however, to some extent went backward on the evolutionary scale, and few achieved fusion. At present Andon and Fonta serve on the first mansion world where with other morontia personalities they welcome ascenders from their onetime home of Urantia.

LATER HISTORY OF THE ANDONITES

The Andonites avoided the forests where their forebears lived, and they avoided the southlands where they might revert to the level of the animals. Human progress has never taken place in the forest, but only in the open lands and in the higher latitudes. The cold and the hardship from living near the ice sheet encouraged invention and resourcefulness, while the primate tribes luxuriated in their tropical paradise.

By *nine hundred fifty thousand* years ago the Andonites had migrated west to England, at that time joined to the continent, and as far east as Java, where their remains have been found in modern times (so-called Java man). The stock of the western branch remained purer than in the east, where the Andonites freely interbred with their animal cousins, then the mongrel offspring again interbred with the once superior Andonites. All trace of earlier teaching about a great Breath Giver was lost, and Urantia's first civilization was failing. The pattern of cultural ascent, then deterioration, was to be repeated again and again in Urantia's history.

Around *nine hundred thousand* years ago arrived in England large numbers of the mongrel tribes from southern France. These peoples were scarcely human and had no religion, but worked some flint and could make a fire. Other tribes occupied Europe, while in the farthest west one branch of the superior progressive Andonites maintained the old traditions. Admixed with subsequent stocks, they moved west of England after a visitation of the ice and became the present-day Eskimos.

THE NEANDERTHALS

Another center of culture persisted in northwest India in the foothills among the tribes of one Badonan, the great-great-grandson of Andon. The *Badonites* were the only early humans who did not practice human sacrifice. About *eight hundred fifty thousand* years ago the Badonites launched a war of extermination against the degenerate half-animal stock of surrounding tribes. After one thousand years of constant war, they had destroyed most or driven them back to more southern regions.

Mixed with some other strands of regressive Andonites, the Badonites emerged as the *Neanderthals*. The Neanderthals were great fighters and traveled widely, spreading gradually from their homes in northwest India west to France, east to China, and south even to North Africa. They dominated the world for half a million years until the migrations of the six evolutionary races of color. Among the Neanderthalers of France originated the custom of giving the best hunter the pick of the women.

The Neanderthals had little culture. They followed the reindeer and made use of every part of its body for food, clothing, and tools. They did improve the working of flint until it approached the level of the days of Andon and Fonta. They reintroduced the flint ax attached to a wooden handle.

Seven hundred fifty thousand years ago the fourth ice sheet was advancing. With their superior flint axes the Neanderthals were able to chop holes in the frozen rivers and spear the fish that came to the surface. Still, they fell back before the glacier's advance. Around *seven hundred thousand* years ago this greatest of glacial advances began its retreat and northern Europe filled with forests, forest animals, and Neanderthals living off them. By *six hundred thousand* years ago the ice was in its northernmost retreat, preparing after fifty thousand years to return once again.

For two hundred fifty thousand years the Neanderthals hunted, fished, and chipped stone, but made absolutely no progress in the evolution of culture, and in many respects went backwards. They were terrified of mists and clouds, and they dreaded nightfall and the dark of the moon. Only human sacrifice could force the moon to shine again. Human sacrifice became the predominant form of Neanderthal religious behavior. The effort to create a new and modified form of intelligent life on Urantia appeared about to fail, once again.

CHAPTER 12. THE EVOLUTIONARY RACES OF COLOR

FIVE HUNDRED THOUSAND YEARS AGO the superior strains of the Andonites were the Badonites of northwestern India. *Suddenly* was born to a single Badonite couple children of exceptional intelligence and ability. This was the *Sangik family*, the ancestors of the six colored races of Urantia.

Not only were the nineteen children of this couple unusually intelligent, but they changed color on exposure to sunlight. Among these nineteen children were five *red*, two *orange*, four *yellow*, two *green*, four *blue*, and two *indigo*. The colors became more pronounced as the children grew older. When these children grew up and mated with their fellow Badonite tribesmen, their offspring tended toward the skin color of the Sangik parent.

THE SANGIK (COLORED) RACES

On an average planet the colored races appear in sequence, widely separated by time. First appears the red race, who rules the world for an age; then the orange race, and so forth. For all six races to appear at one time, and in one family, was an extraordinary experiment. But that is what happened (see Figure 6).

The Sangik races appeared at the time of the fifth glaciation (the third in modern geology), when the advance of the ice sheet was so great that it was not possible to travel from central to east Asia. Before the rise of the Arabian peninsula and the retreat of the Mediterranean, it was also not possible to travel to Africa. The advance of the ice was in general a great and constant trial for these ancient human beings. Hemmed in, for one hundred thousand years the Sangik races spread out in the foothills of northwestern India, multiplying, intermingling but quickly showing strong mutual racial hatred. Not until the ice sheet retreated could these peoples spread out over all the earth.

On Urantia the *red man* was exceptional for his intelligence, a human being superior to Andon and Fonta. The red man first organized tribal government. They were always monogamous. The red men invented the bow and arrow and were fierce warriors, but so weakened by internecine war that, in their long struggle with the yellow man in eastern Asia, they were driven off the continent about eighty-five thousand years ago when the Bering Land Bridge was in place.

Approximately seven thousand mostly pure-line red men, women, and children, divided into eleven tribes, crossed the Bering Strait when expelled from Asia by their yellow enemies. Three small groups of other racial mixes accompanied them, including a group of mixed blues and orange, who later split off and populated Mexico, Central America, and South America, whereas the pure-line red men occupied lands to the north. Later the Bering Bridge sank, and no member of the red race ever returned to Asia, although they left behind strong impressions on the races of China, central Asia, India, and Europe.

The red race at first preserved teachings they had received at the headquarters of the Planetary Prince, but quickly forgot them as they fell into bloody unremitting and suicidal tribal war. It appeared as if the relatively pure line of the red man would disappear altogether. Then about sixty-five thousand years ago appeared the great teacher Onamonalonton, who lived in the redwood forests of northern California. For a while he brought peace to the warring tribes. He taught of the Great Spirit who created the world. His direct line descendants have come down among the Blackfoot Indians of Montana.

But gradually the teachings of Onamonalonton lost force, the tribes returned to mutual destruction, wiping out the best strains and preventing the rise of what might have been a mighty civilization. After around eighty-five thousand years ago the red race never again came into contact with other human groups, except one time with the mostly Andonite Eskimo. About one thousand years ago, European Vikings landed on the Atlantic coast of America, but the red race completely missed out on admixture with the superior Adamic strains that gave to the white man such power (more later on the nature and origins of these genetic strains). They could not rule the white man, and they would not serve him.

The *orange man* loved to build things, even competing with other groups to see who could raise the highest mound of stones. They were not a progressive people, but had sent delegates to the schools of the Prince for instruction. They followed the retreating Mediterranean Sea into the lower Nile Valley, where the green race destroyed them to the last man. The orange race ceased to exist about one hundred thousand years ago, although their DNA survived in the green race and in the later invading indigo race.

The *yellow race* was first to abandon the chase and settle down to an agricultural existence. Although inferior intellectually to the red man, the yellow man was by far superior in the arts of living together with one another in peace, unlike the red, orange, and green races who virtually destroyed themselves until their outside racial enemies moved in to overwhelm them. Through this strength the yellow man overcame the red man and drove him from Asia. The ability of the yellow race more to live at peace than his Sangik rivals has made them the most successful of the Sangik races, so that one of four or five humans alive today is Chinese. They received a small infusion of the blood of Adam.

The red and yellow races had to some extent intermingled in northeast Asia, and later the pure-line yellow man drove the mixed offspring east and south onto the peninsulas and the nearby islands, where they are today. The yellow man clung to central eastern Asia.

The *green race* was one of the least able of early human groups. They quickly dispersed in several directions. Those who went north were subdued, enslaved, and destroyed by the yellow and blue men. Those who went east to what is today India, were absorbed by the peoples there, and important strains of the green race survive today in India. Those who went south entered Egypt, where they exterminated the orange men there, although the two races were fairly evenly matched. Gigantism was common in both races and their leaders were often eight or nine feet tall. Subsequently the indigo race, last to emigrate from the homeland in northwestern India, absorbed the green race in Egypt.

The very able *blue race* had the brain power of the red race and the sentiment of the yellow race. They very early invented the spear, predominant in war until the invention of the light pistol in the mid-nineteenth century AD. The descendants of the race of Adam, the violet race, preferred the blue women to those of the other Sangik races and freely interbred with them.

The blue men had paid special attention to the teachings of Prince Caligastia and his staff at the planetary headquarters and fell into a fatal confusion when the staff began to promulgate the doctrines of Lucifer. The blue men moved in waves into Europe along the same routes used by the Andonites in primordial times. They were bellicose and constantly warred among themselves. Artifacts of the European "Old Stone Age," commonly found, are mostly remnants of cultures of the blue men, who lived in Europe until recently. The so-called "white races" of Urantia are mostly descendants of the blue men, admixed with some of the yellow and red and upstepped by absorbing the greater portion of the violet race, the race of Adam (more later).

In Europe the blue men encountered the Neanderthal descendants of their common ancestor, Andon. The Sangik tribes were more intelligent and in many other ways superior to the Neanderthals, but they interbred with them nonetheless, creating a new Neanderthal race that quickly spread over Europe and as far east as India. The remnant of the blue race left in the Persian peninsula, later amalgamated mostly with the yellow and somewhat by the violet race of Adam, exists today as the nomadic tribes of modern Arabs.

The *indigo race* migrated to Egypt, destroyed the green man and absorbed his remnants and those of the orange men, whom the green had earlier destroyed. Much later the blue men and the mixed races of Arabia drove the indigo man southward up the Nile. The indigo men took possession of the rest of the continent, where they have remained to this day, except when forcibly transported to different parts of the world as slaves. Isolated and cut off from Asia, in Africa the indigo race did not benefit from the cultural advances in Asia and other parts of the world.

Progressive strains of the earlier Andonites at this time became the first seafarers, built boats, and sailed north to escape the Sangik peoples and the recurring incursions of the ice. They reached the extreme northern reaches of Europe, Iceland, Greenland, and northeastern North America.

These were times of intense struggle and unrelenting violence and hatred between the races, except in proximity to the Prince's headquarters in Mesopotamia. There to some extent they lived in peace. From time to time great teachers would arise in each of the races and teach their peoples, in some form, about the one God; then, without writing, the teaching would fade and the Sangik races would sink into their customary backwardness and cultural stagnation. It was a long dark interval between the Lucifer rebellion about two hundred thousand years ago and the time of Adam, about thirty-eight thousand years ago. The days of early humans were dark and relentlessly violent, yet in many cases individuals showed high courage, admirable bravery, and a grandiose heroism in the midst of unrelenting harshness.

The Life Carriers, designers of evolution on each evolutionary world, have good reasons for introducing the colored races, although these reasons may be hard for us to appreciate:

- variability is necessary to sort out superior from inferior strains in a kind of natural selection
- interbreeding among different races may produce a stronger offspring
- a diversity of races stimulates competition
- difference in status between races and within races encourages the development of human tolerance and altruism
- homogeneity of the human races is not desirable until an evolutionary people attains a generally high level of spiritual development

OVERCONTROL OF EVOLUTION

Material life that evolves before connection is made with mind is the formulation of the Master Physical Controllers and the Seven Master Spirits in conjunction with the direct administration of the Life Carriers. As a result of this threefold influence there develops physical capacity to receive mind, material mechanisms that allow for intelligent response to the external environment and, later, to internal stimuli taking origin in the physical mind itself.

The Physical Controllers govern the level where nonteachable organisms respond to the environment. The Adjutant Mind Spirits activate the teachable, nonmechanical mind, the response mechanism capable of learning from experience. The Life Carriers work to control aspects of the environment to favor the progress of evolution up to the moment when the first will creatures appear, creatures capable of knowing God and worshiping him, when the Life Carriers cease to interfere.

Hence evolution is purposeful, not accidental, controlled by the Physical Controllers, the Adjutant Mind Spirits, and the Life Carriers. Natural selection functions to remove life forms that express the interior design encoded in the original life implantations, but are not successful in competition with other species or in responding to environmental crises. But once humans have appeared and the Life Carriers withdraw, further evolution is up to man himself. No more races will ever appear on this planet. Humans on Urantia must work with what they have.

Urantia is a life-experiment world and various new ways of doing things were tried here and progress made in various designs of life processes, especially in how cellular reproduction heals a wound. Ordinarily the Planetary Prince comes at the time of the appearance of the first will-creatures, which would have been in the lifetimes of Andon and Fonta one million years ago, but on Urantia his arrival was delayed for five hundred thousand years. In the interim twelve Melchizedek advisors worked with the planetary Life Carriers. Ordinarily the Sangik races come first, evolving directly from the prehuman animal species, and their appearance is strung out over several generations and they emerge from varied stock. The birth of all Sangiks half a million years after the first humans, the Andonites, and in a single family is unique to this planet.

The Life Carriers' biggest disappointment in the Urantia experiment has been the reversion of certain primitive plant life to the prechlorophyll levels of parasitic bacteria on an extensive and unexpected scale, causing the many dreadful diseases that afflict the higher mammals and especially the vulnerable human species. They somewhat discounted the problem when it arose, because they knew that the admixture of the Adamic life plasm would make the blended race practically immune to diseases produced by the vegetable type of organism. However, the misfortune of the Adamic default put an end to that.

Inherent in life is the striving for perfection, so that all things that live strive to do better in their environment, to improve. Evolution is the unfolding of the life force's urge to be better.

CHAPTER 13. THE REIGN OF THE PRINCE

Delayed for unclear reasons, Caligastia came to Urantia fully five hundred thousand years after Andon and Fonta, at about the same time as the appearance of the Sangik races. He and his staff took up residence in Mesopotamia at the center of the about one-half billion primitive humans living in Europe, Asia, and Africa, where he established schools for the introduction of civilization. Caligastia was a Lanonandek Son experienced in the affairs of the Local Universe of Nebadon and in the System of Satania. On Jerusem he was attached to the counsel that advised the Life Carriers, then elevated to Lucifer's personal staff, where he served well. From early on Caligastia wanted to be installed as a Planetary Prince, especially on an experimental planet, but repeatedly the Constellation Fathers, the Most Highs of Edentia, turned down his request. At last he was assigned to Urantia. Caligastia had a brilliant career behind him and a brilliant one ahead, but he was blinded by the glory of his own personality.

THE CALIGASTIA ONE HUNDRED

The usual corps of assistants accompanied the Prince, including a large number of angels and other beings whose purpose was to assist in the further development of the human races. A special group was the *Caligastia one hundred* that Caligastia chose from 785,000 ascendant morontia personalities on Jerusem who volunteered for the assignment—former mortal humans who had lived in the flesh. Every one of these one hundred was from a different planet and none was from Urantia (see Figure 7, next page).

The morontia forms of the chosen one hundred were brought from Jerusem to Urantia by seraphic transport. On Urantia they were provided with bodies of flesh and blood, but they were also attuned to the life circuits of the Local System. Therefore they took nourishment from two sources. With permission from the System capital of Jerusem and the Constellation capital of Edentia, the two resident Life Carriers, who remained after the others had departed at the maturity of Andon and Fonta, transplanted the life plasm of one hundred superior survivors of the original stock of Andon and Fonta into the bodies prepared for the Prince's corporeal staff.

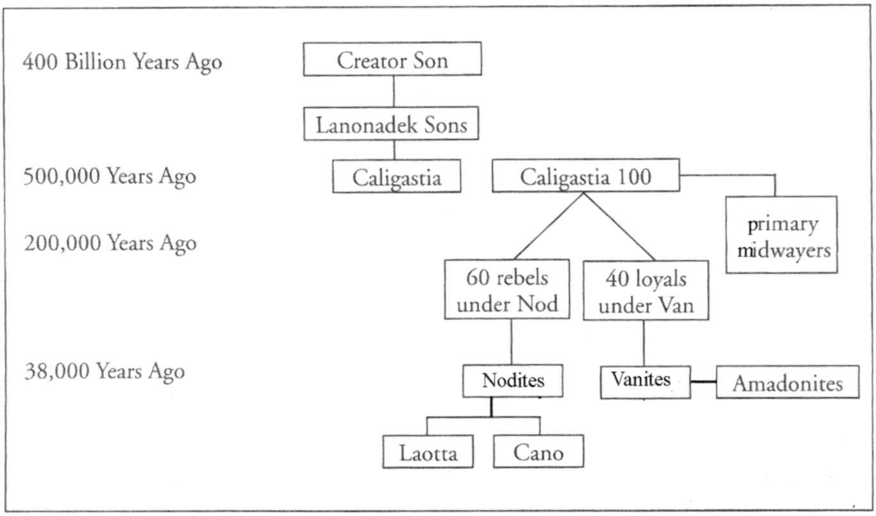

FIGURE 7. Caligastia and the Descendants of His Staff

They selected these fifty male and fifty female Andonites from around the planet. By means of coordinated Thought Adjuster direction and seraphic guidance, the one hundred humans were assembled at the planetary headquarters of the Prince in Mesopotamia. There a highly skilled commission transferred the Andonite plasm to the one hundred new bodies specially prepared for the members of the Prince's corporeal staff. Meanwhile, the volunteers from Jerusem were held in the sleep of seraphic transport. The procedure lasted ten days. When the Life Carriers removed the life plasm from the one hundred Andonites, in a kind of reciprocity they made their bodies receptive to the System circuits. Now they could live on concurrently with the corporeal staff, century after century, in defiance of physical death. These extraordinary events were remembered in distorted form in many Urantian myths and legends: The appearance of the Caligastia one hundred made a tremendous impression on Urantia. They were supermen of a sort, but the tendency was to view them as gods, although they never resorted to superhuman manipulations.

The corporeal staff were sex creatures and capable of reproducing, which ordinarily they did just before returning to Jerusem around the time of the appearance of the planetary Adam and Eve. In this case, however, the Caligastia rebellion overturned the entire regime; after it nothing was normal. The Caligastia one hundred did later reproduce sexually, but by then they had been cut off from the System circuits.

The corporeal staff in appearance resembled the Andonites, whose life plasm they shared. They ate as any human, but because of their unique metabolism did not eat flesh; they subsisted only on fruits and nuts and other vegetable products, the origin of vegetarianism that spread outward from Dalamatia. The Thought Adjusters of the Caligastia one hundred were detached during their service on Urantia, but their souls had been developed through experience on the seven mansion worlds prior to becoming citizens of Jerusem. Truly, these *were* superhuman beings.

DALAMATIA

Five hundred thousand years ago the climate in what was to become Mesopotamia was very different from today, being the most salubrious on the planet. The principal mission of the Prince's colony there, called *Dalamatia,* was to change the Urantia humans from hunters to herders in hopes that one day they would learn the arts of agriculture and the superior social patterns that accompany agriculture.

The Dalamatia headquarters were typical for those on a young and developing planet. The simple but elegant city, divided into ten sections with local administrative centers, was surrounded by walls forty feet high. At the center of the city stood the temple of the unseen God, and in twelve chambers that surrounded this temple Caligastia and his staff conducted their business. All buildings, built of mud brick, were one story high, except for the twelve chambers, which had two stories, and the small temple, which had three.

The first students were recruited from the neighboring primitive tribes. The corporeal staff taught them what they could, then sent the students back to their tribes. The surrounding countryside for as far as one hundred miles around came to be dominated by the superior social and cultural practices introduced by the Prince's staff, including animal husbandry and agriculture—to work the fields was not a punishment for sin, but man's greatest privilege, his most human activity.

The Caligastia one hundred lived in elegant but simple houses as couples. They did not have corporeal children of their own, but as many as five hundred adopted children from the Andonite and then Sangik races, many of them orphans. When grown to adulthood, these adopted children became eligible for marriage and were sent back to their people. The tradition of a monogamous couple living in a settled residence surrounded by their children first came into human society from the model of the supermen and superwomen of Dalamatia.

In Dalamatia time was measured by the day and by a month of twenty-eight days, the origin of the seven-day week, one fourth of twenty-eight.

THE PRIMARY MIDWAYERS

Although instructed to refrain from physical sexual activity, during the thirty-third year of their stay in Dalamatia two of the corporeal staff accidentally discovered that when their morontia bodies (supposedly nonsexual and nonmaterial) came together, the result was the materialization of the first of the primary midwayers that we discussed earlier, beings midway between the morontia and the material levels. This creature was visible to the corporeal staff and their associates but invisible to the human tribes. Seeing what had happened, the Planetary Prince encouraged the entire corporeal staff to undertake the production of similar creatures. In this way the Prince's corporeal staff brought into being the original corps of fifty thousand primary midwayers, of great value in the administration of the planet.

The Caligastia one hundred were immortal and would have lived until the later arrival of a Son of God, a Magisterial Son, because the life circuits to which they were attuned reinvigorated them constantly. When the Magisterial Son arrived, they would ordinarily have returned to Jerusem to resume their universe career. However, rebellion severed their connection with these circuits.

The Caligastia one hundred received the life currents from the fruit of the *tree of life*, an actual shrub from the Constellation capital of Edentia that the Most Highs sent to Urantia at the time of Caligastia's arrival. In Dalamatia this tree grew in the central courtyard of the temple of the unseen Father. While of no value to the evolutionary races, fruit from the tree conferred continuous life on the Caligastia one hundred and on the one hundred modified Andonites who served as their subordinates.

THE TEN COMMISSIONS OF THE CALIGASTIA ONE HUNDRED

The Caligastia one hundred were divided into ten groups, each with a specific mission:

1. The *council on food and material welfare* taught techniques of controlling the flow of water, the manufacture of clothes from animal skins, and techniques of food preservation.

2. The *board of animal domestication* worked to tame various animals (some now extinct) for food production and labor. The blue man had already tamed the elephant, and dogs were long since domesticated. Cattle were tamed and used for butter and cheese. Passenger pigeons carried messages. The board successfully tamed the fandors, gigantic birds capable of transporting humans. The fandors became extinct thirty thousand years ago.

3. The *advisers on the control of predatory animals* taught techniques to trap and destroy the many dangerous animals that threatened human groups.

4. The *faculty on the dissemination and conservation of knowledge* devised the first writing system on Urantia, consisting of twenty-five phonetic symbols, in which enormous numbers of documents were generated on leather, stone, clay, and a sort of paper made of wasps' nests. The library at Dalamatia of over two million documents was destroyed at the time of the Lucifer rebellion.

5. The *commission on industry and trade* encouraged the improvement of manufacture and the desire for trade goods. They created a primitive form of exchange by credit, the use of tokens instead of the actual objects of barter. The same system was in use one half million years later; in modern times thousands of these tokens have been found in the Near East.

6. The *college of revealed religion* made little progress in combating the ghost-fear of primitive humans, but did institute a regular religious service during which prayer was offered to the "Father of all."

7. The *guardians of health and life* devoted themselves to teaching the need to remove human waste from human settlements and sources of water and the necessity

of cooking food to remove diseases from it. They could not persuade the primitive humans to burn refuse, because they viewed fire as something sacred, so instead they taught them to bury it. Early humans would not bathe, except in a ritual context: The guardians introduced a weekly ritual bath. They introduced the handshake to replace the exchange of saliva or blood in proclamation of loyalty. Away from Dalamatia, however, the old practices continued.

8. The *council on art and science* inculcated a taste for beauty and new techniques for making and decorating pottery.

9. The *governors of advanced tribal relations* tried to lay foundations for the emergence of a state. They advocated intertribal marriages and games and festivals that brought together different peoples in peace. They tried to regulate and civilize warfare. But little progress was made beyond the immediate vicinity of Dalamatia.

10. The *supreme court of tribal coordination and racial cooperation* was presided over by the remarkable *Van* (after whom Lake Van in Armenia is still called). Their task was to look after everything that the other nine groups did not and to serve as a court when differences arose.

The Caligastia one hundred were wise from their experience once as mortals and then as students in the mansion worlds. They understood that progress must take place through evolution and not revolution. They did not rob these primitive peoples of the things they valued, but enhanced them. Progress was slow but satisfactory for three hundred thousand years until the outbreak of the rebellion, when everything went backward. The Constellation Fathers deprived Caligastia of his power, but he long remained on the planet and contributed to the default of Adam and Eve around 35000 BC. Machiventa Melchizedek much curtailed his ability to do harm when he incarnated around 2000 BC in the days of Abraham, and Christ Michael at last removed every vestige of the Prince's authority.

SIN, EVIL, AND INIQUITY

The Caligastia betrayal had stupendous consequences on the social and spiritual development of the planet, although it did not affect organic evolution. We live every day with the consequences of this betrayal. Lucifer's assistant Satan, himself a brilliant Lanonandek Son, made periodic inspection tours of planets in Satania. On one tour he divulged to Caligastia the impending "Declaration of Freedom," persuading Caligastia to bring Urantia into the revolt. His betrayal of trust so distorted Caligastia's personality that it never regained equilibrium.

We might think of *sin* as the actions of a personality who knowingly resists cosmic reality and *evil* as maladjustment to universe reality. Sin is purposeful resistance, a conscious choosing to oppose spiritual progress. Evil is dependent on a deficiency of wisdom. *Iniquity* is an open persistent defiance of reality, a kind of cosmic insanity, and the sign of vanishing personality control. While all sin may be forgiven, the iniquitous personality will not ordinarily feel regret for misdeeds or accept forgiveness for sin. One wonders why any personality, a gift of the Father, would freely choose to behave in such ways, but some certainly do.

RESULTS OF THE INSURRECTION

After a long conference with his principal administrator *Daligastia*, a Lanonadek Son, Caligastia summoned a special assembly of administrative personnel, including the ten counsels of the Caligastia one hundred. He informed them that he would soon proclaim himself as the sole power on Urantia and that they should surrender all administrative functions to his assistant Daligastia, pending the reassignment of offices and duties under the new regime.

Van from the corporeal staff, head of the council on tribal coordination, spoke up to oppose the plan, saying that it bordered on planetary rebellion. They should first submit the proposal to Lucifer on Satania for a fair ruling, Van argued. The assembly approved his proposal. Lucifer swiftly replied that Caligastia was in fact Supreme Sovereign and could do as he chose. Van answered the reply with a speech seven hours in length accusing Lucifer, Caligastia, and Daligastia of disloyalty to the universe government. He appealed to the Most Highs of Norlatiadek on Edentia for confirmation.

But already all System circuits were severed. Celestial personalities on Urantia found themselves unable to communicate beyond the planet. For seven years no reply from Edentia was made to Van's appeal. The celestial personalities on Urantia, including seraphim and cherubim, were compelled to choose between Lucifer's plan and the will of the unseen Father. Forty of the Caligastia one hundred refused to join the revolt, but almost half the seraphim and cherubim joined, as did 40,119 of the primary Midway Creatures, all of which Caligastia organized into divisions to do his bidding.

The loyalists of the corporeal corps took possession of the invaluable tree of life and moved it out of Dalamatia to a poorly protected settlement to the east. Seraphim and cherubim and three Midway Creatures protected the tree constantly and allowed only the loyal forty and their modified human associates to eat of its energy-giving fruit and leaves.

Throughout the seven years of waiting for the Most Highs on Edentia to act, Van governed the humans, angels, and Midway Creatures who remained loyal to Michael's regime. At last when every individual had chosen one path or the other, a Most High of Norlatiadek arrived on Urantia with a group of Melchizedeks to seize authority on Urantia. Although Caligastia and his cohorts remained in formal control, from this time began the probationary period of planetary rehabilitation.

The sixty of the corporeal staff who went into rebellion chose *Nod* as leader. They soon discovered that without the tree of life they were degraded to mortal beings. In order to increase their numbers, Daligastia ordered that they engage in sexual reproduction, knowing that they were doomed to die sooner or later. After the later destruction of Dalamatia, the disloyal sixty migrated to the north and the east, where their descendants were known as the *Nodites*, and their dwelling place in the east as "the land of Nod."

These sixty supermen and superwomen, and their modified human associates, all stranded by rebellion, now mated with humans, giving rise to the many stories in later myths of the gods coming to earth and having intercourse with women. "The Nephilim were on the earth in those days, and also afterward, when the sons of God came in to the daughters of men, and they bore children to them. Those were the mighty men who were of old, men of renown" (Genesis 6). They were hardly "sons of the gods," but so they seemed to the peoples with whom they had intercourse. Deprived of the fruit of the tree of life, eventually they died.

All loyal staff members except for Van were removed to Jerusem when the Melchizedek receivers arrived on Urantia seven years after the revolt began. On Jerusem they were reunited with their Thought Adjusters. The Thought Adjusters of the sixty disloyal rebels remain on Jerusem, and the ultimate fate of the sixty no doubt depends on the final adjudication of the insurrection.

FURTHER CONSEQUENCES

For fifty years Dalamatia was in turbulence as Caligastia, Daligastia, and the many who followed them attempted to reorganize the administration of the planet. Revolution was to replace evolution and advanced cultural practices were to be imposed immediately on the hopelessly primitive humans of Urantia. In fact in the immediate vicinity of Dalamatia there was advancement in culture at this time, but in the outlying regions the policy was catastrophic.

Prematurely taught the doctrines of individual and group liberty, of freedom easily mistaken for license, the semisavage peoples of the hinterland banded together and attacked Dalamatia itself. After an energetic defense, the rebel staff and followers abandoned the city and headed north. The efforts to reconstruct human society failed as the Urantia humans sank back to a level close to what it had been three hundred thousand years before, when the Prince arrived. One hundred sixty two years after the outbreak of rebellion a great tidal wave from the Persian Gulf swept over Dalamatia, obliterating every trace of its existence. At the time it was inhabited by renegade Sangiks who had converted the temple to the unseen Father into a shrine to the great god Nog, lord of fire.

Van had refused to return to Jerusem, but led his followers, loyal Andonites once associated with the Prince's staff who had given their life plasma to the one hundred, into the mountain highlands west of India. From there perhaps they could work to rehabilitate the planet. In the first one thousand years of his leadership Van organized over three hundred fifty teaching groups among the descendants mostly of the loyal Andonites, mixed with Sangiks and Nodites. Van sponsored efforts to improve the physical evolution of humans. The offspring of these Andonites were important in sustaining what little culture survived on Urantia in the two hundred thousand postrebellion years.

Van remained on Urantia until the coming of Adam and Eve around thirty-eight thousand years ago, when at last he returned to Jerusem. There he represents the interests of Urantia, reunited with his Thought Adjuster and awaiting the resumption of his long career toward Paradise. For nearly one hundred fifty thousand years he had lived in a material body, sustained by the fruit of the tree of life and the ministry of the Melchizedek receivers resident on the planet.

CHAPTER 14. THE GARDEN OF EDEN

THE CALIGASTIA BETRAYAL AND THE CONSEQUENT cultural and spiritual degeneration had no effect on the continuing biological evolution of humans. Forty thousand years ago the resident Life Carriers observed that biological evolution on Urantia had reached its apex. On their petition, joined by the Melchizedek receivers, the Most Highs of Edentia agreed to dispatch a Material Daughter and Son as biologic uplifters, a planetary Adam and Eve. Because of the Lucifer rebellion, and the isolation of Jerusem, it was the Most Highs' responsibility to make such decisions about the isolated inhabited worlds. The planet was inspected and the plan approved.

One hundred years later a planetary Adam and Eve arrived on Urantia, to do the best they could in conditions of cultural retrogression and spiritual isolation. On a normal planet the arrival of a Material Son and Daughter heralds the dawn of a new and vigorous age, especially of scientific exploration. But on Urantia the tribes lived in backward conditions with little moral development.

NODITES AND AMADONITES

Ten thousand years after the Caligastia secession the planet had lost all the ground established through the influence of Dalamatia. It was as if a Planetary Prince had never come to Urantia at all. Only among two groups, the *Nodites* and *Amadonites*, did any of the culture of Dalamatia persist. Before the arrival of Adam and Eve, they were the most able of all human groups on the planet. The Nodites were the descendants of the rebel corporeal staff, whose early leader was Nod, once commissioner of industry and trade. The Amadonites were descended from Amadon, one of the hundred altered Andonites associated with the Caligastia one hundred and a follower of the loyal Van (see Figure 7). Van's followers were called Vanites, but they quickly blended with the Amadonites. "Nodite" designates a race as well as a culture; the "Amadonites" were really Andonites who shared a certain culture and religion. A powerful enmity existed between these two groups, which surfaced whenever they were thrown together.

After the destruction of Dalamatia by natural causes, the followers of Nod split into three groups. One remained at the head of the Persian Gulf, near where Dalamatia had stood (they built the city of Dilmun) (see Map 1). A second group migrated northeast to Elam, east of the Euphrates at the foot of the Zagros range that divides the Persian plateau from Mesopotamia. The third group migrated northwest to the Syrian shores of the Mediterranean. The Nodites freely intermixed with the Sangik races and left behind a biologically able progeny.

For a hundred years before the planet was inspected by orders of the Most Highs of Edentia, Van had taught his followers to expect the arrival of a Son of God who would advance the biological fitness of the races and promulgate higher teachings, who would in moral leadership replace the treacherous Caligastia. Most had little interest in such a wild prediction, but those close to Van and Amadon prepared for such a Son's arrival.

PREPARING THE GARDEN

Van remembered the Material Sons from his days on Jerusem. He instructed his followers about them and explained how they always lived in simple structures within a garden. Eighty-three years before Adam and Eve appeared, the Amadonites began work on a garden home. Van gathered three thousand volunteers from the communities scattered throughout their highland home. He divided these volunteers into one hundred companies, each with a leader and a subordinate. Amadon remained Van's own subordinate. One of the companies set forth to find the ideal location. Caligastia and Daligastia still had power, however, and they used it to oppose the preparation of this garden.

The company searched for three years, then returned to report three favorable sites: one an island in the Persian Gulf; the second between the Tigris and Euphrates rivers, where they came close together (later the site of a second garden, as we will see); and third, a long narrow peninsula, nearly an island, that protruded into the Mediterranean Sea from its east coast (now beneath the sea, this peninsula seems to have been near Cyprus, in classical myth the birthplace of the great goddess of fertility). The peninsula was chosen and two years spent in transferring their headquarters from near Lake Van, including the tree of life. The native population agreed to give up the peninsula to the newcomers (Map 1).

The project was of immense ambition. Mountains that ran close along the coast of the peninsula surrounded the garden. Protected by the encircling mountains, and with the waters of the inland sea all around, the climate of the peninsula was healthy and equable. A great river, fed by four tributaries (the "four heads that went out of Eden" Genesis 2.10), ran down the center of the peninsula, then east across a plain to Mesopotamia, draining eventually into the Persian Gulf. It rarely rained within the garden, but every night a mist went up from the many irrigation canals that the Amadonites constructed. The garden was to be a glorification of agriculture and horticulture. Eden is named after the glorious horticulture of Edentia, Constellation capital of Norlatiadek.

The neck of the peninsula was only twenty-seven miles wide and easily defensible. On the peninsula gathered the most advanced peoples of Urantia, separated from the savage tribes. In Eden they would make a shining jewel, a model for growth and progress.

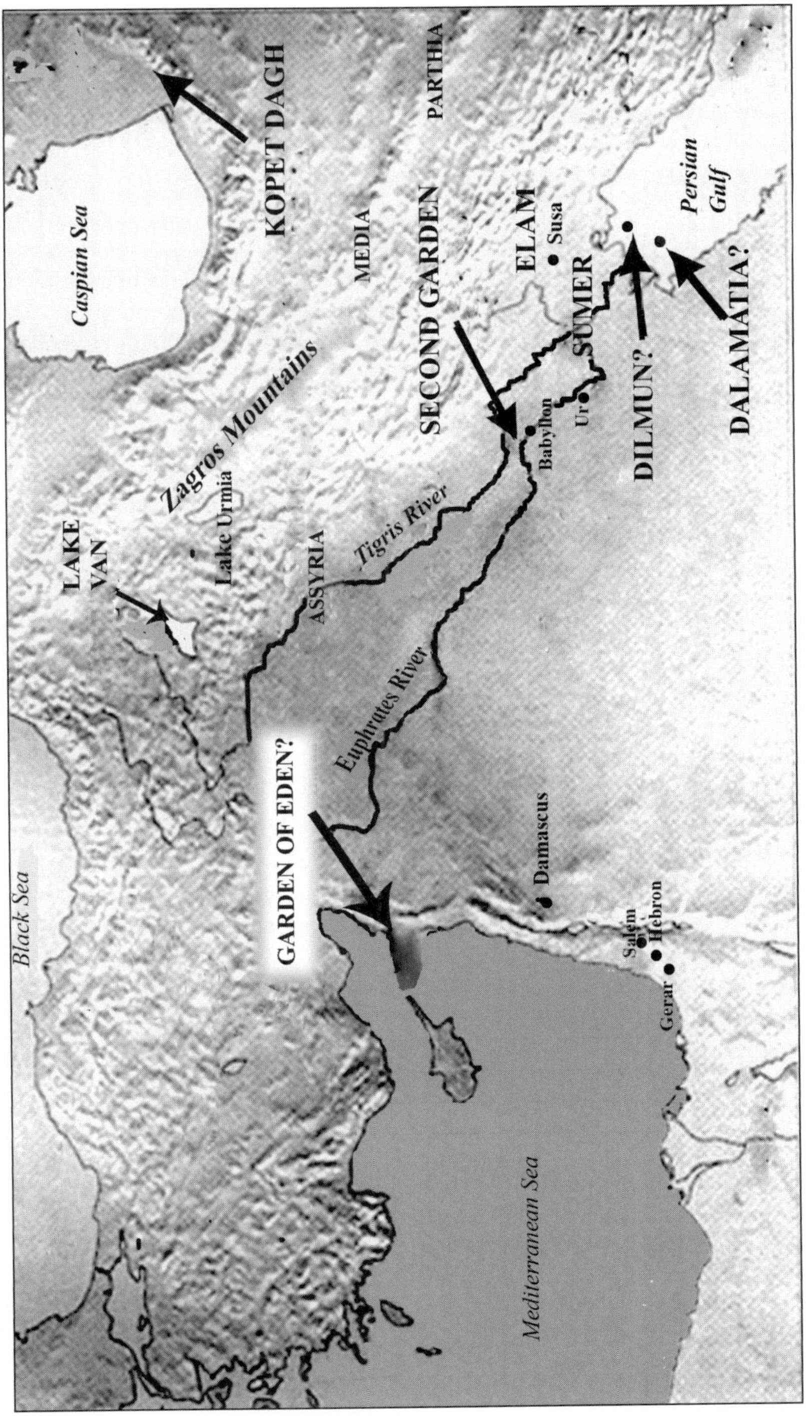

MAP 1. The Near East in the days of the Garden of Eden

Eden was given wholly to the garden. The Amadonites kept animals for husbandry and pasturage on the mainland, under guard, and would import meat from outside, but no animals were ever killed within the garden itself. The first task was to build a brick wall across the peninsula. After that, they turned to the landscaping of the garden and the construction of dwelling places. A second smaller wall was built outside the main defense, divided into twelve sections by twelve raised walkways that led to the twelve gates of the Garden. Within these sections was a zoological garden, and many wild beasts lived there, an additional defense against attack.

Only volunteers worked on the Garden, supported by their herds and their produce, and sometimes assisted by peoples living nearby who believed in the enterprise. Van caused a great disturbance when he suggested that a younger generation be trained to manage the Garden, in case the Material Son and Daughter did not soon come; many believed that he had lost faith and abandoned their work on the Garden, but Van replaced them with younger recruits.

In the center of the Garden stood the simple stone temple to the Universal Father and the other structures were grouped around it, buildings for administration, education, and for the workers to live in. Then "east in Eden" they made a house for the expected Material Son and Daughter and housing for their offspring. All told the workers planned housing for one million inhabitants, with ample land for all.

When Adam and Eve did arrive, the garden was only one fourth complete; yet already in place were twelve thousand miles of paved paths and roads and thousands of miles of irrigation canals. There were more than five thousand buildings made of mud brick, clustered in groups of no more than seven structures. Sanitary arrangements were advanced and no human waste allowed to fall into the water supply, as was otherwise common. At first all waste was scrupulously buried, and later a proper system of sewage disposal installed, a covered brick concourse that ran beneath the walls and emptied into the river of Eden a mile below the walls.

By the time of Adam's arrival good progress had been made in improving cereal plants and fruit trees and other agricultural products, some of which have survived to modern times. Knowledge of others was later lost. Five percent of the garden was under intense cultivation, another fifteen percent partially cultivated, and the rest left in a natural state until Adam's arrival.

THE TREE OF LIFE

Van planted the tree of life, which had kept him alive for so many generations, in the middle of the Father's temple in the garden. Adam and Eve, too, would depend on the fruits of this plant from Edentia. They do not require its fruits while living on the System capital, but their physically reconstituted bodies on another world require the fruit. The biblical "tree of the knowledge of good and evil" is a literary invention to describe an aspect of human life, a moral tale, but the tree of life was quite real, which the Most Highs had delivered, by means of the Melchizedeks, at the time of Caligastia's installation.

Such plants grow not only on Edentia, but on the capitals of the Local Universe and Superuniverses and on the Havona spheres, but not on the System capitals. The tree absorbed certain energies that defied the aging mechanisms in animal life. It was of no use to ordinary humans but essential to the bodies of the Prince's corporeal staff and to the one hundred Andonites who contributed their life plasm, as we have seen. The tree, at first transported from Dalamatia to the Armenian highlands and regrown from its original core, had supported Van and Amadon for one hundred and fifty thousand years.

When Adam and Eve were expelled from the Garden, they were not allowed to take the tree. When the Nodites invaded the Garden, they found the tree unguarded, to their surprise. Believing that by eating the tree's fruit they could be immortal, they ate it for many years, but continued to age and die. They lacked the complement in their bodies that would enable them to receive the energy from the fruit. At last during an internal dissension the temple of the Father was burned and the tree destroyed. Adam and Eve, their children, and their children's children all grew old and died, bound to the mortal ascension career like other mortals of the realm.

Around four thousand years after the death of Adam and Eve, the peninsula of Eden began gradually to sink beneath the waters in connection with nearby volcanic activity and the submergence of the Sicilian land bridge to Africa. Its destruction was a wholly natural event, the sad end of a glorious chapter in the history of Urantia.

ARRIVAL OF ADAM AND EVE

The Material Son and Daughter arrived on this planet in the year 35914 BC, in the height of the season. At noon, unannounced, the two seraphic transports settled to the ground near the stone temple to the Universal Father. Accompanying agents from Jerusem rematerialized the bodies of the couple within the temple in a process that lasted ten days. They awoke simultaneously. Each was over eight feet in height.

The Material Sons and Daughters always work as a pair, never alone. For fifteen thousand years before being chosen to come to Urantia, this Adam and Eve had worked in a laboratory that experimented with the application of certain kinds of energy to evolving life forms. Earlier, they taught in the schools for new arriving mortal ascenders on Jerusem.

After taking oaths to Michael, to whom they had remained loyal, and to the Most Highs, this Material Son and Daughter were instructed in the difficult circumstances of life on Urantia. They were to be under the authority of the Melchizedek receivers then in charge of Urantia, although in practice they would be the virtual rulers of planet 606, Satania system. As they departed for Urantia, their one hundred children born on Jerusem, fifty male and fifty female, accompanied their parents to the dematerialization chamber to bid them farewell.

When after seraphic transport Adam and Eve awoke in the Father's temple in Eden, a huge crowd, and the now legendary Van and Amadon, were there to greet them. The story of Eve being made from Adam's rib comes from a confused conflation of the seraphic arrival of the couple around 36000 BC and the surgical implantation of Andonite plasm into the bodies of the Caligastia one hundred, which had taken place more than four hundred sixty thousand years before but was never forgotten.

The speech of Eden was an Andonic dialect. Van had devised an advanced alphabetic writing for this speech, which he hoped would spread over the planet and become universal. This was the second system of writing on Urantia after that introduced under Caligastia, lost at the time of the Lucifer rebellion around two hundred thousand years ago. Adam and Eve learned the Andonic dialect while on Jerusem and were therefore able to speak immediately to the crowd in their own tongue.

Carrier pigeons were released to outlying communities to announce the arrival of the couple, which made plausible teachings about the Universal Father that Van and Amadon had long promulgated. Thousands of pilgrims arrived in Eden to behold the rematerialized Son and Daughter. After they awoke, they were escorted to a high mound behind the temple prepared for this purpose. There a local committee received them made up of representatives from the Sangik races, a female representative of the Nodites, the acting chief of the midwayers, the son of the architect of the Garden (who had died), and the two resident Life Carriers.

The senior Melchizedek administered an oath of loyalty to Michael of Nebadon and to the Most Highs of Norlatiadek. Van then gave up his titular rulership of the planet that he had held for over one hundred and fifty thousand years by authority of the Melchizedek receivers. Weaving, introduced in the days of Dalamatia, was still practiced and Adam and Eve donned regal robes. At this moment was heard the voice of Gabriel of the universe of Nebadon, the Bright and Morning Star, who proclaimed the second dispensational resurrection of sleeping survivors (the first was when Caligastia arrived, around five hundred thousand years ago). The third epoch in the history of humans on Urantia had begun.

ADAM AND EVE INSPECT THEIR SURROUNDINGS

Straightaway Adam and Eve understood their isolation. Their companions on Jerusem had gone to worlds where the Planetary Prince and his staff were well established and in a position to assist the biologic uplifters. On Urantia the universe broadcasts were silent and the Planetary Prince, although shorn of most power, was still in a position to oppose their mission. They were stranded in a distant and lonely place. Deep into that night they discussed the realities of their situation. On the second day they met with the Melchizedeks, who instructed them in greater detail about the Caligastia secession and its consequences. Caligastia's ill-conceived plans for accelerating social evolution, departing from the established plan for evolution, had collapsed.

On the third day they mounted on the backs of the giant birds called fandors and from the air inspected the vast reaches of the Garden, the most beautiful place on earth. The day ended with a banquet to acknowledge the efforts of all who had made the Garden possible. Again, Adam and Eve discussed late into the night the immensity of the problems facing them.

On the fourth day the couple spoke to the people about plans to regenerate the world from its rude and undeveloped condition. It was astonishing to behold a female acting with the same authority as a male, unknown since the days of Dalamatia. Women, too, sat on the councils now formed to administer the planet.

On the fifth day they organized a temporary government designed to function until the Melchizedek receivers would leave the planet.

On the sixth day they inspected the kinds of animals that live on Urantia, walking along the wall in the east of Eden overseeing the zoological park. To the surprise of his guides, Adam immediately recognized and easily named the nature and function of the thousands of different animals; for on Jerusem he was a leading expert on animal anatomy, and Eve was similarly trained. With his dual nature Adam could also discern myriads of living things invisible to his guides, and he astonished them with his descriptions.

On the seventh day they planned to rest. Adam's speech to the assembly and his deep understanding of animals had already convinced many followers that these beings were gods. Even on the night of the sixth day a plan was devised to bring them before the Father's temple and to fall to the ground in adoration. Van protested, but the crowd wished to hail him as a god too, who seemed immune from death. Van sent a midwayer, with whom he could communicate, to warn Adam and Eve of this unhappy plan.

Even as on the morning of the seventh day fandors winged their way to fetch the couple to be presented as gods, the midwayers transported them to the temple of the Father. In an impassioned speech Adam explained how the Material Sons and Daughters were not gods. Only the Universal Father and those he appointed were appropriate objects of worship. If they loved him, they must go to the Father's temple and bow down in worship there. From this act came the tradition in Eden of setting aside the seventh day as a time of worship and for the cultivation of physical, mental, and social skills, the origin of the seven-day week and the legend that the world was made in six days.

LIFE IN THE GARDEN

In the seventh year the twelve Melchizedek receivers turned over the full administration of the planet to Adam and departed for Jerusem by seraphic transport. Van and Amadon, too, left the planet in their company.

For a while all went well. Adam fostered the arts of manufacture in order to establish trade with outlying tribes and strong communication with them. But few groups outside the Garden itself were ready for social and cultural advance. Adam hoped to establish a world government, and within Eden set up a federation of representatives from the different human groups, but outside the Garden he made no progress. Caligastia and Daligastia, invisible to human eyes, contrived in various ways to frustrate the Adamic plan and to strengthen those who advocated unbridled personal liberty. Some of Caligastia's followers lived within the Garden, where they caused constant disturbance.

Adam had hoped to replace monarchical government with representative government, but there was no government worth the name anywhere on the planet. He abandoned this plan. By the time of their expulsion from the Garden he had established one hundred outposts outside the Garden where his own lieutenants ran affairs. The practice of sending ambassadors from one tribe to the next began at this time.

The compound of Adam and Eve within the Garden encompassed five square miles. Beyond its bounds was planned housing for three hundred thousand of their offspring, but the Edenic adventure was aborted before even one third of this housing was complete. Their first child on Urantia was called *Adamson*, then a girl was born, then a second son called Eveson. Eve was the mother of five children before the departure of the Melchizedeks, and bore sixty-three children, thirty-two daughters and thirty-one sons before the default. When expelled from the Garden, their family consisted of four generations and 1,647 pure-line descendants (including Adamson). Eve bore an additional forty-two children after leaving the Garden, and two children fathered by a human.

Cooking was practiced universally in the Garden outside the rulers' compound, but within the compound Adam and Eve and their children ate only fruit, nuts, and cereals. They ate once a day, soon after noon. Adam and Eve also received vital energy from certain space emanations in conjunction with their eating fruit from the tree of life. Their bodies gave off a shimmer of light, which they covered with the same clothing as worn by others. From their heads, however, came a glow, the ultimate origin of the halo common in artistic representations to represent individuals of high spiritual attainment.

Adam and Eve could communicate with one another and their children by thought transference over a distance of about fifty miles through the agencies of delicate gas chambers near their brains. Their children attended school until sixteen, the older students working with the younger. Time every day was set aside for play, activity for the sheer joy of it. Earlier humans had no sense of humor, but the children of the Garden had a fine sense of humor and a developed appreciation for music, qualities passed on to modern times.

In the schools the children studied such practical topics as horticulture and agriculture, but also received moral instruction in the golden rule and the relationship of the individual to the group. They engaged in competitive play as a substitute for war and learned how to keep their bodies healthy. They studied means for advancing trade and learned the history of the Urantia races and of the planet.

Unarmed visitors were welcome at any time, but had to be "adopted" by the Garden and to acknowledge the rulership of Adam and Eve and the supernal authority of the Unseen Father. At age eighteen the children of Adam and Eve were betrothed. They married at age twenty after two years of instruction, then began their life's work or specialized training for it. Traditions of brother-sister marriage in a royal family go back to Eden.

Adam opposed the use of fixed prayers in worship and of animal sacrifice, but made little headway on either topic. He strongly promulgated equality of the sexes, and the people saw how Eve was always at his side. He declared the truth about human biology and the contribution of the female to the creation of the fetus; the female's role had been viewed as solely a nurturing one. But Adam made little progress in advancing the culture or wellbeing of the Urantians.

Distorted memories of these far away events gave rise, around 500 BC, to the composition of the biblical account of creation, which combines misinformation with misunderstanding and falsely attributes the document to Moses. The Hebrews did not adopt a Local System of writing until around 900 BC, long after the death of Moses. Moses had claimed, however, in a simplified account to his rude followers that Adam was the first man (although Cain, to find a wife, went "to the land of Nod"). Such confusions led to the myth of the Golden Age, the widespread notion that once everything was utopian and has since steadily declined. In fact there never was a Golden Age, but there really was an Eden, the Garden home of Adam and Eve.

CHAPTER 15. THE DEFAULT OF THE MATERIAL SON AND DAUGHTER

After one hundred years Adam could see little progress outside of the Garden itself. The realization of race improvement seemed an impossible dream. The situation was desperate. The Material Son and Daughter were loyal, but isolated and without advice. Only indirectly, through the angelic orders, was it possible to communicate with any being off the planet.

Ordinarily the first task of the Material Son and Daughter was to blend the races, but on Urantia all the races held regressive and degenerate strains. It was pointless to proclaim the brotherhood of man to barbarian cannibals. There were hundreds of languages and religious behavior was primitive in the extreme. Hopes for religious unity were preposterous. As the spirits of Adam and Eve drooped, their courage weakened. They were noble beings, but impatient, especially Eve. Because they could not wait, they brought catastrophe on themselves and on the planet.

BREAKING WITH THE PLAN

Frequent visitors to the Garden, Caligastia and Daligastia often spoke with Adam and Eve. They presented various of their own plans for planetary improvement along the lines of policies that had failed disastrously. Caligastia realized the weakness in Eve, mother of the violet race. She had become close friends with one Serapatatia, a leader of the Nodites and descendant of Caligastia's corporeal staff. His line had always held authority among the western Nodites. He was brilliant, brown-tinted, and charming. They spent a lot of time in private conversation.

Serapatatia was a fervent believer in Adam's mission and had brought over nearly in a body the powerful Syrian tribe of the western Nodites, who pledged to help in the cause. He and his subordinates established direct liaison with Eden, and Adam and Eve entertained them in their own home. Serapatatia soon became one of Adam's most able and trusted lieutenants, scrupulously honest and reliable in every way. Serapatatia never understood how he was becoming a tool of Caligastia's plan.

The Edenic committee on tribal relations, of which Serapatatia was an associate chair, planned to pursue a more aggressive policy to persuade the outlying tribes to follow the cause of the Garden. Serapatatia was long in conference with Adam and

Eve, and in one conversation with Eve he wondered if a good tactic would be not to wait until Eve had born thousands of offspring to establish the violet race, which on some faraway day might emigrate outward into the tribes. If they could right now produce a great leader of the Nodites, a blend of the violet and Nodite strains, then they could straightaway establish permanent ties of blood between the Nodites and the Garden. This special child, reared in the Garden, would bring great advantage to his father's people, the Nodites.

For more than five years they pondered the plan. A secret meeting was arranged with a Nodite named Cano, the most brilliant of the Nodites living nearby, and also an enthusiastic supporter of Adam's mission. Eve had never before met the handsome Cano. She was flattered by him, charmed, and persuaded. The deed was soon done.

Adam immediately knew that something was wrong and called Eve aside. She explained to him the plan for world improvement that she and Serapatatia had devised. While they spoke, they heard "the voice in the garden," really a seraphim named Solonia (author of several papers in *The Urantia Book*). They had disobeyed the instructions of the Melchizedeks and broken their oath to the supreme sovereign. The archangel custodian of the tree of life had warned them, every time they ate from it, "In the day that you commingle good and evil, you shall become as the mortals of the realm. You shall surely die." While sin is the deliberate conscious transgression of divine law, evil is the distorting of divine plans such that confusion and disharmony result. Eve had told Cano of the archangel's admonition, but he assured her that for those whose intentions are pure, no evil is possible. Eve had also admired Cano's good looks.

In despair at what had happened, Adam on the next day called out one Laotta, a prominent Nodite woman in charge of some of the Edenic schools. To join his wife in folly, he slept with her. Thus he and Eve could not be separated but would share a similar fate, united in the love that had always joined them.

The news spread swiftly through the camp. The enraged inhabitants of the Garden poured out through the gates and descended on the unsuspecting nearby Nodite settlement. They killed every man, woman, and child, including Cano, whose unborn child, *Cain*, lay in Eve's womb. The remorseful Serapatatia drowned himself in the great river of Eden.

Adam wandered in solitude for thirty days. Eve was nearly beside herself with grief and fear; she never fully recovered from the mental anguish of those thirty days, against which all later suffering seemed inconsequential. Had Adam followed Serapatatia and destroyed himself, or been removed from the planet because of her act? The innocent children of the couple were swept up in the tragedy.

Seventy days after Eve's intercourse with Cano, the twelve Melchizedek receivers returned to the planet to resume control of world affairs. In the meanwhile, news of the massacre at the Nodite settlement had reached Serapatatia's peoples in the north, who now prepared to invade the Garden. So began a long and bitter war between the Adamites and the Nodites that went on long after Adam and Eve had migrated to the Second Garden in the valley of the Euphrates.

Adam sought advice from the Melchizedeks, but they refused to interfere in his affairs. He held an all-night conference with twelve hundred of his loyal followers, who agreed to follow him out of Eden no matter where he should go. On the next day they departed. Adam did not wish to war against the Nodites who were descending on the Garden.

DEPARTURE FROM THE GARDEN

On the third day out from the Garden, seraphic transports appeared and, while the parents waited, the children of Adam above age twenty were allowed a choice: to remain on Urantia with their parents or travel by seraphic transport to Edentia, where they would be under the protection of the Most Highs. Two thirds elected to go, where they are today. The departure of their children struck deep sorrow into the hearts of the Material Son and Daughter.

While they were halted, Gabriel himself appeared to inform Adam and Eve officially of their default. They had now lost their immortal status and must live out their lives as humans of the realm. Guilty not of rebellion, but of poor judgment, they must join the mortal races in the ascension adventure. They had been dual-circuit beings, but the second circuit, fed from the tree of life, had now ceased to function. Like all humans, they would grow old and one day die.

There never was a "fall of man," a myth built on the Adamic default. The descendants of Adam and Eve did a great deal to upgrade the native human stock. We very much feel the effects of their contribution today. Although under strict command not prematurely to admix the superior strains of the violet race with the native stock and the descendants of the Prince's staff, she hoped through her ill-considered plan to improve the seemingly hopeless conditions of life on Urantia. Such mistakes are inevitable in an imperfect universe, where all strive for perfection, where the governing force is personality and not a blind, mechanical energy.

THE SECOND GARDEN

Only the way to the east lay open: to the west was open water, to the north were the Nodites assembling to attack, and the highlands to the south were infested with hostile tribes. Thus they journeyed to the then pleasant land between the two rivers in Mesopotamia.

Cano's child, called Cain, and Sansa, the daughter of the Nodite Laotta, were both born before they reached the land between the rivers. Laotta died in childbirth and Eve nearly so. Eve reared both children. Sansa grew up to be a woman of high ability, who married the chief of the northern blue men, one Sargan.

It took a full year for the caravan to reach the Euphrates, then in flood. They camped for six weeks near the bank until the river subsided, then crossed into what would become the Second Garden. Adam knew about this site because it was one of the three that Van and Amadon had chosen as possible locations for the original Garden.

The local inhabitants fled at their approach. They camped south of where the two rivers came closest together. By building a wall fifty-six miles long, they could protect themselves from attack from the north, while the rivers themselves provided protection from the east and west. The first Garden was partially prepared when they arrived. Now they had to wrest a living from the raw ground "in the sweat of their brows."

The first child born to Adam and Eve after they left the first Garden was called *Abel* (see Figure 8). When twelve years old, Abel chose to be a herder. Cain preferred agriculture. The boys hated each other from the beginning and their enmity grew worse. Cain was in any event the object of popular (although unexpressed) resentment, as a symbol of what had caused the default. Adam was so preoccupied with defense, construction, and economy that he turned over the religious and educational duties to the Nodite priests who had lived in the first Garden, who already slid back into practices from the days before the Garden. They accepted blood sacrifice from petitioners. Abel offered the preferred blood sacrifice (which Adam had earlier forbidden), and Cain the fruits of the field.

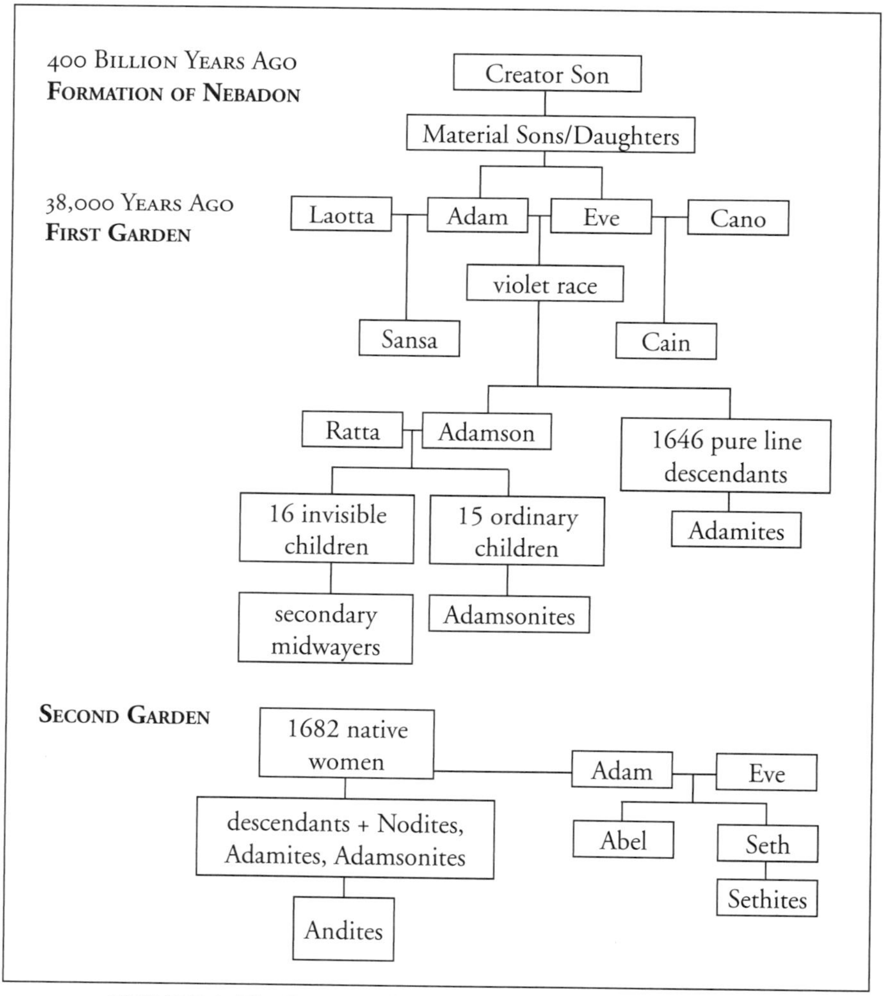

FIGURE 8. The Descent of Adam and Eve and Their Children

CAIN AND ABEL

One day when Abel viciously taunted Cain, reminding him that his father was not Adam but a Nodite of mixed descent from the Andonite, blue, and red races, Cain turned on him and killed him. Cain had not been indwelt by a Thought Adjuster, but when in sorrow he sought advice from his mother, an Adjuster indwelt him.

The murderer Cain left the community and traveled east to the land of Nod, where he fulfilled the plans of his mother and father by bringing together these Nodites and the Adamites: The blood of both groups ran in his veins. He married a distant cousin and their son Enoch became leader of the Nodites of Elam (southeast of Mesopotamia, see Map 1), who for a long time maintained friendly relations with the Adamites in Mesopotamia.

The civil rulers of the Adamites traced their lineage to the children of the first Garden. Adamson, the first born, traveled north and established an independent community of the violet race. Eveson, the second child, became a great leader in the Second Garden, and his son took over the Adamites at the death of Adam.

A new Adamite priesthood originated with *Seth*, the oldest surviving son of Adam and Eve of the children born in the Second Garden. Seth was born 129 years after Adam's arrival on Urantia. The Sethite priesthood was concerned with education, health, and worship. They served at religious rituals, were physicians and sanitation inspectors, and taught school.

The Adamites had brought seeds and bulbs from the first Garden, and herds of domesticated animals, resources that gave them an enormous advantage over all outlying peoples. Although the eating of flesh soon became common in the second Garden, Adam and Eve themselves, and Adamson and Eveson, were never flesh eaters.

THE ADAMITES

The Adamites were by far the most advanced group on the planet. Adam introduced a new system of writing, the third on the planet (after the systems used in Dalamatia and in the first Garden). They practiced metalworking, pottery manufacture, weaving, and a kind of architecture superior to anything found for many thousands of years thereafter.

Home life among the violet race was ideal. Children were trained in agriculture, animal husbandry, and the crafts; or if they chose to follow the Sethite priesthood, they learned medicine and religious matters and would teach school. The Sethite priesthood was a superior organization whose teachings about divinity and the universe were roughly accurate. Their methods of education have never been surpassed.

Adam and Eve were blue-eyed and fair-haired and so were their children. Neither Eve nor the earlier evolutionary races experienced pain in childbirth; only when mixed with the Nodites, then the Adamites did women experience pain in childbirth, as did Eve herself at the birth of Cain. Adam and Eve had had dual energy systems, but their children were born as children of the realm, with a single system. At first they lived a very long life, but their lifespan dropped closer toward the human norm with each generation. The special senses of Adam and Eve enabled them to see the midwayers, the angelic hosts, the Melchizedeks, and Caligastia, but in their descendants these special senses diminished more and more.

Most of the Adamic children were of survivor potential and were Adjuster indwelt. The cells of their bodies were disease-resistant, and so would ours be, had we received more of the blood of Adam. Adam designed, nonetheless, to leave as much of his life plasm as possible on Urantia, and a commission over which Eve presided selected 1682 of the highest type of native women, all of whom Adam impregnated. One hundred and twelve did not reach maturity, but the surviving 1,570 offspring formed the beginnings of the mighty so-called *Andite* race, a mixture of Adamic, Nodite, and Sangik blood. Although the mothers were chosen from all the evolutionary races, the majority came from the Nodites. Such children were reared in the mothers' tribal surroundings.

Adam lived for five hundred thirty years, then died of old age. After his reduction to mortal status, and his sincere regret, he was Adjuster indwelt, as was Eve. Hence they became candidates for the ascension career. Eve died nineteen years earlier than Adam, from a weakened heart. They were buried in the floor of the temple of divine service, the origin of burying important people in places of worship.

On the third day after Adam's death, pursuant to the orders of Lanaforge (the System Sovereign who replaced Lucifer), sustained by the Most Highs of Edentia and a Trinity personality on Salvington, a special roll call was held of the distinguished survivors of the Adamic default on Urantia. In accordance with this mandate of special resurrection, Adam and Eve were repersonalized and reassembled in the resurrection halls of the mansion worlds of Satania along with 1,316 of their associates in the first Garden.

Adam and Eve quickly passed through the mansion-world school and returned to Jerusem again as citizens, but this time as a different order of being: no longer Sons of God, now the sons of man. They serve today on the advisory committee of twenty-four who keep watch on the Urantia situation.

CHAPTER 16. NODITES, ADAMSONITES, ANDITES

It was not possible to maintain the high civilization of the second Garden because of the early dilution and eventual submergence of the violet race. People make a civilization, a civilization does not make a people. For four hundred and fifty thousand years there had been a direct, physical connection between the universe administration and the evolutionary races, first through the Prince and his staff, then through the loyal followers Van and Amadon, then Adam and Eve. Now that was over. With the death of the Material Sons all physical connection between the universe administration and the evolutionary races came to an end. Not until the arrival of Machiventa Melchizedek about 2000 BC in the days of Abraham was any plan put in place to further the welfare of the wretched inhabitants of planet 606, Satania system.

THE NODITES

The advent of a divine Son on an evolutionary world always has curious, not always explicable effects on the relationship between the material and spiritual realms. The production of Midway Creatures in two phases, once under Caligastia and once again under Adam, was just such an effect. We have seen how the primary midwayers came into being through supermaterial, nonsexual contact between male and female members of the Prince's corporeal staff, once morontial ascendant beings resident on Jerusem without reproductive prerogatives; on Urantia, the corporeal staff were descendant sex creatures with material bodies, capable of sexual reproduction. Each couple was capable of creating 1,000 midwayers in this fashion, but for reasons completely unknown, no more than that number. Eventually a full corps of 50,000 was reached, who served as an intelligence service, ranging far and wide, bringing in information and acting to further the Prince's plans among remote peoples. Four fifths of these midwayers joined the rebellion, but the remaining one fifth went with Van, with whom they associated until the arrival of Adam. We will have more to say about the secondary midwayers in a moment.

The Prince's corporeal staff, morontia ascenders from Jerusem, were constituted sex creatures on this planet in order to follow out the Life Carriers' plan to admix the offspring of the staff and the evolutionary peoples with the first-generation offspring, one distant day, of Adam and Eve, the violet race. The Life Carriers hoped that descendants of these unions would become the teacher-rulers of the evolutionary races.

The offspring of the sixty rebels on the Prince's staff, when they interbred with the evolutionary races, were children of exceptional physical and intellectual ability, and with deep spiritual power too. Mutation of the DNA had taken place in the germ plasm of the materialized staff, who furthermore received powerful life-maintenance energies from the System, rather as Urantia scientists today use X-rays to cause mutations in laboratory experiment. The whole situation was extraordinary, but the changes that brought forth the superior race of Nodites had not been expected, was not part of any plan. The situation was further complicated by sexual relations between the Nodites and the forty-four members of the Andonic staff loyal to Caligastia, who in return for offering their germ plasm were enabled to absorb energy from fruit of the tree of life. Their offspring in turn blended with the Nodites. In this way the "Nodites" became in effect the *eighth human race* to appear on Urantia, an unforeseen development.

The Nodites were a magnificent people, but they interbred with the evolutionary races and gradually merged with them. After ten thousand years their offspring lived no longer than evolutionary humans. On Sumerian tablets appear the names of kings going back several thousand years. The further back you go, the longer the lives of the kings, in some cases reflecting accurate information (and in other cases, wildly inaccurate).

THE TOWER OF BABEL

After the submergence of Dalamatia around two hundred thousand years ago, followers of Nod migrated northeast where they founded the new city of *Dilmun* near the head of the Persian Gulf, mentioned on surviving cuneiform tablets as the paradisiacal land from where the Sumerians (descended from Nodites) came (but the Sumerians were thinking of Dalamatia as much as the real city of Dilmun). In Dilmun the Nodites reproduced rapidly and began to intermarry with the surrounding Sangik and Andonite peoples. About fifty thousand years after the death of Nod, about one hundred fifty thousand years ago, a council was called to address the issue of the races' diffusion among the evolutionary peoples. After much discussion the plan of one Bablot, a direct descendant of Nod, was adopted.

Bablot proposed that they build an enormous tower, unlike anything ever seen, as a symbol of Nodite greatness, but outside Dilmun in case it too were one day engulfed by the sea, as had been Dalamatia. Around the tower they would construct many other buildings to serve as a center of Nodite culture and greatness.

The plan was adopted and construction begun, but after four and a half years of labor controversy intensified about how the new construction was to be used, and even what was the purpose of so much work. Actual war broke out between two factions, who fought one another until most were dead. After this incident this region was long known as the "land of Babel" (really named after the man Bablot but explained in

Semitic as meaning "gate of God"). About twelve thousand years ago, the *Andites*, the mixed-raced descendants of Nodites and Adamites, attempted to rebuild the tower on the ancient ruins, but nothing came of the project.

The war over the tower of Babel much reduced the number of Nodites, who now dispersed from the ruins of Dilmun in several directions. For one hundred and twenty thousand years they declined until the days of Adam, with whom they were closely associated. Even in those distant days the Nodites were an advanced and able people.

DISPERSAL OF THE NODITES

Some Nodites went north up the Tigris River, where they blended with Andonite peoples to become the ancestors of the Assyrians. Some went east to Elam in southwest Iran, "the land of Nod," where other Nodites had earlier emigrated. They intermingled with them and with the mixed Sangik peoples.

A third group, the smallest, settled at the mouth of the Tigris and Euphrates rivers, where they later blended with the Adamites of the second Garden, who lived north of them between the two rivers. Descendants of this group would much later appear as the historical Sumerians, who come as if from nowhere about 3400 BC with a full-grown superior culture embracing temples, metalwork, agriculture, animal husbandry, pottery, weaving, commercial law, civil codes, religious ceremonial, and a system of writing. To modern scholars the Sumerians are the "first and oldest civilization."

A fourth group of Nodites were really the Vanites or Amadonites, the followers of Van and Amadon after the time of the planetary rebellion, who had lived in Armenia north of Mesopotamia long before the Bablot conflict. Some of Van's early followers settled around the lake still called Lake Van. Nodites had traveled there and interbred with them. For the Vanites, Mount Ararat was sacred, much as was Mount Sinai to the Jews: There Van, they said, had received the seven commandments of proper behavior from the gods and, while worshiping, was carried away to heaven (in fact Van was taken to Jerusem by seraphic transport when Adam and Eve arrived). Ararat is mentioned in the biblical story of Noah and the Flood.

ORIGIN OF THE SECONDARY MIDWAYERS

The secondary midwayers partly descend from the Nodites and partly from the Adamites. Adamson, the first-born son of Adam and Eve, chose to stay on Urantia with his parents. He remembered the stories that Van and Amadon told of their highland home to the north of Mesopotamia, and when he matured he was determined to go there. Adamson was 120 years old at the time and had fathered thirty-two pure-line children, but his wife and all his children had accepted seraphic transport to Edentia, Constellation capital of Norlatiadek, where they became wards of the Most Highs. Adamson wished to help his parents in the second Garden, but was unhappy there, and at last bid goodbye and journeyed north with twenty-seven followers. In three years they found the settlements described by Van and Amadon east of the Caspian Sea, near the ridge today called the Kopet Dagh that separates Iran from Turkmenistan (see Map 1).

Adamson married a woman named Ratta, last of the pure line descendants of the rebels of the Prince's staff. They produced sixty-seven children, ancestors of important leaders in later times. Inexplicably, however, and without precedent, every fourth child of the couple would sometimes become invisible. In fact both Adamson and Ratta were superhuman. Knowing about the primary midwayers, the astonished Adamson and Ratta wondered if something similar were happening.

Sixteen of the peculiar invisible children were born in this way, equally divided between male and female, and when they came to maturity and paired off, each couple was able to produce a secondary midwayer every seventy days by a combined technique of sexual and nonsexual liaison. Thus did each couple produce 248 offspring, engendering the original core of 1,984 secondary midwayers, invisible to ordinary human eyes. Such an event had not been possible on earth before that time, nor has it occurred since. The sixteen strange invisible children lived and died as mortals of the realm, but their electrically energized offspring live on and some are on the planet today. They, not the short-lived humans, are the permanent residents of Urantia.

ADAMSONITES AND ANDITES

Adamson and Ratta stayed in touch with the second Garden through the secondary midwayers, and every seven years they traveled there themselves for a visit. Their home in the Kopet Dagh became a center for the teaching of righteousness and truth. The corps of midwayers assisted immeasurably in the mission, whose benefits have never altogether disappeared from the world. Adamson lived for 396 years.

His descendants were the *Adamsonites*. They maintained a vigorous and advanced culture in the lowland valleys of the Kopet range, where archaeological remnants of the headquarters of the violet race can be found today. There arose four successive cultures. Some belonging to the second group migrated west to Greece and the Balkan Peninsula. Others migrated north and west into Europe, while still other remnants of the Adamsonite stock, combined with the Nodites and some Sangiks to create the Andites, later entered northwest India, remembered as the "Aryan invasion." The Andites appear as a distinct group around fifteen thousand years ago. The Andites were the protoIndoEuropeans of modern European linguistics, their speech being a blend of the Andonic dialect of Turkestan with the speech of the Adamsonites (all modern European languages with small exceptions belong to the IndoEuropean family).

LOYAL AND REBELLIOUS MIDWAYERS

The loyal primary midwayers had worked with Adam and Eve in the first Garden, but after the default submitted to the authority of the Melchizedek receivers. The secondary midwayers worked with Adamson until his death, when a group of their leaders went over to the Melchizedek receivers while the remainder joined rebel primary midwayers to become a strange, disorganized, and mischievous lot without organization or control until the time of Machiventa Melchizedek around 2000 BC. These midwayers were responsible for many unusual goings-on up until the present planetary dispensation that began with the incarnation of Michael of Nebadon in 7 BC. Such troublesome midwayers, who could reveal themselves on occasion to mortal eyes, were usually designated as "unclean spirits" or "demons" by those who had dealings with them. Possession by an outside personality is not possible after the dispensation of a Paradise Bestowal Son and the pouring out of the Spirit of Truth, but before Michael lived as the mortal Jesus of Nazareth, some midwayers actually did take possession of the weak or feeble-minded and they directed their actions.

Of the original 50,000 primary midwayers, 40,119 joined the rebellion. Of the original number of 1,984 secondary midwayers, 873 refused to align themselves with Michael during his sojourn here. Both groups of rebellious midwayers were interned at the time of the adjudication of the planet at Pentecost in 30 AD. They have now been removed from Urantia and await an uncertain fate.

The 10,992 loyal midwayers, primary and secondary, have since Pentecost joined together as the *United Midwayers of Urantia.* They work for the spiritual and semispiritual wellbeing of the planet and are indispensable to the work of the seraphim who serve as personal guardians. They serve as messengers and enable contact between spiritual personalities and material ones (for example they were important in the very communications received as *The Urantia Book*).

The midwayers are nonmaterial beings and do not sleep, but they share many personality traits with humans, including an appreciation for humor and play. They do not procreate beyond their original numbers. The primary midwayers are closer to the angels than to material realities, but the secondary midwayers are just outside the range of mortal vision and are able to make contact with the material world. For example a midway creature, not an angel as reported, freed Peter from prison after the execution of James the brother of Jesus (Acts 12). Their chief work today is as unperceived agents of liaison with the human reserve corps of destiny.

The Midway Creatures might depart from the planet by traversing the energy circuits, but only one midwayer has ever done this, the leader of the primary midwayers, who now serves on Jerusem in the council of twenty-four advisors on Urantia affairs. Midwayers have a culture of their own, but unlike humans they forget nothing; they are the repository of the truth about what has happened on this planet. Midwayers provided the information about the life of Jesus, which makes up fully one fourth of *The Urantia Book.* The rebellion-tested corps of the United Midwayers has pledged to remain on Urantia until in the distant future it is settled in light and life.

OVERVIEW OF RACIAL DISTRIBUTION 35,000 YEARS AGO

For thirty thousand years the Adamites lived in Mesopotamia, the cradle of modern civilization, which first appears clearly in the archaeological record about 5,000 years ago with the invention of the fourth form of writing on Urantia, later called cuneiform; traces of the earlier three systems of writing (in Dalamatia, in the First Eden, in the Second Eden) have disappeared completely. As the Adamites journeyed out from Mesopotamia to the farthest reaches of the earth they mixed with Nodites and Sangiks. Adam's contribution to the biologic status of the evolutionary races, in spite of the partial failure of the plan, therefore enormously improved the peoples of Urantia. Adam and Eve also contributed to the social, intellectual, and moral advancement of those who lived in and near the second Garden, although in the planet as a whole little culture of any kind existed. The human race continued to languish in dismal savagery—fruits of the tree of Caligastia and his friends.

Around 35,000 years ago the general situation was as follows:

The Adamites lived in the second Garden in the triangle formed by the Tigris and Euphrates rivers, near where Babylon later rose, and modern Baghdad. A second center of the violet race was east of the southern shore of the Caspian Sea near the Kopet Dagh mountains, where Adamson had gone.

Also in Mesopotamia, at the mouths of the rivers, remnants of the ancient culture of the Dalamatians lived on, Nodites who became thoroughly blended with the Adamites living to the north. These were the preSumerians. In preDravidian India was the most complex mix of races on Urantia, every race being represented but especially the green, orange, and indigo.

On the Saharas, then green and fertile, lived the most progressive of the indigo men. Deep strains of the green and orange men also lived in the DNA of these peoples.

Around the Mediterranean lived the most highly blended of races, except for those in India. Here came together blue men from the north, indigo from the south, and Nodites and Adamites from the east.

Such was the situation before the violet race began to infiltrate outward. Truly, in Mesopotamia lay the best hope for the future in the midst of what seemed hopelessness and endless chaos.

The aboriginal Andonites, descendants of Andon and Fonta, lived in several conclaves north and east of the headquarters of Adamson and were otherwise dispersed in Turkestan and in the highlands of northern Eurasia, and in Greenland and Iceland; but the blue man had driven them out of the plains of Europe and the yellow man had driven them out of the lowland valleys of China (Map 2).

The red man occupied the Americas, driven out of Asia by the yellow man about 85,000 years ago.

The yellow man was in secure possession of eastern Asia. The main centers of power were to the northwest of modern China near Tibet.

The blue man held Europe. He was the most aggressive, adventurous, and exploratory of the evolutionary races in Eurasia, but had slowed down a good deal after interbreeding with the Neanderthals.

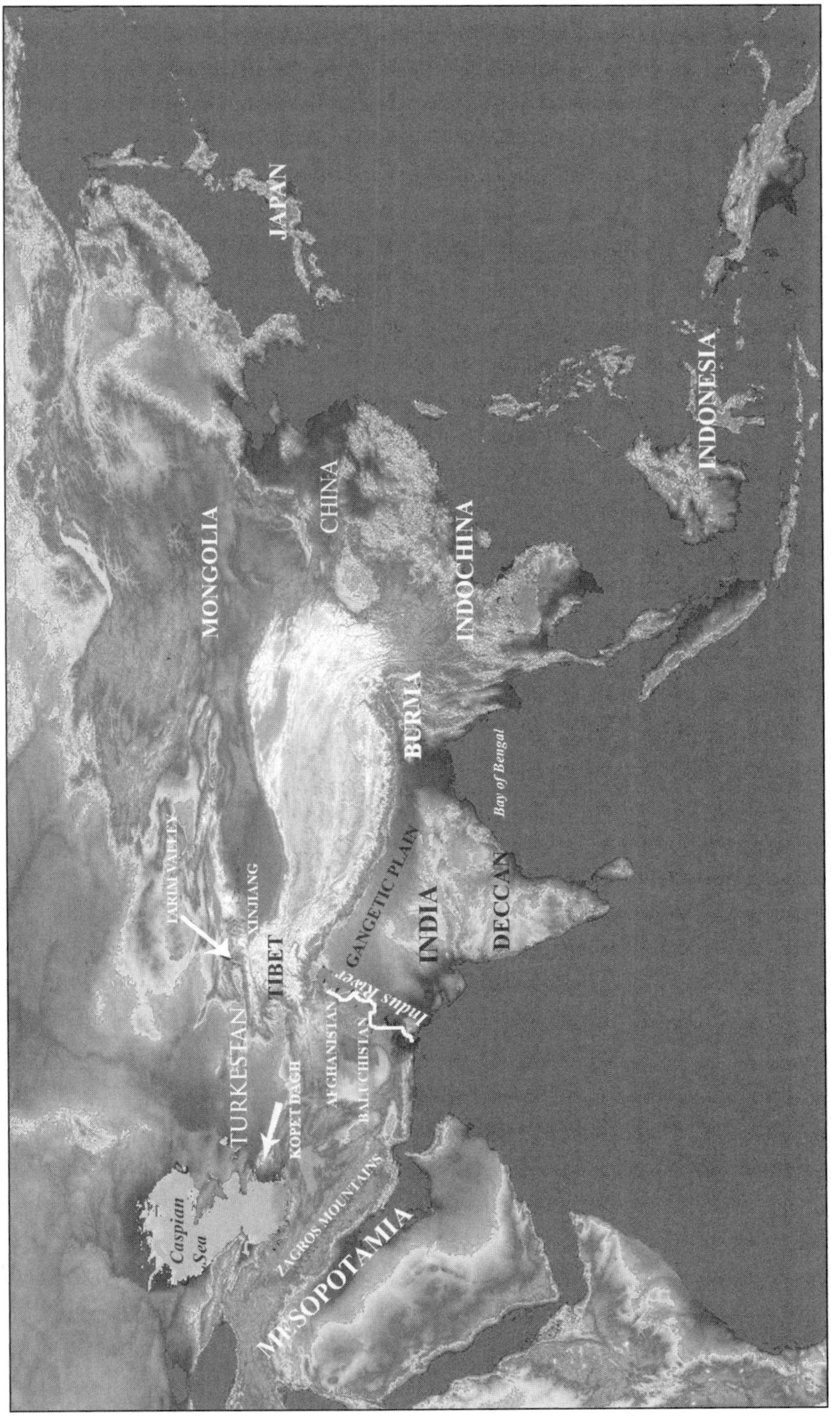

MAP 2. Lands of the postAdamic dispersions

CHAPTER 17. DISPERSALS AND WARS OF THE RACES

In the second Garden was the only real culture on the planet. Elsewhere only among the Adamsonites of the Kopet Dagh did the descendants of Van and Amadon preserve something similar. The Adamites of the second Garden never forgot the initial mission of the errant Adam and Eve, but the civilization there was implanted from outside: It had not *evolved* and was therefore destined to sink to surrounding levels. Still, by 19000 BC the Adamites of Mesopotamia were a cohesive people, numbering about four and a half million, and by this time millions of their progeny had gone forth into the world.

ANDITE DISPERSIONS

Some of the Adamites went west to the valley of the Nile, and some east, but largely because of the logical route of migrations Europe inherited more of the ancient Mesopotamian culture than any other part of the world, the ultimate antecedents of modern Europeans. The relentless thrust of migration was north along the east coast of the Caspian Sea, then west above the Caspian Sea into Europe where they blended with the blue men already there.

Few pure line members of the violet race, or Adamites, ever reached Europe; by the time they reached Turkestan they were already thoroughly blended with the Sangik and Nodite peoples; they were Andites. Nonetheless by 15000 BC there were more descendants of Adam in Europe and in Turkestan than any other part of the world, even in Mesopotamia. The Andites were about one eighth to one sixth violet, a far higher percentage than found in any human group today. They were the best human stock to appear on the planet since the days of the pure line violet people. The general appearance of the so-called Caucasian or white races derives from their Andite inheritance.

There was strong sexual attraction between the Andites and the blue men and women, who aspired to breed with them, a high honor. In this way much of the genetic damage suffered from the blue man's earlier breeding with Neanderthal strains was reversed. As for the Neanderthalers themselves, the immigrant Adamites and Andites hunted them down in a war of extermination. From the interbreeding of the Andites and blue men came a dozen or so virile and progressive groups, one of which is called CroMagnon in modern anthropology.

At first preserving the peaceful behavior taught by Adam, the Nodites and then the Andites became ever more aggressive. Andite leaders learned to conduct effective military campaigns of conquest. The restless Andites had a roving disposition, and their descendants in historical times moved out from Europe in bold expeditions until they had explored and described the whole planet. You can trace the movements of the last wave of Andite emigration by the spread of the IndoEuropean languages into northwest India, the upper reaches of the Mesopotamian plains, Russia, and Europe.

By 12000 BC sixty-five percent of all Andites lived in northern and eastern Europe. Other groups of Andites migrated to northwest India, the so-called Dravidian and, later, Aryan conquerors. Andites migrated to northern China over Tibet and Xinjiang (= Chinese "New Frontier," in central Asia) and greatly influenced the culture there. In the Tarim Valley in Xinjiang they established a civilization rivaling that of Mesopotamia that lasted for fifteen thousand years. Today the valley is a desert, but then it was fertile and open land. The Gobi to the east, also now a desert, was well-watered and fertile then. Remnants of these ancient Andites are the so-called Tocharians, who spoke the easternmost of the Indo-European languages. The civilization perished when the rain-winds shifted direction.

Many individual Andites traveled to the ends of the earth as missionaries, traders, and teachers, and small groups traveled to Japan, southern China, and Indonesia. A group of 132 Andites sailed from Japan across the Pacific ocean to South America. They interbred with the native red men living there and founded the ancient ancestors of the rulers of the Inca. Their crossing was leisurely, and they left behind many descendants on the Pacific Islands, populated by the "brown man," a mixture of the red and yellow Sangik races. Easter Island was long a center of the culture of the migrating Andites, but only 132 of them ever came to the shores of South America.

The final dispersions of the Andites, and their knowledge of agriculture, animal husbandry, crafts, and art, took place between 8000 and 6000 BC. Their migrations strengthened outlying peoples, but progressively enfeebled the homeland. Subject to constant attack by cavalrymen descending from the Zagros Mountains to the east, assisted by their large numbers of tamed horses, and plainsmen from the open spaces to the west, the Andites left Mesopotamia in three massive emigrations.

Sixty-five percent entered Europe by the route over the Caspian. Ten percent, including most of the Sethite priests, went east over the Zagros Mountains behind Elam up onto the Iranian plateau, then northeast to Turkestan, the land of Adamson. Later they, and remnants of the Adamsonites, were driven into India. Ten percent went over Xinjiang where they mixed with the blended Andite-yellow peoples to form ancestors of the later Han Chinese. Ten percent crossed the desert and entered Egypt. Five percent remained at the mouths of the Tigris and Euphrates, preserving a high culture that reached all the way back to Dalamatia.

NOAH AND THE GREAT FLOOD

The same geological shifts that caused the sinking of the first Garden beneath the Mediterranean raised the mountains from which the Tigris and Euphrates rivers flowed, increased the snow pack, and made constantly worse the floods that had always devastated the Mesopotamian plains during the spring runoff. Around 5000 BC the situation became so bad that most inhabitants left the valley altogether to live in the eastern highlands.

Nearly 5000 years later, Hebrew priests, carried away in captivity to Babylon (586-538 BC), were attempting to make sense of traditions about their ancestry, which they wanted to trace back to Adam. It occurred to one of them that the easiest way to cut through the muddle was simply to wipe out the whole earth in a great flood, then trace Abraham back to one of the sons of Noah. Stories of a great flood, found all over the world, were in any event very old in Mesopotamia. In fact there never was a great flood that engulfed the whole world, except in primordial times before the emergence of land.

Noah, however, was a real man, a wine merchant, who lived in the early third millennium BC in a city called Aram, near Erech, in southern Mesopotamia. He kept a written record of the annual height of the flood and constantly urged his fellow townsmen to build their houses of wood and be prepared to put their household animals inside in case a huge flood came. They laughed, but when a great flood did come, all were lost save Noah in his houseboat.

The ever worsening floods finished off what remained of the second Garden. Soon all trace was gone, save some remnants in the south among the Sumerians who by 6000 BC were mostly Andite. By this time the Andites were gone from the other parts of the valley, now occupied by the wild and uncouth peoples who surrounded it. The germs of civilization went west to Egypt and to the Mediterranean islands.

END OF THE ANDITES

The invaders into Mesopotamia, although little cultured, were often themselves of superior Andite stock, some coming from as far as Turkestan, and once settled in the valley they revived many of the cultural forms of the earlier Andites. These were to become the Semitic Mesopotamian peoples of historical times. They could not take over the Andite Sumerians, who defended themselves through superior arts, intelligence, and a system of military canals. Coming to respect and admire the Sumerians, the wild northern peoples took teachers, administrators, and governors from them, as did the Egyptians to the west and the Indians to the east.

At first confederated, the Sumerian city-states came to be ruled by individual Sethite priests who called themselves kings only after conquering a neighboring state. Each city-state considered its own god superior to all others. In the late third millennium BC, Sargon the great, priest of the city of Kish in south-central Mesopotamia somewhat east of Babylon, ended this era by conquering and uniting Mesopotamia and outlying lands into a single administrative structure. The system soon collapsed, Sumer fell, and by the time of Hammurabi around 1800 BC the Sumerians were absorbed into the ranks of the northern Semites. The Andites were no more.

Asia is the homeland of the human race. On a southern peninsula of that continent (the geography is different now) Andon and Fonta were born. A descendant named

Badonan created a center of Andonite culture in the highlands of Afghanistan that lasted for 500,000 years. Here appeared the Sangik races. In southwestern Asia flourished the cultures of Dalamatia, the Nodites, the Adamites, and the Andites. From these regions have flowed the currents that became modern civilization.

DEVELOPMENTS IN INDIA

Only in India did all the Urantia races mix and blend. The Sangik races arose in the highlands northwest of India and their descendants went down into the peninsula where Andonites also lived and, later, the Andites. An earlier mixture of red and yellow and Andonite absorbed the disappearing green race and large strains of the orange and indigo races.

Although traces of the Adamic blood had entered northwest India by 20,000 years ago, not until 15,000 years ago did large numbers of Andites come down from Turkestan. For 1500 years they poured through Baluchistan into the river valleys of the Indus and the Ganges, then pushed south toward the Deccan (Sanskrit "south") Plateau, driving many of the inhabitants into Burma and southern China and others even further south onto the ever-narrowing plateau surrounded by the sea. Fearing racial obliteration in an ocean of conquered peoples, the Andites put in place rigorous rules for intermarriage. Nonetheless, by 10,000 years ago they had disappeared into the native population, producing the people now called Dravidian, who produced one of the most advanced civilizations on earth at that time, whose ruins can be seen today at Mohenjodaro and other sites in Pakistan. ("Dravidian" is of unclear meaning and refers both to the peoples of the southern Deccan and to the family of languages that they speak.)

Many of the Sethite priesthood entered India, first with the conquering Andites, then with later waves of Andites, the so-called Aryans. The monotheistic teachings that permeate Indian religious traditions go back to Adamite teachings in the second Garden. Teachings about the Trinity degenerated into the triune symbol of the fire god found in India today.

The Dravidians were among the first to build cities and by 7000 BC carried on extensive overseas trade with Sumer by land and sea and by crossing the Bay of Bengal as far as the East Indies. The later-appearing Aryans did not recognize their cousins submerged in the Dravidian population, but admired the civilization that they found there. These Aryans, the last of the Andites to evacuate Turkestan, entered India over a period of five hundred years in the middle of the third millennium BC. Their centers of power were in the north, especially the northwest.

On the Gangetic plain the mixed Aryan and Dravidian peoples produced a high culture, reinforced by cultural forms coming down from China. To prevent absorption, the Aryans created the complex caste system still vigorous today, but their numbers were too small to prevent amalgamation with the Dravidians of the south. The premier class, the teacher-priests, were descendants of the Sethite priests of the second garden. The modern Brahmans descend from them, whose teachings, corrupted by enslaving Dravidian superstition, bear little resemblance to those of their illustrious ancestors.

EAST ASIA: RED MAN, YELLOW MAN

In India successive Andite invasions conquered the native highly mixed peoples, then were absorbed by them, but in East Asia the story was very different. There two of the primary Sangik colored races, the yellow and the red, escaped the debasing admixture with the subhuman Neanderthal strains that so retarded the blue man in Europe.

The red man was first to migrate from their central Asian homeland. They traveled northeast around the highlands of India to the far East, where they reigned supreme for 100,000 years. Then around 300,000 years ago the yellow men entered China from the south. As their population increased, they pushed ever northwards until they encroached on the hunting grounds of the red man.

Fueled by racial antagonism, the struggle for control of further Asia had begun, one of the greatest of all struggles in the history of humans on this planet. For two hundred thousand years the red and yellow men engaged in bitter unremitting war. At first the warlike red men were successful, but gradually the yellow men improved their own arts of war, enhanced by their disposition to live at peace with one another: through unity, strength. In the meanwhile, the red men continued ferociously to fight each another even as they struggled against the yellow men. In historical times, the inability of the red man in North America to unite against the invading Europeans insured their destruction.

Around 100,000 years ago the decimated tribes of the red men were fighting with their backs to the retreating ice sheet. When land passage over the Bering Strait emerged, they did not hesitate to abandon Asia for the Americas. The last of the red men crossed the strait 85,000 years ago, but they left much of their genetic heritage with the northern Chinese peoples and the Andonite Siberians. Thus separated from the rest of humankind, the red man never had the benefit of the Andite inheritance, being dispossessed of their homeland fifty thousand years before Adam's arrival. Not until around AD 1000 did the red man have any contact with other races, except for the Andonite Eskimos who lived in the far north, and traces of Andite blood that reached Peru.

CHINESE CIVILIZATION

After driving the red man from Asia, the Chinese destroyed or drove out the Andonites living in the river valleys into Siberia. In Burma and Indochina the Indian and Chinese cultures blended. A greater percentage of the submerged green race is found there than anywhere else. Highly mixed races occupied the Pacific islands, with a large percentage of green and indigo in the racial makeup of occupiers of the southern islands, and Andonite mixed with red and yellow in the northern islands. Ancestors of the Japanese were not driven off the Asian mainland, by northern Chinese, until around 12000 BC.

Andite immigration from the Tarim Basin and Tibet strengthened the yellow man, now settled along the Yellow and Yangtze Rivers, where they built a superior culture. The value placed on peacefulness and social cooperation led to a massive population increase so that between 25000 and 5000 BC the largest centers of population on the whole planet were in northern and central china. By 15000 BC there were twelve million Chinese speaking a common language and sharing a common culture. Their devotion to the family, the basis for all high civilization, was superb and unparalleled, encouraging the preservation of wealth, the pooling of experience from more than one generation, the education of children in the arts and sciences of the past, a strong sense of duty, and a devotion to moral behavior.

The yellow man was the first human group to achieve a large-scale social, cultural, and political civilization. Following the teachings of a prophet named Singlanton, the Chinese had an advanced religion that recognized the One Truth. Were it not for the mountain chains to the West, inhibiting communication with the Andite centers of Turkestan, China would surely have created a culture that dominated the world. Even so, numerous Andites did travel over the Gansu corridor north of the Himalayas to the upper reaches of the Yellow River, whose culture and peoples to this day are different from those of southern China.

The Chinese suffered a further disadvantage from the thoroughness of their conquest of East Asia, which left them protected in a remote region without dangerous enemies and the impetus to growth that danger always encourages. Their devotion to agriculture and the pacific and cooperative values that further agriculture generally led this great people more and more to venerate the past, to worship their ancestors instead of the One Truth, to ignore the unknown in favor of what was established, and finally to a cultural stagnation that belied their magnificent beginnings. The bizarre complexity of their system of writing restricted learning to a tiny class and forever shut out the populace at large, although the Chinese first discovered printing with movable type. Although the mechanical and religious developments of the so-called white races have been of a high order, they have never excelled the Chinese in family loyalty, group ethics, or personal morality. Millions upon millions of human beings have lived lives of happiness in the embrace of the superior Chinese culture, which after sliding into a temporary abeyance, is poised once again to forge ahead in the struggle for human progress.

CLIMATIC AND GEOLOGIC CHANGE

The so-called European white race is a combination of the red and yellow and especially the blue races, with a strong Admixture of Andonite stock and later mixtures with the Nodite and Andite strains. The blue men in the south were mixed with Saharan peoples. The European blue man did not himself produce a great civilization, but when combined with such rich outside strains he became the most aggressive and progressive of all races on Urantia since the days of the violet race in the second Garden.

Before the last of the Andites were driven from the Euphrates Valley, their brethren had entered Europe in large numbers, as we have seen—as traders, teachers, and warriors, mostly over the Caspian Sea. At this time, however, land bridges across Gibraltar and across Sicily from Italy to Africa formed two large inland lakes, scene of the earliest maritime experiments, and on the shores of these lakes also met and intermingled the blue men from the north, Nodites and Andites from the east, and Saharans from the south.

The early and successful advance of the violet race into Europe was followed by a sudden shift in the climate when the water-laden winds that had blown from the west shifted to the north. The once populated grasslands of the Sahara turned into the burning wasteland they are today, and its small, dark-eyed long-skulled inhabitants were dispersed. Purer elements of the indigo race penetrated southward into the forests of Africa, where they have remained to this day. Like the red men, they became physically isolated as North Africa dried into desert.

In western Africa, the more-mixed inhabitants crossed into Spain to become the ancestors of the later Mediterranean long-headed brunet races. Those living southeast of the great African plateau migrated to Arabia, then through Mesopotamia and Iran to faraway Sri Lanka, while those living in the northeast of Africa migrated into Egypt and Palestine.

At about the same time as the winds shifted to the north, desiccating northern Africa, England separated from the continent, Denmark rose from the sea, and a great earthquake broke the Gibraltar land bridge separating the inland lakes of the Mediterranean from the Atlantic Ocean (Map 3).

MAP 3. Places important to the Andite dispersions in Europe

The ocean rushed in and swiftly raised the water level, overwhelming the Sicilian land bridge, uniting the two inland lakes, and forming the rough shape of the modern Mediterranean Sea. In this catastrophic inundation thousands of human settlements along the Mediterranean shores were submerged. More humans died in a single flood than ever before or after.

The engulfment and enormous expansion of the Mediterranean Sea now hindered migration to the west. The Sahara had been a grassy plain, where the Saharans herded flocks and practiced agriculture, though they did not build. Now great numbers of Saharans poured into Egypt and Mesopotamia to escape rapid desiccation of northern Africa. Adamite/Nodites had long exerted a strong presence in Egypt where they introduced domestic animals, pottery-making, and metal-working. For thirty thousand years a constant stream of Mesopotamians had entered Egypt, bringing with them the arts of the second Garden, but the influx of the undeveloped Saharans depressed the culture.

The Adamites continued to head north, as Adamson had done long before. However, an enlarged Caspian Sea held them for generations in Turkestan, where they hunted, herded, and tilled the soil. Gradually, in spite of the obstacles, they again drifted west, where they found the blue men admixed with their earlier Adamite relatives, long isolated and having by now slipped far behind the cultures of Mesopotamia.

DEVELOPMENTS IN EUROPE

The blue men lived along the course of the rivers, which then flowed in different channels (except for the Somme). Along the Atlantic and in present-day Russia there was heavy admixture with the Andonite stock. A great array of racial types lived in Europe 15,000 years ago. The CroMagnoid Andite/blue men were a people of great vigor, courage, and industry, and inclined to artistic expression (they left behind their art in caves in southern France). However, they greatly diluted the religious traditions of the immigrant Adamites, who lost the blue man's respect through their racial contempt for the blue man and their practice of debauching his women. By contrast to the mixed Adamites' taste for sexual adventure, the blue men were honest and ordinarily married to a single woman.

As the winds shifted to the north, dense forests covered the once open grazing lands of Europe, encouraging more or less permanent villages along the European rivers, herding instead of hunting on the open plains, and some agriculture. Instead of killing inferior surrounding peoples defeated in war, as had always been the custom, the mixed CroMagnoid people began to take slaves in whom ran strains of the Neanderthalers, and gradually through interbreeding with these inferior peoples they slipped backward. The retrogression in culture continued until the arrival of the final wave of Andite immigrants from Mesopotamia, who swiftly absorbed the CroMagnoid racial type and initiated what became the modern cultures of the so-called white races. The "white man" is not Andite, but a blend of the Andite with the highly mixed peoples already living in Europe.

While the mixed Adamite/Nodites called the Andites poured into Europe in a steady stream over a very long period, there were seven major migrations. In the last migration many rode on horseback, the first mounted cavalry, now penetrating Europe across the Aegean islands and up the Danube Valley, or more commonly across the steppe through which flows the Volga and the Don, north of the Caspian (see Map 3). Only in Lapland (and to a small extent in Brittany) did the ancient Andonites retain a sense of identity.

The earlier migrations, where the pacific Adamites were stronger, were more peaceful than the later ones, dominated by the aggressive and belligerent Nodite stock. The key to the success of the later invaders was their mastery of the horse, which enabled them to move swiftly north of the Caspian and into Europe without losing their group and racial solidarity. In earlier migrations the migrating Andites moved so slowly that they lost group identity and the cultural traditions of the second Garden.

By the end of the Andite migrations and invasions, more of Andite descent lived in Europe than in any other part of the world. For thousands of years their headquarters were in Denmark, but gradually they dispersed to blend and disappear into the white races. The final battle between the "white man" and the highly organized unassimilated blue man took place in the valley of the Somme in northern France. The valley runs for about one hundred fifty miles from south to north to empty into the English Channel, and in AD 1916 witnessed one of the greatest battles ever to take place on this planet: One and a half million men died or were wounded for no military advantage. In primordial times, in the same place, for five hundred years the blue men defended their northern territory, then succumbed to the superior military strategy of the white men, who married some of the blue women, but killed all the rest.

In the last battle on the Somme a man named Thor, later thought to be a god, led the white men to victory. The white men subsequently systematically, and swiftly, exterminated or absorbed all other peoples in Europe. By 5000 BC they controlled all of northern Europe. Only in the Danube Valley did some Andonites hold out and, in central Europe, remnants of the blue man.

DEVELOPMENTS IN THE EASTERN MEDITERRANEAN

As the Andite migrations continued, culture in Mesopotamia declined. The immediate center of advanced culture on Urantia shifted to the Nile Valley. By 5000 BC, seven distinct groups lived in Egypt, all of them, except one, of Mesopotamian origin. The immigrants were familiar with river life, enjoyed the protection of the Nile Valley, and understood irrigation agriculture. The Egyptians early fashioned their multifarious local deities into an integrated system served by an elaborate and burdensome priesthood. These priests formed a complex theology; attempts to return to the earlier simple teachings of the Sethite priesthood made little impression.

Many of the best Andite artisans, skilled in the working of stone and metals, went to Egypt. In the third millennium BC the Andite architectural genius Imhotep designed and supervised the construction of Egypt's first stone building, the step pyramid complex at Saqqara, never surpassed on earth for sophistication in stone construction, an amazing site still today.

About 12000 BC an advanced group of Andites crossed the Mediterranean to Crete, but not until 2,000 years later did their descendants spread out to inhabit the Aegean isles. These early Cretans were the narrow-headed, smaller-statured Andites who had intermarried with the Vanite division of the northern Nodites. They were skilled in the manufacture of textiles, pottery, metalworking, hygienic plumbing, and building with stone. They early devised a system of writing by which they managed production from their herds and agriculture. The Cretan religious cult of the "great mother" glorified the memory of Eve. Images of her were everywhere, in thousands of public shrines throughout Crete and Asia Minor. This mother cult persisted to the times of Christ and was later incorporated into the early Christian religion as the worship of Mary the earth mother of Jesus, a cult still vigorous today.

About the same time that the Cretans occupied the Aegean islands, a group of 375 Nodites under the leadership of a direct descendant of Adamson and Ratta crossed the northern islands to Greece, coming directly from their highland homes in the lands north of Mesopotamia. These were the distant ancestors of the historical Greeks. These newcomers were of a high intellectual order and the most handsome of all humans since the days of the first Garden, the inspiration for later traditions that the Greeks were descended from the gods.

By 6500 BC the ancient and superior culture of the Andites had steeply declined as they lived amidst the other more numerous human races; a spiritual life by then scarcely existed among them. By 5000 BC the centers of greatest future potential lay with the Andites of Sumer (south of Mesopotamia at the head of the Persian Gulf), northern Europe, and Greece. In Scandinavia the mother cult was predominant, brought there by Cretan missionaries, but in Spain and southern France the people worshiped the sun and set up stones in honor of the sun's progress. Stonehenge in England was later set up in connection with worship of the sun.

PART III

CULTURAL CONSIDERATIONS

CHAPTER 18. THE EVOLUTION OF CIVILIZATION

ONCE OUR ANCESTORS WERE ANIMALS; now we live in a kind of civilization. How did we get from there to here? Civilization is not biological in origin, but is a cultural achievement. Therefore its values, bases, premises, and practices must be taught afresh to every succeeding generation.

Humans may be gregarious by nature, but just as much they are suspicious and resentful and enjoy enmity and hate. Only the harsh school of reality teaches that in unity there is strength. In early times a man alone, without a mark indicating his group or tribe, was doomed, the "mark of Cain." Civilization provides protection against random death, but you pay a high price and you must obey the law.

Nor is cooperation, and its enormous benefits, a natural form of behavior; it must be learned. Cooperation brings success against nature and against one's fellow man, and its survival value encourages social progress. By lessening pain and increasing pleasure, social organization progresses ever forward in spite of reversals and setbacks that come when the goal of self-maintenance is replaced by the goal of self-gratification.

THE ARTS OF CIVILIZATION: OVERVIEW

Despite the bitter setbacks consequent upon the defection of Caligastia, and the relative failure of Adam and Eve, the organic evolution of the human races proceeded unimpeded. Evolution is a principle of the universe and although it can be delayed it cannot be stopped. Although the violet race, the offspring of Adam and Eve, were far fewer in numbers than had been envisioned, nonetheless under the influence of the descendants of this race human civilization has progressed more in 37,000 years than in the preceding one million.

The first four great advances in civilization were: (1) the making of fire; (2) the domestication of animals; (3) the enslavement of captives; (4) the institution of private property. Andon had discovered how to make *fire* by striking a flint, but the theory of natural causes was unknown to early humans. Everything out of their everyday experience had a supernatural cause. Hence fire came down from heaven, or something similar, according to tales told all over the world. Still today many view the world in this way, but science has grown from the fearless pursuit of true causes.

Only *tools* make civilization possible, because they release humans to turn their attention to things other than just keeping alive. In early times humans had no time whatsoever for reflection on who they were and where they came from. *Domesticated animals* are living tools. Without them humans must perform every task themselves. Domesticated animals make agriculture and transportation possible, and without them civilization cannot progress. Most animals suitable for domestication lived in central and southwestern Asia, one reason that modern civilization began there. Most of these animals had been tamed before, then reverted to a wild state, except for the dog, tamed by the blue man in the remote past and ever since a companion and protector to man.

The Andites of Turkestan were first to tame the *horse*, one reason for their long ascendance, and by 5000 BC horses were raised in the civilized and not-so-civilized areas. They pulled wheeled vehicles, carts, and war chariots. Other animals tamed by 5000 BC in Mesopotamia, Turkestan, and China were sheep, goats, cows, poultry, camels, and elephants. For beasts of burden these peoples used the ox, camel, horse, and yak, so that humans no longer needed to carry everything.

Civilization depends on *mechanical inventions*, because early humans were always slaves, slaves to nature. Agriculture brought an institutional slavery and the private ownership of land. *Human slavery* raised the master's standard of living and created leisure for social and intellectual development. Scientific discovery in modern times has made slavery unnecessary and exalted the freedom of all humankind.

Before the coming of Adam, humans lived where they could, under rock ledges and in caves. Of the modern races, the Andites in Turkestan were first to build independent structures of wood, similar to the log cabins of the American pioneers. On the plains, *houses* were made of mud brick, then much later of fired brick. The older river races had from the earliest times lived in houses made of a circle of stakes whose tops were brought together to make the frame for a kind of inverted basket; daubed with mud and dried, such huts made a rainproof dwelling. *Basketweaving*, one of the primordial arts, was discovered in this way independently in different parts of the world, as *pottery*-making was discovered when one of these huts caught fire and hardened the clay encased on it.

Many transforming discoveries were made accidentally in this fashion in human history, at least until the time of the coming of Adam. In fact the Prince had introduced techniques of potterymaking 500,000 years earlier, but they were lost. Only the gulf-dwelling preSumerian Nodites continued the art, until revived in Adam's time when it spread out east and west from Mesopotamia.

Around 12000 BC began the great drought that ruined the open grasslands of Turkestan that had been so ideal for herding and hunting. The Andites living there had no choice but to adopt new forms of behavior, some becoming agriculturalists and others manufacturers or traders. The superior tribes from the Nile in Egypt to the Hindu Kush in Afghanistan, from the Ganges River in India to the Yellow River in China all adopted the new way of life, *agriculture* with *trade* on the side.

Trade is a transforming activity, because humans at this time lived in tribal groups—that is, an extended family—and had little contact with other human groups. Trade brought them together, made a virtue of peace, and encouraged cultural cross-fertilization. Traders and roving *explorers* were the single most potent force in the spread of historic civilization, the developments that one would study today in a university class in the "history of civilization." Military conquest, colonization, and missionaries were also important in spreading the arts of civilization.

Around 10000 BC came the first *cities*, which arose as centers of trade and manufacture, but always surrounded by zones of agriculture and the raising of stock. By any modern standard these city-dwelling humans lived in squalor and filth. They generated so much garbage that the height of the mound they lived on rose by one or two feet every twenty-five years. They were steeped in excrement. The mud brick houses, too, would decay, then be rebuilt on the same site. Today the Near East is covered with high *tells*, "hills," remnants from these early settlements.

These early city-dwellers worked with *metals*—gold, copper, and iron—but in no sense can we divide early ages into stone, bronze, and iron ages: They all existed at the same time. Gold was the first metal worked, used for decoration. Copper was easily worked but not widely sought until an Adamsonite in Turkestan accidentally discovered how to combine copper and tin to form the much harder bronze. His highland copper mine happened to be next to a deposit of tin.

Humans are not born into cultural society as equal members entitled to equal benefits. Only the nobility of those who work tirelessly to make the world a better place for themselves and their descendants are admitted to this exclusive guild of civilization. Entrance requires high admission fees, imposes rigorous discipline, and imposes dire penalties on dissenters and nonconformists while conferring few personal privileges except enhanced security against common dangers. You give up your freedom, but in return are safer than before, insulated somewhat from the terrible and constant violence that afflicts human beings. The mission of government is to define what is right, just, and fair and to enforce equality of opportunity under the rule of law. But every human right brings with it social duty; liberty without restrictions is a dangerous fantasy. The rights of groups, as well as of the individual, must be protected, including the right to regulate sexual behavior.

ORIGIN OF THE FAMILY

Early human groups were held together by hunger for food and for sex, desires they shared with the animal kingdom. Other factors bringing them together were vanity—the lust for approval—and fear, especially of the ubiquitous ghosts. The home was founded on the male's desire for frequent and convenient sexual gratification, combined with the female's innate and instinctual love for the children that came, an emotion women share with the animal kingdom. She needs to be in one place as much as possible to care for the children, and where the woman is becomes home.

The woman gathered foods and, later, tilled the soil, contributing directly to the supply of food. She was a beast of burden and a constant companion willing to take tremendous abuse without violent resentment and at the same time an object to provide sexual gratification for the male. The function of the home is not to provide personal happiness or gratification, but to ensure the viability of the race. The aggressive pressure of sexual desire had just this result. Most values of civilization take root in the family, where two humans first learn to live at peace in spite of constant differences and antagonisms; and they teach such values to their children.

THE GHOST CULT: INTRODUCTION

Desire produced the earliest societies, but *ghost fear* bound them together. The ghost cults that have so afflicted human civilization, and which are strong today, began in the ghost dream, common or universal today, when a close relative or intimate friend dreams vividly of someone who has died. Such dreams inspired abject terror in early humans and threw the living together in an effort to combat the dangers of the unseen dead.

The ghost dream marked the division between humans and animals, because animals do not visualize the "spirits of the dead." Through ghost fear, and its inexhaustible power, greater and more complex social units came into being. Such baseless fears are part of social evolution, because later awe for Deity inspired by revelation will replace them.

Hunger, sexual desire, vanity, and ghost fear created early societies, but the same forces smashed them to pieces when allowed to rein unchecked. Humans are not by nature peaceful, but must learn peaceful behavior from revealed teachings, the experience of progressive communities, and especially from the teachings of Jesus of Nazareth.

The ghost cult established the reality of an invisible world whence might emanate rules for life and for progress. The belief that the dead will resent any change in the customs by which they themselves lived their lives, and the fear that they will punish such innovation, gave force to customary behavior and established the authority of traditional mores, "the way things are done." Customary behavior gives social stability, but also imprisons the individual and opposes creativity. Advancing civilization frees the individual from such fear and bondage to superstition. A society can only survive when its mores evolve.

CULTURE AND SOIL

The earliest human cultures arose along the rivers of the eastern hemisphere. They passed through four stages: (1) collecting food, when lines of gatherers up to ten miles long combed the countryside for anything to eat; (2) hunting animals, dependent on the repeated reinvention of the stone hammer used as a weapon; (3) herding domesticated animals, when the male, who once wandered as a hunter, now easily obtained meat, milk, wool, and even sexual gratification from his animals, while the female continued to labor long and hard at obtaining vegetables and, like an animal, delivering and caring for the young; (4) growing domesticated crops, the highest development of material civilization.

As a hunter the males were kind to the females, but with herding began the dark chapter of the male's shameful and brutal treatment of the female on this planet. Growing plants ennobles the spirit, by contrast, and places a value on peace and good relationships between the sexes. Caligastia and Adam and Eve tried to inculcate agricultural practice. Now industry has to some extent replaced agriculture, but agriculture will remain the basis for progressive social development.

The mores always codify man's relationship with the soil. Humans come from the soil, are of the soil, and return to the soil. Through human intelligence the soil has been made more productive at the same time that humans have learned to control their rate of reproduction, thus creating the wealth and leisure required for civilization.

Supply and demand establish the value of both humans and land: When the land is wide open, human life is highly valued; when people live on top of one another, human life has far less value. Dense populations bring out the worst in human nature. Cities, wherein art and science can flourish, multiply the power to do good and do bad. The size of the family depends on the standard of living: the higher the standard, the smaller the family, up to the point of gradual extinction of the line. Intense competition gives rise to the formation of castes, to keep separate different levels of standards of living.

Early peoples worked in many ways to reduce their populations and so increase the ratio of land to the individual and increase personal wealth. All primitive tribes killed sickly and deformed children. Girl babies were commonly killed. From a set of twins, one was ordinarily killed as the product of voodoo or infidelity. Many early peoples understood techniques of abortion, which became widespread when the taboo spread against unmarried girls giving birth. If the abortofacient did not work, the girl would kill the infant (or sometimes give it to her mother). Through widespread abortion and infanticide many clans dwindled away and ceased to exist.

In the past, world overpopulation has not been a problem, but advances in medicine and the diminishing practice of war have made overpopulation a grave issue in parts of the world today.

EARLY INSTITUTIONS

Human emotions differ from the emotions of animals in their capacity to enjoy humor, art, and religion. Socially, humans are superior to animals in their ability to make tools, to communicate in complex ways, and to build institutions. Most institutions in some way save labor while they provide for security. Civilized humans revere their institutions, but in fact they are the aggregations of conventional ways of doing things built up over time, sanctioned by irrational taboos and dignified by religion.

Institutions are of three kinds: (1) institutions of *self-maintenance*, that is, ways of doing things that relate in some way to the production of food, which means how relationships to the land are handled, including war for gain; (2) institutions of *self-perpetuation* that grow from sex hunger and maternal instinct and preserve the home and family, including marriage customs and war for defense; (3) institutions of *self-gratification* that grow from vanity and glorify the self through display, costume, adornment, and war for glory. In practice the three kinds of institutions are intimately intertwined and function as a single social reality.

Man's natural estate is abject, horrifying poverty, and early humans lived in a state of constant destitution, in direct competition with the animals for something to eat. The lesson that a group can do more than individuals was learned early, which led to a rough organization of society and a crude division of labor and the greater efficiency that such behavior brings in the contest against death and starvation.

But early humans did not value labor or ever do real work cheerfully or willingly. The notion of completing a certain task within a certain time is wholly modern. Advancing standards of living brought the necessity for real labor. Labor separates human from the animal, whose activity is always instinctual. The necessity of labor is the human's greatest gift as they lead this earthly life. Still, the old ways of hoping for the big break, of manipulating charms and chanting spells to get ahead, long combated foresight, self-denial, and industry.

Specialization of labor depended on nature and on custom. Sexual difference forced specialization. Women bore the children, then were instinctually attached to them more than were the males. Women became the routine laborers, doing such daily repetitive work as gathering wood while the males hunted and fought or did nothing. The males took the most interesting and challenging activities and left the drudgery to the females for much of the history of the human race; yet in making the home, male and female worked together.

Master and slave created its own categories as did age and disability. Cripples and the old would stay at home and make weapons. Religion early generated its own specialization in the medicine man or shaman, the first class of humans exempt from work, the first professional class. The earliest industrial specialist was the chipper of stone artifacts and the stone mason, followed by the smiths, the rock-salt exporters, and the potters.

Shamans bitterly competed with the smiths, the first nonreligious group to enjoy special privileges, whose ability to refine and shape metal inspired awe. Smiths were early associated with magical practice and were often hated for their power and manipulations. The competition and hostility between the medicine men and smiths was the first contest between religion and science.

At first, exchange of goods took place through raids, when one group attacked another and took their possessions. Then barter began, when goods were placed in the open between hostile groups. Women were the first traders because they carried the burden and thus produce to a market while the males hunted. But piracy continued for a long, long time between the early practice of silent barter and modern methods of exchange. The trading counter, still standard today, began as a barrier wide enough to prevent the buyer and seller from killing one other with their weapons.

Markets were sacrosanct and immune to theft, protected always by a fetish, an object imbued with power. They became places of refuge where you could be safe from violence. Merchants within a tribe were scrupulously honest with one another, although one was of course expected to cheat an outsider. The life of the trader combined adventure with profit, and traders, the great civilizers, spread new ideas and better methods far and wide.

FIRE

The possession of fire separates the human world from the animals. Fire enabled early humans to sleep on the ground and not in the trees, for all wild animals fear fire. Fire enables altruism because the giver could give and still have what he gave. The fire was the center of the first home, around which the family gathered. The need to maintain the fire encouraged watchfulness and dependability.

Andon discovered fire but never venerated it, as did many of his followers, who saw fire as a fetish or spirit. Early humans did not reap the sanitary benefits of fire because they would not defile fire with refuse. Everything associated with fire was sacred, including the flints used to start it. Sacred fires burned forever in special locations. Women priestesses tended them because they tended the fire in the home.

Fire makes cooking possible, which heightens caloric intake by reducing the energy required for the digestion of raw food. The "raw-eaters" were viewed with contempt and derision. Fire made possible the working of metals and, eventually, the discovery of steam power and electricity.

PROPERTY

At first all animals were the enemy to man, competing for the same food and through their superior power and high intelligence preying upon humans. Then humans killed and ate the animals and later tamed them by corralling herds of grazing animals, an early form of property and wealth. The dog was tamed earliest to protect this property, serve as food, and offer companionship.

Capital allows labor in the present to be preserved for the future. It begins in food hoarding, of grains and animal parts, which teaches self-control and places the food hoarder in a superior position (so long as the food is not stolen). The first bankers were those brave enough to protect the wealth of a whole tribe. When attacked, all gathered around the hut of the banker, a need that led to military organizations that were themselves used to raid the wealth of other tribes.

Hoarding led to gross distinctions in wealth and status. Material possessions were used to wield power over others, including the taking of bond slaves when loaned wealth was not returned with sufficient interest. A common way to establish prestige was to destroy large quantities of wealth, including the killing of large numbers of animals and slaves. Capital is the basis for modern industrial society, in spite of its famous abuses. Capital together with invention has liberated humans alive today in ways unimaginable to earlier generations.

Those taken in battle were property and in early times were eaten, sacrificed to the spirits, tortured to death, or set to fight one another to the death. Slavery was a great advance over such primitive practices and a basis for advancing civilization. The hunter-gatherer, like the American Indian, rarely took slaves, but either adopted captives or killed them, usually in prolonged and exquisite tests of pain. Flocks were a form of capital off whose interest herdsmen lived, the yearly increase of the flocks, and herding was not the work of slaves. In the early agricultural societies, by contrast, male slaves taken in war were forced to till the land.

Slavery was indispensable to the evolution of civilization because it imposed order on the chaos of indolence and incompetence. Those who were poorly organized, or who could not do things well, fell victim to those who were well organized and in this way provided their conquerors with the wealth and leisure that made social advancement possible. Slavery requires strong regulation and inspires the birth of government to provide and enforce those regulations. Ironically, in time slavery becomes a cancer within the very society that it first made possible. Modern machines have, of course, made slavery obsolete. Returning to fire as a source of energy and power (as in coal-fired electric plants), where humans first began, man is discarding slavery and the animals that first raised him up from savagery.

In early times all things were held in common, a sort of communism. The women belonged to the group, but were controlled and allocated by the chiefs. There was no theft and no prostitution and no beggary. All tools and weapons were also held in common. Because the indolent and shiftless depended on the industrious and imaginative for their survival, innate forces guaranteed an end to the communist communities.

The evolution of the family depended on the human desire to acquire property and to bequeath this property to offspring, not have it absorbed into the community at large. The religious belief that only property can assure a pleasant afterlife enhanced the desire to accumulate wealth, then bury or burn it with the corpse. The urge for security and power led progressive members of society to find ways to hoard their wealth away from the others.

When communism passed, the father gradually took control of the family and monogamy began to replace polygamy, a relic of the days when all women were slaves. Only monogamy makes possible the fulfillment of the high ideal of one husband and one wife united in childrearing and homebuilding, sharing a common culture of self-discovery.

Private property began with the notion that certain places and things were imbued with someone's personality. The sleeping place was the first private property, often assigned by the tribal chief. A fire pit or well could belong to someone. Charms to ward off ghosts, the first jewelry, were private. Real estate was marked by private insignia, the origin of family crests. A crop was considered to belong to its planter, and successive crops gave title to the land. Agriculture brought the first private ownership of tracts of land, the basis for all modern government, law, civil order, civil rights, social liberty, peace, and happiness. The American red man could never understand the concept of private land in the long and unhappy war he fought with the European invaders.

CHAPTER 19.
THE EVOLUTION OF GOVERNMENT

Antagonism between peoples and individuals on an evolving planet is inevitable. The demands placed on humans just for staying alive are so great that chaos in the social order brings wretchedness, starvation, and untold misery. Thus does government evolve, to impose a modicum of order and lawfulness and relieve the terror of being alive.

WAR

War is the natural state of early humans; peace is the measure of the progress of civilization. Humans are naturally individualistic, resentful, quarrelsome, and spoiling for a fight. War is simply the collective expression of these innate tendencies. War is an expression of man's animal nature. The Sangik races were all bellicose (though the yellow man came to understand the advantages of peace). When social groupings began, the resentments of the individual became those of the group. Hence peace came to the group while hostility was directed to the outsider, the other tribe. Tribes defined themselves by their collective hatreds (as do many modern states). Even so, to dispel antagonisms early tribal leaders would periodically organize a free-for-all between different factions within the tribe, to "blow off steam."

Although based in man's animal nature, war is provoked by numerous other factors. *Hunger* provoked raids for food so that peoples with peaceful inclinations were soon exterminated. *Scarcity of women,* that is, domestic labor, has provoked many wars. *Vanity,* the desire to impose ones' own superior values on inferiors; the *need for slaves* in developed societies; the *desire for vengeance* when having suffered a wrong; and the *need for recreation,* because war is a kind of sport, have all caused uncountable conflicts over time. The desire to make *converts to a cult* can also cause war, and early priests were always aligned with the warriors. All early religions were bellicose; only the separation of church and state has made peace possible, at least in theory.

Early wars were always at the bidding of the gods as interpreted by tribal chiefs and were atrociously violent. War underwent various refinements. Sometimes two heroes, like David and Goliath, would decide the outcome. The taking of prisoners, instead of the wholesale slaughter of men, women, and children, was a great advance.

Soon women were excluded from battle, although they egged it on and cared for the injured. Military castes emerged who would not fight on certain holy days, or in certain places. Thus did war evolve to its present conditions of delimited barbarity.

The horrors of war are a great civilizer, forcing in a generation changes in society that would not occur in peace after ten thousand years. Unfortunately, the medicine sometimes kills the patient. Nonetheless, many and profound are the debts that society owes to war: War imposes discipline and demands organization; war exalts fortitude, courage, and self-sacrifice; war destroys the weak and unfit; war dissolves the illusion of communal equality.

Although a great civilizer, like slavery war must one day be abandoned. In our day the advantages that war once brought have turned to self-destructive disadvantage. Industrial development today provides the same advantages that war once brought, although it brings its own dangers of materialism—the mindless pursuit of useless wealth—and the slavery of drudgery.

Still, peace can never triumph until it is shown to have material advantage and until appropriate substitutes are found to express the animalistic instinct to fight, struggle, and destroy. Once war taught the need to subordinate rampant individualism to a central authority; war chose the best men to lead, but today this is no longer so.

GROWTH OF GOVERNMENT

Government has survival value and so it becomes traditional. In the most primitive society, there is *no government* and the horde is all. Children belong to the horde and are held in common. There are no chiefs (as among African bushmen today), but leaders are informally chosen and only for the occasion. The evolution of the *family*, based on sex hunger and mother love, placed children under the care of their mothers, as the evolution of the superfamily *clan* and the *tribe* were the beginnings of forms of government that have evolved into the modern territorial nation state (which still prepares constantly for war). The native Australians never advanced beyond the clan even to the stage of tribal organization.

The clan was an extended family, supported by blood ties. Closely knit and internally controlled, the clans long resisted the formation of the state. The headmen of the clans gathered together in a council of elders, the first form of government, discharging what would later be recognized as judicial, legislative, and executive functions.

When tribes later emerged, each a collection of clans ruled by a *chief*, the chief retained a council of elders for advice and control. A tribe lived in a territory and warred against its neighbors. The tribal leader was also a war leader; tribal organization always defeated organization by clans. The need for commerce and trade brought what peace there was. Trade, not highminded ideals, brings peace, as money is a universal language, transcending other difficulties in communication.

Early tribes lived at the center of concentric circles of terror. Every stranger was killed. If for some reason an outsider was befriended, the man was adopted into the clan and his membership confirmed by the mutual drinking of blood. Among Semitic tribes the person adopted placed his hand on the genitals of a tribal member and swore an oath; he was reborn into the tribe.

Later, saliva was drunk, the origin of social kissing, or blood was mixed with wine, then just wine drunk, the origin of touching cups for good luck. The first peace treaties were sealed by representatives from either side drinking one another's blood to create a "blood bond." One side would bring their best-looking girls to sleep with leaders of the other side, who would reciprocate, leading to intermarriage between the tribes as a means of maintaining peace. Every adoption or marriage to a woman from outside the tribe was sealed by a feast.

From the tribal chiefs evolved *kingship*, where all effective power is invested in a single personality. Rulership depended on the authority of a wealthy family, and the early kings were often called "father of the people." Later their power became hereditary, supported by the notion that kings were descended from the gods (some believe this today). Hereditary kingship curtailed the ferocious and bloody struggles that otherwise followed the death of a king.

As a being of divine origin, the king himself became a fetish, and everything connected with him was treated in a special way. He spoke a special language, was kept hidden except on certain days, and his touch was thought to cure all disease. He appointed representatives to do the actual work of government, the origin of the elite classes. His wife was respected as queen. Nonetheless assassination was always near, and the medicine men and shamans regularly opposed his power, as did the elite who served him.

A step toward *higher forms of government* took place with the formation of private clubs devoted at first to religious practice, then to sociopolitical aims. At first such clubs were confined to males, then females too formed secret societies. Such societies controlled initiation into manhood for the boys, who at first gathered vegetables with the women, then after initiation hunted with the men.

Male initiation would last around five years and involved a wide variety of torture, mutilation, and privation. In this way the young males learned to be strong and to resist the pressures of life. Circumcision began as an initiation rite, and the carving of a society's emblem into the body is the origin of tattoos. Not all candidates passed such tests, a personal catastrophe. Women's societies prepared young girls for married life, or from an early time formed groups that refused to marry.

Such societies performed many functions, including worship of the ancestors in secret rites, when members wore terrifying masks. The secret societies imposed restrictions on sexual behavior, especially of adolescent youths. They made it a value to keep away from other men's wives. Later the societies inaugurated commercial prostitution by hiring girls from another tribe.

The societies functioned as vigilantes when crimes were committed, and as spies and secret police. They were the first political parties, putting forth usually with violence their own candidates to rule after a leader's death. They collected taxes, often one tenth of the spoils given to the house of the king, or disguised as a religious offering. From them grew later religious organizations, and some became intertribal.

CLASS AND RIGHTS

The natural mental and physical differences between humans guarantees the formation of social classes, except among the most primitive humans before classes have emerged, and among the most advanced, where differences are absorbed in spiritual achievement. Social classes derive from many factors: proximity in birth to the king; personal achievement; wealth; political dominance, especially after conquest; social worth in the popular imagination (priests are better than laborers); race—when two races intermingle, one is held to be inferior; age, the old being in a different class from the young. Flexible social classes are indispensable to an evolving society. When class become caste, greater social stability is the result but at the high price of stifling individual initiative and an original response to life.

Nature gives humans a place to live but confers no other rights on the individual, even the right to live, as will quickly become clear when an unarmed man meets a tiger in the forest. Only society provides protection from life's dangers. Only society assures a source of food, protection from outside enemies, peace within the social order, the control of sexual behavior and the protection of the family, the right to own property, and education for the young. When rights are very old, they are often called "natural rights," but they are entirely social in origin: arbitrary, shifting, and relative. The right of one age is the wrong of another. The weak and inferior have always supported rights that would recognize them as equals and give them the fruits of others' labor.

There is no natural justice either. In nature the only justice is the inevitable conformity of results to causes. In primitive society, justice, or the enforcement of perceived rights, was enforced by ghosts working through the priests and medicine men. Such leaders devised crude tests of pain and torture to determine guilt or innocence. If administered poison, the innocent person would in theory vomit it up. In the Bible the accused adulteress was made to drink a repulsive concoction from sweepings off the temple floor. If guilty, "the water that causes the curse shall enter into her and become bitter, and her belly shall swell, and her thighs shall rot, and the woman shall be accursed among her people" (Numbers 5:27).

Vengeance, retaliation of an eye for an eye, was the basis for all primitive justice. Suicide was commonly practiced so that as a ghost the dead person could persecute the guilty. Hunger strikes in modern times are a descendant of this practice. Because everything has its price, including human life, blood money could be accepted in place of pursuing a vendetta; in early times such fines were paid in women and cattle. Eventually the practice of vengeance was taken from the control of the family wronged and placed in the hands of the state, as in modern times. Lynching and dueling, common until recently, are holdovers from the days when private parties managed the affairs of revenge.

Every grant of rights or liberty to the individual is an infringement on the rights of others. Law is at first entirely negative, a list of prohibitions based on taboo, something you must not do. The taboos, always claimed to have supernatural origin and to be sanctified by the oversight of dangerous ghosts, became fixed as the mores and later codified and enforced as the law.

But it was never the purpose of primitive courts to be fair or just in our own sense; rather, to settle the dispute so that it would not continue to fester within the society. Contestants were allowed to fight it out with the judges as umpires, then, later, to argue it out. The sense of injustice so strong in modern times did not exist: It was taken for granted, and hardly resented, that those in power would twist matters to their own benefit. Yet a society is itself judged by the honesty and impartiality of its judges.

Primitive authority was based on physical strength. Better is the *representative system of government* wherein leadership is based on ability, but in the early days of barbarism constant warfare did not permit such government to function effectively. Although a divine and ideal form of government exists, it cannot be revealed: The humans of every inhabited planet throughout the universes of time and space must discover it for themselves.

THE EARLY STATE

The state is entirely evolutionary in origin, growing from the pressure of endless war. The modern state is the institution that survived in the long struggle for group power. The state proved a superior form of organization that overwhelmed other forms of group power. The state lives by the myth that the citizen has a responsibility to live and die for it, a moral pretension that gives the state its power. The state has no divine origin and does not depend on the intelligent volition of those living within states, but is a wholly automatic development.

States are territorial and regulate the social behavior of those who live within a territory. In the strongest states the citizens share a common language and customs. Early states were small and always the result of conquest, never of voluntary association. Many were founded by nomads who would descend on herders and agriculturalists, enslave them, and live from the fruits of their labor. Hence division into classes and the specialization of labor accompanied state formation.

Emerging states depended on a technology of writing; the possession and inheritance of private property that writing makes possible; cities supported by agriculture and supporting industry; helpful domestic animals; descent in the family through the male, not the female (as was true of the American red men); a defined territory; a strong executive; the enslavement of captives; and decisive military conquests. Such were the features of the great Roman state and of the imperial Chinese state, whose achievements laid the foundations for the modern nation state.

The state was possible only through the diminution in power of the primitive blood bond and the organization into clans and tribes, from which the state emerged after a long and bitter struggle. The basis of the state is not blood, but territory. Nonetheless the power of the clans and tribes within the state remains strong and always a point of contention. When the territorial state collapses, as it did in western Europe in the fifth century AD, society reverts to organization by tribe and clan, as happened in the feudal society of medieval Europe.

Democratic rule is an ideal, but subject to profound disabilities: the glorification of mediocrity; the choice of base and worthless leaders; refusal to recognize the basic facts of life; power placed in the hands of the uneducated and indolent; slavery to public opinion (which so often is wrong). Public opinion retards social evolution, but it preserves civilization. Only through educating public opinion is progress in civilization possible. Forced advancement works only temporarily.

Representative democratic government, to be successful, requires: freedom of the individual; freedom of the mind, which depends on education; the reign of law; the security of property; the right to petition a grievance; universal suffrage; the control of public servants; intelligent and trained representation. Only on the basis of these conditions can democracy flourish through representative government. No state can transcend the moral values of its citizens as exemplified in their chosen leaders.

CHAPTER 20. THE EVOLUTION OF MARRIAGE AND THE FAMILY

Marriage results inevitably from there being two sexes of opposite polarities, strongly attracted to one another. It is nonetheless not inherent in biological evolution. Marriage is the basis of social evolution, as we have seen, without which society cannot progress. Marriage created the home, the crowning glory of social evolution. The family is the master civilizer, where the child learns most of what that individual will ever know about life.

Although men and women have different kinds of personalities, their overwhelming mutual attraction brings them together to engage in sexual activity and so to reproduce the species. In the guise of personal satisfaction, sexual desire ensures that humans place the perpetuation of the race above individual ease. Such attraction long preceded any feeling of love, devotion, or fidelity.

RESTRICTIONS ON SEXUAL BEHAVIOR

Among early peoples, there was little interest in the refinements of sexual behavior; it was something simple and natural, taken for granted. Early humans had ordinary animal desire, but no sense of sex appeal or appreciation for beauty in the opposite sex. Affection was based not on sexual behavior but on the fact that male and female lived together. Only the mixed races, and those with the largest proportion of Andite and Nodite blood, embellished and refined sexual behavior, and not always with a happy result.

Once the need for food is satisfied, the predominant concern of early humans, the sexual impulse may become dominant and require rigorous social control. No human emotion, when unregulated, brings more unhappiness and sorrow. Nature has no morals and cares nothing for the happiness of the individual, but insists on reproduction. Marriage rides on the tension between basic instincts and evolving ethics. The story of the evolution of marriage is the story of control imposed on the sexual instinct by social, religious, and civic power.

Among the early races there was little or no regulation imposed on sexual behavior. Hence prostitution did not exist. "Free love," however, has never enjoyed a good reputation above the level of basic savagery. Once societal groups form, rules appear that govern mating behavior. From a primordial and primitive sex license, where there are no restrictions, the institution of marriage has evolved to its present form, where all is restricted. Marriage was originally a group affair, then a matter for the family, and only in modern times an adventure of the individual.

In the earliest stages of development, systems of taboo evolved to keep the sexes apart in a crude way and so began to delimit the range of sexual liberty, inadvertently creating the concepts of vice, crime, and sin. Yet on days of high festival, like May Day, it was common to suspend all sex regulations.

Women were always subject to more taboos than men. Before marriage they were as free as the males, but after marriage a wife could not sleep with other males. His activities were not so restricted. The female's married condition was always indicated in some formal way, by dress, seclusion, or adornment.

There has always been, and will always be, two realms of marriage. Externally there are the rules governing the marriage, but internally there are the secret thoughts and personal behavior of men and women. The individual resents the sex regulations imposed by society, so there is an eternal tension between the group's self-reproductive need to govern the individual's behavior, and the individual's selfish need to give in to desire. Those races that placed effective control over the marriage mores produced more children that survived, became more numerous, and overcame their rivals.

EVOLUTON OF THE MARRIAGE MORES

Marriage has fulfilled very different social roles, at first a mark of prestige, then a fulfillment of religious obligations, then a means to procreate citizens for the state. In early times, often the bride was prize in a contest, or winnable only after a feat of theft, or the solving of a riddle. Later, the women themselves decided what groom was worthy by imposing tests of various kinds, or by having the groom live for a year in the family of the bride. As for the women, they needed always to be strong and capable of hard work and able to bear children. If the woman had already born a child, so much the better, because she would probably bear more. Without children, a woman was ruined. She would be divorced and returned to her family, an object of shame.

It was sinful not to be married because only the married could enter spiritland after death. From this conviction came the practice of child marriage, and even marriage before birth, if the sexes worked out right. The dead, too, needed to be married, and special matchmakers might make such arrangements for a fee.

Many tribes allowed the power-possessing males in the tribe to have intercourse with a bride before her wedding night. It was the custom for these men to give the bride a present in exchange for her favors, the origin of the modern custom of giving gifts at a wedding. In some tribes women earned their dowry in just this way. Thus were morals in many ways the opposite from modern fashion: Wives took pride in their husbands' affairs, a sign of his attractiveness, and chastity in young girls was a hindrance to marriage.

Widowhood was a fate to be feared. Widows were often blamed for their husbands' death and summarily executed or burned alive. A decent wife was expected to kill herself over her husband's grave, to accompany him to spiritland. If she survived her husband's death, her life was in any event one of bitterness and suffering, as all reproved her and no one would marry her again.

Marriage was always a business deal, closely tied to the mores governing property. Marriage existed for the advantage of the group, which therefore supervised the details. As the notion of holding property in private entered the evolving mores, theft became the greatest crime. Adultery was a form of theft, not a sexual crime as such. The woman was the private property of her father, who delivered title to her husband, who had rights over his property. He could lend her to a friend or guest, and such customs were common in some tribes. The Hebrew Old Testament teaches that women are property, and the Quran teaches that women are inferior beings: These are recent documents in the history of human custom.

Sex jealousy is not innate but is a modern development. Women were held to a closer restriction on their sexual behavior because issues of descent and inheritance were at stake. At first adulterous women were killed, but later and for a long time the husband or responsible father had the right to kill the male trespasser too. Such "crimes of honor" are common today in some parts of the world.

Because the taboo requiring married women to remain chaste depends on the mores governing property, it did not govern the behavior of unmarried girls. Later the girls' father came to enforce her chastity because that way he could get a better price for her. Once the demand for female chastity took hold, it spread with a vengeance; sometimes girls were kept in cages before marriage to ensure their purity. Prostitution came from the high value placed on chastity because those girls found to be unchaste by the grooms' mothers had no other choice in life.

INMARRIAGE, OUTMARRIAGE

There is no biologic instinct against any degree of in-marriage, and all such rules are a matter of taboo. Early humans noticed that the offspring of inbred couples were weaker than the offspring of outbred couples and gradually placed taboos on inbreeding; those who did not restrict inbreeding would weaken and disappear. On the other hand, inbreeding by no means always led to inferior offspring if the parent stock was strong. Inbreeding had the advantage of preserving the knowledge of certain crafts within a tribe and of preserving wealth within the family. For over one hundred fifty thousand years the Nodites practiced consanguineous marriage. Early members of the violet race perforce practiced consanguineous marriage, and traditions of their behavior long persisted. Brother-sister marriage was common in the Egyptian royal family in an effort to keep the blood pure, and practiced also in Syria, Mesopotamia, and Persia. Abraham himself married his half-sister.

Nonetheless outmarriage won out in the end because the men preferred it. Outmarriage lessened the influence of in-laws. Also the males preferred a mate who was not so familiar as a sister so that marriage outside the tribe became the rule. The practice promoted peace, encouraged alliances, and was a first step on the long road to the building of nations. Only kings and important leaders preserved the right to marry their close kin, in order to preserve their bloodline. Kings have always been allowed wide latitude in sexual matters.

WINNING A WIFE

The institution of marriage has been threatened many times in history, but it will always survive because of the fact that men and women simply will not live without one another, whether they are the most primitive savage or the most modern sophisticate. The irresistible sex urge is the basis for progress in civilization because through it, in the issue of children, everyone must learn the reality of self-denial and the technique of altruism. Because of sex, humans are forced to think and through sex they are led to love.

Early marriage was based on coercion, not affection. Common was to capture a woman from another tribe. Sometimes the woman connived at the capture, wishing to escape the supervision of her own male relatives and to be with someone her own age. The mock capture succeeded the real thing, and then much later the capture by charming the lady's interest. The modern custom of carrying the bride over the threshold descends directly from primitive customs of wife-capture.

The wife could also be bought by the payment of an agreed-on sum. Later, not wishing to appear to be selling their daughters, fathers gave gifts to the newly married couple of a value similar to the bride-price. As the mores became more refined, the bride price was abandoned but the gifts remained, the origin of the dowry, sign that the wife was not a slave or just a piece of property. A man could not then divorce his wife without repaying the dowry fully, so the dowry reflected the increasing power of women within the social hierarchy.

The wedding itself was a public affair, never private. Because there were no records, it was important that many witness the couple's union in order to clarify the legitimacy of children and their rights to inherited property. Because marriage without children was not a marriage at all, early humans did everything they could to assure a fertile union, choosing a favorable day, and a favorable phase of the moon. The modern casting of rice or grain on the married couple goes back to such magical behavior.

The great threats to a successful marriage are the envious ghosts and hostile spirits, but the bride's veil conceals her identity from them and the carpet from the coach to the altar prevents a discharge of protective energy into the ground. At one time a human sacrifice always accompanied marriage among the well to do; the blood would feed the hungry dead who then would leave the couple alone. Best to have some priests at hand to defeat the envious ghosts, the origin of the wedding in a church still common today.

POLYGAMY

At first the unmarried women of the tribe belonged to all the men, then the woman belonged to one man at a time. Although the man could break the relationship at any time, the children belonged to the mother. Then came the practice of group marriage when, for example, a group of brothers would marry a group of sisters. The Jewish custom of requiring a man to father children on the wife of his dead brother descends from this practice. From group marriage evolved polyandry and polygyny, although the rare polyandry was mostly restricted to queens and women of high social prestige. Even so, the woman would usually "marry" just one man, the others being considered as "uncles."

In polygyny the wives were always of different status depending on how they were acquired. There was always a "status wife" who was ceremonially married to the man and who through her position dominated and controlled the other wives. Only her children could inherit property. The status wife was ordinarily not the "love wife," or sweetheart of the man. This system led to formal concubinage, where the husband was permitted to maintain sexual relations with concubines, but they were no longer "wives." Often the concubines were the slaves or servants of the status wife, and as among the Jews all children born to the man were considered to be the children of the status wife.

Sex taboos against having intercourse with pregnant or nursing women fostered polygyny, because women tended to be pregnant all the time, leaving the man without opportunity to satisfy his sexual desire. Sex jealousy is not found in polygyny because the older wife, worn out through constant child-bearing and endless hard work, welcomes the assistance and relief of the younger wife or wives. The very short lives of early women were a constant misery, and they only survived at all because for one week out of every month, during menstruation, the woman rested in isolation.

The man could have as many wives as he could support. If he was wealthy, he would have many, because only in this way could he insure that his children survived the murderous rate of infant mortality. Many such wives were slave wives, mere laborers.

MONOGAMY

Monogamy is a kind of monopoly: It is fine for those who achieve it successfully, but it shuts out in the cold those who are unsuccessful, even if they have tried hard and played by the rules. Monogamy originates in poverty, where a man can only afford one wife, but monogamy always makes a stronger economic unit of the family. Apart from the suffering monogamy imposes on the individual, it is the best system for raising children. Monogamy is unnatural for the evolutionary races (whereas the Nodites and Adamites were always monogamous), but it is the ideal form of marriage and of great social value to those who adopt the practice.

By the time of Christ all of the civilized world had accepted monogamous marriage in theory if rarely in practice. Monogamy requires powerful self-control, a virtue hard to attain, and its failure always traces back to a lack of self-control. Monogamy refines one's sentiment, enhances moral character, and encourages spiritual growth, all impossible in polygyny where a woman's need constantly to compete for her husband's affections guarantees that she will not be much of a mother.

DIVORCE

In primordial times a man simply dismissed a woman when he tired of her, as still today in some parts of the world, or if there were no children, always blamed on the woman. Any children went with the mother where the instinctual bond is not cultural or social, but inherent. About half of all marriages ended in separation, but the rise of bride purchase and dowry worked to keep couples together, as today economic considerations often have the same effect.

The rise in modern times of romantic love, not property, as the basis for marriage has resulted in a high rate of divorce that is certain to remain high until the young are educated in the realities and responsibilities of pair-mating. The steep rise in very recent times of woman's status, so long denied, has placed extreme strain on the pair-marriage and the home it sustains. Nonetheless, marriage in modern times stands at a high position when compared with mores of the past.

FAMILY LIFE

In spite of the contradictions in the lives of early men and women, and the looseness of their attachments, marriage improved the chances for survival in a relentlessly harsh world, and the division of labor improved the happiness of all. Society is the combination of all its family units, through whom culture and education is passed. The individual has no continuity and in this way counts for little, but the family goes on. Through the family social evolution takes place. The propagation of new life will always be the basis for the family and for all social life. The home combines the three basic human instincts for self-perpetuation, self-maintenance, and self-gratification, while sexual behavior sets the family off from all other kinds of social behavior.

The male in early societies did not need marriage for sex; it was presented to him everywhere he turned. Who needs to be encumbered by wife and child? The woman, by contrast, was held to her children by physical and emotional attraction. Mammalian mother love is the special endowment of the Adjutant Mind Spirits of the Local Universe. In strength and devotion such mother love is directly proportional to the length of the helpless infancy of the species, which in the case of humans is very long.

The woman's mother love has always handicapped her in the contest for power with the male, who does not feel this love. Ambition, selfishness, and religion can overthrow and thwart this all-powerful emotion, but ordinarily the woman wants the security of marriage, an environment in which she can raise her children. Not sex, but food hunger brought in the male, who enjoyed the rude shelter the woman provided for her children.

The connection between sexual intercourse and female pregnancy was completely unknown and unsuspected. A virgin could easily become pregnant when struck by the evil eye or because of something she ate. A baby came from spiritland, a ghost working its way out. Ghosts lived in the water, so a woman could get pregnant by bathing at the wrong time. Deformed or premature babies came from an animal's spirit attempting to penetrate the woman, and all such babies were destroyed. Later, when the connection between sexual intercourse and pregnancy was made, the theory was common that the penetrating penis opened a path for the ghost to enter. The child always came from the father; the mother was the soil in which the seed grew, but not the source of the seed itself.

MOTHER-FAMILY, FATHER-FAMILY

The woman's periodic discharge of blood that stopped at pregnancy, and the discharge of blood at childbirth, suggested that the child was made of blood and led to the notion of the blood-bond. In early times this bond passed down through the woman, the only certain agent in the child's descent. Many peoples never gave up the mother-family, based in biology, including the red man of North America. The father family, by contrast, is social, economic, and political and made possible the formation of the nation state.

In the mother-family the wife's mother is all-powerful and her brothers and nephews can exert more authority in the household than the husband. The child belongs entirely to the mother. In the father-family, the situation is reversed, and now the child belongs to the father alone. When his wife is due, he takes to his bed and awaits the delivery, then from his bed accepts congratulations for the delivery, while his wife is sent back to the fields.

Under the mother-family, the husband goes to live with his wife's mother; in the father-family, the bride comes to live with the groom's family, where she submits to the absolute authority of the groom's mother. The father-family was superior to the mother-family because the husband/father was able to concentrate more power than the wife's male relatives, among whom the power was always diffused. The wife herself was too busy with child-rearing to acquire and maintain power effectively. Wife-stealing and wife-purchase fit with the father-family. Societies based on the father-family always overcome societies based on the mother-family when they come into direct conflict, as they did when Europeans settled in North America.

The shift from mother-family to father-family was one of the greatest about-face alterations in human social life ever to have taken place, and its consequences were gargantuan. The shift paralleled and accompanied the shift from hunting and gathering to the herding of flocks, when the patriarchs held absolute power. As reflected in the biblical narratives, the male held scant respect for the female during the herding stage of social evolution. After all, the women were burdened constantly by the demands of child-rearing, they were physically weak and unable to defend the tribe from marauders, and they behaved poorly in a crisis. Women only worsened their position in the social order by praising and glorifying the strongest and most brutal males, successful in the very tasks they could not succeed in.

SEX ROLES

Throughout human history the woman has been the workhorse, toiling at vegetable plots and gathering wood and water, tasks unworthy of the male, who gloried in his freedom as a hunter, warrior, and herdsman. In agricultural society a great advance for women came with the enslavement of captives, who could be forced to work the fields, leaving the woman free to run the house. When the male eventually came to accept field labor as an honorable endeavor, he applied methods of organization he had learned as a hunter and warrior to the increase of the food supply, whereas female labor had been loose and disorganized.

As a rule the status of woman within marriage measures the progress of the institution, and the progress of the institution measures the progress of the whole society. Although the woman often had the status of a slave, a piece of property, through her sexual relationship with the more highly sexed male she was always able to exercise subtle power and through her charm and wiles get some of what she wanted. The male did not appreciate this power, when he understood it, and saw the female as crafty and duplicitous, although an object of irresistible fascination.

The sexes simply could not understand one another. The male's fear and mistrust is clear in the many stories, such as those of Eve and Pandora, that attribute to the female the introduction of evil into the world. In the male imagination the female was weird, probably had no soul, and could do harm even with her shadow. Childbearing, accompanied by a pouring out of watery fluids, blood, and excrement was overwhelmingly polluting, and the male would have nothing to do with it. The expectant mother was forced to remain alone and bear the child by herself, until later old women might attend, the first midwives. No male would come anywhere near a mother who had given birth until she was purified by elaborate ritual. The mother's death in childbirth was a clear result of spirit adultery, as was the bearing of twins.

So-called feminine modesty with regard to clothing comes from the fear of being observed during menstruation. From the earliest times the menstruating woman was quarantined for one week a month, a time of blessed relief from the onerous daily life. Anything she touched at this time would be polluted, ruined. Even her glance would pollute.

This attitude toward the menstruating woman had a great deal to do with woman's traditional lowly status. But primitive women did not pity themselves or complain, or dare to hope that anything could be different. Not religion, but science is today setting woman free in a world where brain power, not physical might is decisive. Not revelation, but evolution is setting woman free.

THE ADVANTAGES OF SEX COMPETITION

Lust brings men and women together, but only the parental instinct and the social mores keep them together. Sexual intercourse is biologic, natural; marriage is sociologic, social. In monogamous marriage one learns to accept antagonism and go on in a practical working harmony. Such antagonism is inevitable between individuals in the nature of things. In the family life that goes with marriage one must learn to adjust one's own temperament to that of others, who live intimately with you. This is growth in the spirit. The parental love and affection is the child's first lesson in the Father's love for all his children.

Each sex will remain forever supreme in its own domain, determined by biologic difference and by mental dissimilarity. The man and woman are distinct varieties of the same species compelled by biology to live in close and intimate association. Female and male viewpoints and their entire approach to life are essentially different. The sexes are opposites and can never understand one another. But woman has always been the moral standard-bearer.

Humans are fortunate in their sexual bipolarity because a recurring principle in universe organization is the bipolar pair, which always multiplies variability. Men and women need each other, not only in this mortal life, but in the morontial and spiritual lives to come where the division remains real. Even in Havona the beings who once were male or female will continue to help one another, and maleness or femaleness never wholly disappears. From their irreconcilable points of view comes an ever-flowing richness.

CHAPTER 21. THE EVOLUTION OF RELIGION I: NATURE AND GHOSTS

Primitive religion has a biologic basis. Animals fear but have no illusions, hence no religion. Humans build religion on their fears by means of illusion. Hence they have at some time worshiped everything on earth, above the earth, beneath the earth, including man himself.

The objects of early worship are based entirely on suggestion. If something impressed early humans, they imagined that it was filled with a special power, a spirit, and the things of nature made powerful impressions upon them: volcanoes, earthquake, storm, flood, fire, cold, infestations, disease. Still today we call inexplicable destructive events "acts of God."

THE POWERS IN NATURE

Humans have a deep affection for stones, and they were the first objects of adoration. Still today the meteorite in Mecca is the object of unquestioning adoration for the world-wide religion of Islam, and within the ancient mosque on the temple mount in Jerusalem is the stone from which Mohammed supposedly entered heaven, or where Abraham nearly sacrificed Jacob, according to other legends. In the center of the church of the Holy Sepulcher, most sacred site in Christianity, is a stone thought to have been part of the tomb of Jesus; for a few dollars the tearful tourist can touch it and even kiss it. Sacred stones, which often resemble animals, are everywhere. The wearing of stones as jewelry began as devotion to the powerful spirit in the stone, sure to bring good luck. A hole in a stone always increased its power.

From stones, the human imagination turned to large stone formations that suggested something or other, then to whole mountains: Up there lived the gods, as the demons of death lived beneath the earth, where the bodies went, into whose realm penetrated deep caves.

The intoxicating powers of some plants made them obvious objects of veneration; clearly a spirit lived inside them. Intoxication was coming close to the divine, even as we say that liquor contains "spirits." The sprouting grain was a subject for awe and quickly became a symbol and magical agent for rebirth. The moistened grain was placed in tombs in a shallow tray, there to sprout and give life to the deceased. Paul of Tarsus (died AD 65) famously said, "Take a grain of wheat, only if it die, may it live again."

Tree cults appeared very early. Virgins embraced a special tree when they wished to become pregnant. Different peoples imagined the tree spirits in different ways: Either they were tricky, cruel, or kind. When the cult declined, copses remained the site of later religious cult. There was a universal cult of the Tree of Life, except in China, descending from memory of the real tree of life in the Garden of Eden that had sustained the mortal bodies of the Material Son and Daughter and their coworkers. The Maypole, the Christmas tree, the practice of rapping on wood to avert evil, the use of divining rods to detect water or metals are inherited directly from ancient tree cults. Such early religious behavior was activated by the Adjutant Mind Spirit of worship and was strong long before humans were susceptible to higher spirit influences.

Primitive humans, as many moderns, had a strong feeling of kinship with animals. Their ancestors had lived with them and even mated with them. They held animals in awe for their special powers of speed, keen sight, and strength. Such powers must depend on spirit assistance. The souls of some humans came back as animals, they thought. Some human group has worshiped every animal on earth at some time. Snakes, especially, have attracted attention; the Chinese dragon descends from the ancient snake cult. Winged creatures were always popular.

Rivers and springs, with their mysterious bubbling and rushing waters, must be where spirits dwell; for this reason a virgin must be careful when bathing so that the spirit does not enter her. When we toss a penny into a fountain today and make a wish, we behave just as did our ancestors when they cast humans and other valuable objects into the waters to satisfy the power that resided there.

In the sky the rainbow was a great being. Earning devotion among many groups was wind, rain, hail, and especially lightning, the voice of a terrifying god. Fire worship has had a long life, ranging from adoration of the flame itself to taking fire as a symbol for something else. Still today worshipers burn candles before altars and, just as there burned a sacred fire guarded by vestal virgins in ancient Rome, a perennial fire burns beneath the *arc de Triomphe* in Paris and on the American tomb of the unknown soldier in Washington, DC.

The prominent spheres in the heavens attracted religious cult—the sun, the moon, the stars. Many believed the stars to be the souls of the mighty dead. Worship of the moon preceded that of the sun, being popular among hunters. Agriculturists preferred cults of the sun and imagined that their kings descended from the sun god. The powerful cult of the god Mithras, strong in the Roman armies at the time of Christ, began as a Persian cult of the sun.

Exceptional humans were always regarded with awe, whether they were idiots or geniuses. When kings died, it was easy to worship their spirits, now made into gods. Some kings claimed to be gods and demanded worship from their subjects. Humans' worship of other humans, alive and dead, is the highest that religion can go through evolution alone. Then revelation is required, the teaching of higher truths, as in *The Urantia Book*. You will never reach these truths on your own, but must be told them.

Humans make gods in their own image, but revelation makes humans into the image of God.

THE DUAL ORIGIN OF EVOLUTIONARY RELIGION

Evolutionary religion, which takes form in illusion based on fear, finds its origin in two sources, in nature cult and in the ghost cult that we introduced in the last chapter. Spirits of nature and spirits of the dead easily turn into gods, where they may become mixed up. For example, Thor was a real man who once lived, but in the religious imagination of the later German tribes he was a god who controlled the lightning. The sixth Adjutant Mind Spirit of worship, bestowed on early peoples as a directing influence, constantly stimulated the urge to worship, motivated by animal fear. It is logical that early cult focused on things in nature and imaginary ghosts because feeling, not thought, guided the evolutionary development. To primitive humans there was little difference between fearing, shunning, honoring, and worshiping.

When the seventh Adjutant Mind Spirit adds wisdom to the urge to worship–meditative thought that absorbs the lessons of experience—then appears what we might think of as real religion. The religious human turns away from nature and ghosts to the eternal creator of all things natural and of humans themselves.

GOOD LUCK, BAD LUCK

The theory of good and bad luck was critical in the evolution of early religion. The primitive male was a hunter, where success always varies. One day the hunter has "good luck," another day "bad luck." The consequences are dire to humans living on the edge of starvation, always in fear of the next atrocious thing, dreading inappropriate behavior that might anger the spirits and bring still more bad luck.

The worst thing that could happen to anyone was a sequence of fortunate events because a catastrophe was sure to follow. The ever-present fear of bad luck paralyzed action. Why work hard when bad luck will soon overturn it all? Why not simply wait for good luck and not work at all? Early humans alternated between a passion to get something for nothing and the fear of getting nothing for something.

Existence was a gamble and the early human was a gambler. Not only hunting, but herding and agriculture too appeared to be playthings of good and bad luck. Sometimes crops were good, but in other times came drought, hail, floods, and pestilence: good luck, bad luck. "I returned and saw that the race is not to the swift, nor the battle to the strong, neither bread to the wise, nor riches to men of understanding, nor favor to men of skill; but fate and chance befall them all. For man knows not his fate. As fishes are taken in an evil net, and as birds are caught in a snare, so are the sons of men snared in an evil time when it falls suddenly upon them" (Ecclesiastes 9:11).

Because bad luck constantly threatened, primitive humans lived in constant fear. But pain and suffering are essential to progressive evolution. When in modern times an individual is constantly anxious, wracked with tension, that person has only returned to the ordinary state of mind of all primitive humans, when life was constant suffering. In pain humans cried, "Who is tormenting me?" Obviously the powerful spirits in the natural world were to blame, and of course the maleficent ghosts of the dead.

Early humans were logical but knew nothing. What we call superstition was really rank ignorance about reality. If one thing follows another, the relationship must be causal, they reasoned: after this, because of this. They could not think abstractly. They personalized everything. The forces of nature were persons, spirits, ghosts, then gods. They dealt with them in the only way they knew, as one person to another.

The notion of chance—good luck and bad luck—depended on the primitive belief that nothing in the world is accidental. Everything happens from intention. Most primitive humans died violently or from disease, so there was no expectation of a necessary death. You do not have to die, and many cultural traditions told tales of men who never died. The unusual death from natural causes was therefore quite mysterious. Life was natural, death came from the outside, from the invisible world of spirits and evil magicians.

THE BREATH-SOUL

As we saw earlier, the belief in a supermaterial existence of the individual human depends on the ghost dream. How else can you explain that several members of a tribe will dream in the same night of their dead chief, awakening in a cold sweat and shaking with emotion? But the notion of the ghost's survival alleviated the terror of death, so disturbing to the biologic sense of self-preservation.

Obviously the breath is in some way connected with the life of an individual because the absence of breath distinguishes the dead from the living—the breath has escaped, is out there, and probably not happy about what has happened. While alive the individual had a dual nature, body and breath. At death, the body falls away but the breath goes on. They thought the breath could also escape temporarily, as in a faint or a coma and during sleep. For this reason we say "God bless you" to a sneeze because the breath-soul came close to escaping. The breath-soul goes out at night in dreams and has various adventures; the dream world is as real as the waking world. The breath-soul is a double, the same as the material person except you cannot touch a breath-soul. After death, the breath-soul lives in the spirit world and although it had a human origin, it is supernatural. Certainly these unhappy breath-souls are causing all this bad luck.

Such views were not a belief in immortality, a concept beyond the reasoning powers of humans who could not count beyond twenty. The breath-souls did not stay in spirit land forever, but would come back as another human, or an animal, or something else. Reincarnation appeared to explain hereditary similarities within a family line. From this belief comes the common practice of naming a child after his grandparent or other deceased family member.

At first there was no notion of heaven or hell; the spirit world was just like this one, except no more bad luck! Because you entered the spirit world just as you left this one, few wanted to become old and enfeebled. Much better to die young and in the full flush of life. Later appeared the belief that good souls and bad souls went to different places, a kind of heaven and hell.

Still, different peoples held very different views about the fate of the breath-soul. The Egyptians and Chinese thought that it remained near the body, which for this reason they carefully preserved. Others thought the soul returned to the ancestral lands. In the Hebrew tradition the soul went to a place called Sheol, like the Greek Hades, from where it could never return. That was an advance in thought.

Humans are born into a natural environment; they acquire a social environment; and they imagine a ghost environment. The state evolves in response to the dangers of the natural environment, the home evolves in response to the social environment, and religion in response to the fear of bad luck and the illusory environment of gods and ghosts. Evolutionary religion can go no further than this, and in some parts of the world today has not progressed beyond this stage. Only revelation can turn religion in a different direction. Yet long before revelation, the evolving religion based on chance and ghost fear established in human consciousness the moral categories of right and wrong: The good spirits do right, and the bad spirits do wrong. Humans, too, must operate within these categories.

It is easy to condemn the folly that grew from illusion based on fear of the ever displeased spirits—the gigantic tombs, the killing of billions of animals and humans in bloody sacrifice, the sheer oceans of blood, the destruction of property, the building of pretentious temples, the support of expensive and gluttonous priesthoods. It is rather depressing to witness the overwhelming power that the ghost cult has held on the Urantian imagination. This worldwide all-pervasive cult has been entirely negative, attempting to prevent disaster and never aspiring to a higher knowledge or a devotion to deity. Yet in its insistence on the reality of supernatural beings, this ugly cult opened human apprehension of what is real in the spirit world. The creation of the consciousness of relative right and wrong, human ethics, was alone worth the terrible cost. Humans learned that they simply must regulate their conduct: There is a supermaterial world out there that is in control of human destiny, as in fact there is.

Primitive humans paid insurance against the certainty of bad luck through their evolutionary religion. The costs were high but the danger was great. In modern times we pay money for protection against catastrophe based on scientific actuarial tables. Removing religion from the insurance business has released it to speculate on less material things, although it has shown little inclination to do so. Primitive religion prepared the human mind, by the overwhelming force of fear, for the bestowal of a real spiritual force of supernatural origin. The divine Adjusters have ever since labored to transmute fear of the divine into love of the divine. Although evolution is slow, it works in the end.

MORE ON GHOST CULT

In ghost cult, the most dangerous time is just after death, before the ghost has a chance to enter ghostland. Although supernatural, ghosts are not very smart and they can be tricked, forestalled, and blocked by various means. When someone got sick and might die, they were removed from the family home and placed outside or in a special small hut. No one would live in a house where someone had died, for millennia an obstacle to the development of larger and permanent buildings and of the establishment of villages and towns. Still today the Navajo destroy the hogan in which someone has died.

The corpse is dangerous and anyone who touches it needs to undergo purification. When someone died, the other members of the tribe would stay up all night, talking; to fall asleep was to risk dying too. The corpse must not be left in the dark, where all manner of mischief is possible, but lighted, the origin of the burning of candles at funerals.

It is important that the ghost be kept from returning to its old haunts, so after burial the mourners take a different path from the one they came on. Men and women dress as each other to confuse the stupid ghost, who might take the opportunity of its release from the flesh to settle old scores, or just generally to take pleasure in the suffering of those still alive. Special mourning clothes are worn, later explained as showing respect for the dead, but in origin their purpose was to confuse the ghost.

To calm and satisfy the ghost was the initial purpose of ghost cult. Later evolved the coercion of ghosts and the supplication of ghosts. To remove the ghost from this world and human contact, to "lay" the ghost, required precise ritual. Omissions or errors will invite ghost anger and reprisal. The funeral was in origin a ritual to accomplish the laying of the ghost, to get it out of this world and into ghostland. Naturally the ghost will enjoy food and clothing for its journey to ghostland, placed with the corpse in the grave. It might take from three days to a whole year to break the ghost's connection with the corpse and the grave, but then you can be free ... until someone else dies.

One can treat the corpse itself in various ways to prevent the ghost's return: break its legs, cut off the head, remove the testicles, tie the arms with rope, place a large stone on top (one of the origins of the tombstone). You can also burn the corpse in the fire, throw it into the sea, or expose it for wild animals to eat.

Loud noises of bell, drum, or voice, and foul smells, and ugly masks will all frighten away the ghost. Otherwise silence is good (after the noisy wake), because ordinary voices may attract the ghost. You can tell when the ghost is near: The dogs will howl and the cock will crow. Fire will drive them away. Ghosts also hate water, especially the holy water in which the priest has bathed his feet. Ghosts will not cross a body or stream of water. Bathing among early tribes is never for hygienic purposes, but always a ritual act; baptism still washes away ghost pollution.

Self-immolation will delight the ghost—cutting the hair, chopping off a finger joint, scratching the face, fasting, wearing black cloth—because such behavior assures the envious spirit that you are not enjoying life now that he or she is gone. Long periods of mourning will prove how sorry you are—such periods of worthless idleness have been a great obstacle to human progress. The names of the dead must never be spoken: You could attract the ghost.

The best strategy for ghost-placation is to provide the ghost with things it had in life: servants, a wife, a mother. In some early tribes a slave wife was customarily buried alive with her dead husband. Later, a good wife would kill herself on her husband's grave (still practiced occasionally in India today). If a valued child died, the mother, aunt or grandmother would be strangled to care for the child in ghostland. Such human sacrifices are always voluntary; it would hardly be possible to behave otherwise when ghost fear and public disapproval would make an already miserable life unbearable. Many died at the funeral of a great chief, to serve their master in ghostland.

Ghosts enjoy the smell of food, and funerals always had food offerings for the dead. A pinch of food tossed in the fire, or bit of wine, will calm any hungry ghost. To transfer tools, weapons, and other objects into ghostland, mourners would break them, burn them, or bury them, "kill" them so they could join their master. In this way untold quantities of wealth were destroyed, greatly retarding human advancement. Later substitutions of fake objects (or as in China today, fake money) alleviated the practice's harmful effects, but nonetheless still today vast quantities of wealth are squandered on funeral display. Ghost fear led inevitably to ancestor worship, the predominant religion in China today and always throughout history of enormous power and importance.

EVOLUTION OF THE GHOST CULT

Every human race and people at one time believed in ghosts, which do not in fact exist. At first and for a very long time all ghosts were the same, but very gradually, and as an independent parallel development in different societies, came the recognition that some spirits were "higher" than others. These were the "good" spirits, able to bring good luck, as the other ghosts were the "bad" spirits, far more numerous than the good ones and the bringers of bad luck.

The evolution of a dual spiritism was a radical change in the human religious consciousness because, instead of the earlier altogether capricious ghosts who might jump this way or that, now was recognized a supermortal guidance of human affairs by beings who were consistent and logical in their behavior. The down side of the evolving dual spiritism was the tendency to regard the forces of good arrayed against the forces of evil, when in reality the universe is ruled by the cosmic good. Most modern religions preserve in some form this primitive dualism, which grew from the ghost cult.

The higher and benign spirits, no longer ghosts but gods, received their priests and their cults, but most religious activity was still directed toward foiling the hostility of the lower and malevolent ghosts, thwarting them through bribery and tricks. Humans were the pathetic and helpless victims of ghost persecution, that was obvious.

When notions of higher "good" spirits evolved, then devotees of these spirits turned more toward persuading them to grant good luck, and then actually to compelling them to do so. Exorcism was driving out one spirit by means of another. Primitive cursing was a means of driving back the lowly ghosts. Self-torture and the austere denial of sexual activity also helped one to control the ghosts, who could hardly envy someone living in conditions of suffering and deprivation. Thus did humans begin to take some control of their destiny, even if they were only indulging their illusions.

The ghosts' principal trick was to bring harm through the evil eye. One maleficent glance will allow penetration of the ghost and certain disaster. Elaborate procedures evolved to meet the danger of the evil eye, the single religious belief that all humans have held (the Quran devotes a whole chapter to the evil eye; "fascinate" meant originally "to fix with the evil eye"). Women first were veiled to protect them against it (as still today in Islamic countries), children were not let out after dark, and prayers would end with "and deliver us from the evil eye." Phallic emblems were effective against the evil eye (the origin of giving an aggressor "the finger"), as were representations of an eye (opposing fire with fire) or of a hand (stop!).

The best strategy in such a world, beset on every hand with goblins determined to get you, is to strike a low profile in life, not to appear too wealthy or too prominent, to be physically ugly with an ugly name, and if anyone compliments you, or you are involved in a happy occasion, to turn away the compliment and deflate the value of the happy occasion by insults or other deprecating behavior. For the ghost is envious, like living humans. One's success in life never depended on effort, initiative, or imagination, but on how able you were in defeating the malicious spirits, to which you devoted your whole life. Fortunately you could learn what the ghosts intended through oracles and prophets, or by inspecting the stars. Dreams were valuable channels of communication, too (in reality higher beings rarely use dreams to communicate with humans). The flight of birds is informative, as is any "chance" occurrence. Still today humans seek omens and "signs" everywhere.

THE FETISH

Corollary with the ghost cult was the belief in the *fetish*, an object, animal, or human in which a ghost resided. A fetish can be almost anything—a peculiar stone, a comet (but not a star), a volcano, a tree. For the Hebrews, the stone altar was a fetish in which they believed that the spirit of Yaweh lived. When animals became a fetish, a taboo on eating them soon followed. The snake as fetish has been widespread. Days and numbers can be fetishes, as Friday and the number 13 both bring bad luck. Saliva was potent and could drive out evil spirits. The umbilical cord was a powerful fetish and, set with stones, became the first necklace. The fingernails of powerful people were coveted and thought to be imbued with power.

Chiefs, kings, priests, prophets, and church rulers, as well as men of genius, were thought to be infused by a higher spirit, and they have always wielded great power. Their bones, infused by their spirit, were held in reverence and awe and preserved in special places. Commonly today people journey to the tombs of powerful people, and churches and shrines commonly preserve in glass cases relics of holy men and women, recipients of the candles of the devout.

Idols were a form of fetishism. The first idols were made to preserve the memories of the illustrious dead. At first they were matter and form, but the spirit could be compelled to enter the idol. Even so the holy man may bless an ordinary object, which then becomes magical and special, a charm.

Moses explicitly condemned the cult of fetishes in the second commandment: "You shall not make a graven image or any likeness of anything that is in heaven above, or on the earth beneath, or in the waters of the earth"—that is, a fetish. The commandment retarded the development of artistic expression among the Hebrews, who made no artistic accomplishments, but lessened somewhat the power of the ancient fetish cults.

Nothing is more dangerous than the fetish of the word. No doubt for this reason Jesus was careful to leave behind no writing of any kind. Repeatedly in modern religion books have imprisoned the spiritual imagination. Such books, thought not only to be entirely true but to contain all truths, have led humans into bigotry, fanaticism, superstition, intolerance, and atrocious and exquisite cruelty. To swear on a book is primitive fetishism, as it is to open a sacred book at random, let the eye fall upon it, and there read the answer to a question. Even so, such fetish books are superior to the veneration of a chief's toe nails.

Fetishism pervaded virtually every aspect of religious behavior and still does today. One development was the totem, when the clan chose usually an animal to represent the clan. This animal was given special respect and not eaten. Totemism was a way of socializing the otherwise personal fetish. Modern flags, as emblems of a people, descend directly from the practice.

Common were fetish or medicine bags, in which were kept a collection of ghost-impregnated objects. The pouch, like the flag, could never touch the ground without losing its power, as by electrical grounding; for the impregnated ghosts might slip away into the ground.

Kings had fetish scepters, emblems of their power, which they "inherited from God." Such thought has long dominated political structures. Even an idea can be a fetish, as in the notion that democracy is a God-given form of government: The views of one man on any topic have little weight, but gather them together in a mass display and suddenly they are the arbiter of justice and truth. With a fetish you do not question its power, but accept it without question. That is where its power comes from.

CHAPTER 22.
THE EVOLUTION OF RELIGION II: MAGIC, PRIESTS, SACRIFICE

In modern civilized society, problems in the world are attacked through science, through knowledge of facts achieved through experimental means. In earlier times, the real problems of an imaginary ghost environment led to *magic*, a means to control the spirits through persuasion or force and so control one's own destiny. Fetishes and other spirits were the magician's agents.

MAGIC

Magic and science have the same goals, to learn the future and to influence the environment, but the transition from magic to science has been long, bitter, and drenched in blood. Only now have Urantians backed into the truth through long experience and have turned around to face the world with the technology of science. Primitive humans were driven by the same curiosity and fear, to know about the world and to control it.

The inability to understand the concept of natural death encouraged the belief in magic, the deliberate manipulation of ghost power to kill. Commonly ten people would be killed in retribution for the death of one man who died naturally, a custom that greatly held back the growth of population. Before death, the victim usually confessed. The later notion of original sin, that humans were born to die, was therefore a great advance over recurring mass murder.

When everyone believes in magic, the fear of death through magic can actually cause death, which of course reinforces the theory. If the magic does not work, then there is need for more magic. Anything having to do with the body can be manipulated to harmful effect. For this reasons early humans buried their excreta in secret and covered their spittle and left no food in the bowl. None of this behavior had anything to do with the practice of hygiene, which did not exist.

Magic was public when performed by the shaman or priest, and private—to harm one's enemies or seduce one's beloved—when performed by the sorcerer or wizard. Wands and incantations were common, and the magician was often naked. There were many women magicians. Magical incantations, often danced or acted out, evolved into public prayers. Mimicry had magical power. Sexual license at May Day is imitative magic that appeals to the sex lives of plants. Dolls bring fertility.

Because there are two kinds of spirits, good and bad, there must be white and black magicians. It is dangerous to do too well in life—you could be practicing the black arts and therefore must be killed, as continues in Africa today. Half the world still languishes in the grip of ghost-fear, fetishism, and the belief in magic. And even those living in the civilized world are *spellbound* by the one they love, *ill-starred* when they lose at the track, covetous of *possessions*, longing for *inspiration*, fearful their children might be *spirited away*, hopeful for *ingenuity* (when the spirit "is born within"), *entranced* by beauty, *thunderstruck* by levels of ignorance (because a spirit lives in thunder), and *astonished* ("thundered," when a spirit enters) at the whole thing.

TABOO AND SIN

Early humans realize that they are in debt to the ghosts, they owe them something. They may suffer from the ghosts' persecution, but they might suffer a lot more. They should be thankful. You need to buy back your life from the ungrateful dead. From such thinking emerged the notion of sin, a condition of having done something wrong to excite the ghost's anger, and salvation, a condition that comes from having paid the penalty for wrongdoing.

Scapegoats are useful for this system of necessary repayment, one of the origins of headhunting: Not my head, but the one hanging from my belt will take the punishment, in fact has already done so. You must give something to the ghosts, you must win their favor not in hopes of future bliss, but to forestall immediate calamity.

The ghosts are offended in two ways, by commission and by omission. Sins of commission come from violating taboo. The taboo is always negative, "Thou shalt not" do this or that. The taboo, a system of right and wrong based on ghost fear, was the first form of the regulation of society and is still very strong. Law is at first a codification of taboo. Taboos were the ancestor of self-control.

The common food taboos originated in fetishism or totemism. The Egyptian taboo on pork was taken over by the Hebrews and Muslims. To the Hindus the cow was taboo. The famous ten commandments of the Hebrews are all taboos—don't do this, don't do that—as were the seven commandments of Dalamatia and Eden. They were an enormous advance over the thousands of negative commands found among most peoples because they generalized the terms of avoidance and they promised something in return for obedience.

The emerging systems of taboo, making possible the concept of sin, at last offered a logical explanation for natural death. "The wages of sin is death." If you violate the taboos, you will die. Therefore death is not necessarily the result of black magic, or ghost manipulation, although necessarily an offended spirit remains responsible for the sinner's death.

Such manner of thinking was supported by dim memories of the days of Adam and the two Gardens and before that of life in Dilmun. Once humans had lived in paradise, but because of their sinful behavior, and their transgression of the taboos ("Do not eat of the fruit of the tree ..."), they had fallen into their present sorry plight, their world of sin and suffering. If you are a sinner, it is good to confess your transgressions of taboos in a public ceremony, then receive a public purification. That will lessen the danger that your deeds will bring misfortune to the whole people, as it often does.

So long as you are doing poorly in life, the spirits will be delighted, and hence appeared the custom of the fast (because spirits want food), often accompanied by sexual abstinence (whose loss the spirits regret) and the renunciation of pleasureful things in general. The religions of the Hebrews, Hindus, and Buddhists have made much of physical humiliation. Followers of the mother goddess in classical times refined such techniques by castrating themselves in honor of the goddess. Both Philo and Paul thought that material possessions were harmful to spiritual growth, and their influence has been great. Paul personally followed a cult of sexual continence and did great harm to Christian theology when he pronounced that "It is good for a man not to touch a woman"(1 Corinthians 7.1) and "I would that all men were even as I myself" (that is, celibate: 1 Corinthians 7.7). If society followed Paul, there would be no souls left to save. Jesus did not teach these things, as Paul knew full well.

SACRIFICE

This complexity of thought about ghosts and their demands led logically to *sacrifice*, whose value was always measured by the amount of pain it caused. Renunciation is the negative form of propitiation; sacrifice is the positive form. At first sacrifices were simple forms of self-mutilation—the knocking out of a tooth, the cutting of the flesh, the amputation of a finger—then came to embrace anything at all of value, including other human beings. There were two kinds of sacrifice: one a gift to the power or god to create good will; the other a buying back from the ghost the pain you already owed it, redemption.

Animal sacrifice was common and to early men a far greater loss than to ourselves. They knew well their animals, loved them, and did not clearly distinguish a difference between animals and humans. At least they could eat the animal's flesh, because its ghost had fed the spirit to whom it was given. Hence arose the communal meal, or as we would say, the communion service.

CANNIBALISM

Modern theories of cannibalism are mistaken. We abhor cannibalism, but it was part of the mores of primitive humans. The forebears of all of us were cannibals. They did it for one reason or another, military, economic, social, or religious. Its origins and functions are therefore complex.

The Sangiks were all cannibals, and the Nodites and Andonites when mixed with them became cannibals too. Women and children were raised and fattened and bartered to be eaten. The Nodites encouraged cremation as a way to forestall the digging up of the bodies and eating them; for early humans enjoyed human flesh and ate it often. For this reason they liked to kill and eat humans in honor of the spirits. After all, being once human themselves, the spirits enjoyed human flesh.

Once one has tasted human flesh through revenge, friendship, religious ritual, duty, or hunger, the taste for it grows. It satisfied the intense emotions of early men, who believed that by eating their enemies they destroyed the persecuting ghost, or at least blended its power with their own. In this way they absorbed their enemies' power. As if in contradiction, they also ate the bodies of their friends and parents as a sign of respect and endearment.

Cannibalism disappeared through the following steps. At first it was a communal exercise, as in the eating of condemned criminals. It was a public ritual governed by priests. Then, instead of eating the whole body, only certain parts were consumed, those containing the soul, especially the liver (so-called because it was thought to be the seat of the soul). Because cannibalism was a special rite, women were excluded. Then ordinary men were excluded and the rite restricted to chiefs, priests, and shamans. This made the practice taboo to most, and eventually to everyone, as it is in modern times. Human sacrifice ended cannibalism because the sacrificed person was to be food for the gods and not for men, unless perhaps they ate just a small portion of some special organ.

HUMAN SACRIFICE

Human sacrifice has been universal among all peoples, even up until recent times and occasionally today, a logical extension of the theory of sacrifice. Humans were commonly killed as foundation deposits for an important building, to provide a protective ghost. The Chinese immersed a virgin girl in the hot metal cast for a bell so that her spirit would improve the tone. It was long the practice to bury living slaves in an important wall. Moses spoke against the foundation sacrifice, but they continued vigorously after his death. Still today we place trinkets in the foundation holes of a new building.

There once existed a worldwide conviction that a human sacrifice must be offered when something extraordinary happened. Common in the ancient world was to kill one's first-born son in a time of crisis, as in the story of Abraham and Isaac. The Phoenicians kept up this rite down through their wars with the Romans in the third and second centuries BC, which weakened their power. Moses condemned the practice, recommending substitutes be offered for the first-born son, and the abandonment of this practice quickly placed the Hebrews in a superior position to those around them who continued it.

A woman taken to sacrifice could ransom her life by sexual surrender, the origin of one kind of temple prostitution. Instead of being killed, the woman dedicated her body to the goddess and through sexual relations with devotees purchased her redemption. It was thought an honor to have sexual relations with such a woman and certain to bring the benefits that sacrifice itself might have entailed. Many women of high status earned their dowries through sexual service in the temples, and their future husbands thought that they had benefited from this experience. As temple prostitution was a substitute for human sacrifice, so was the partial sacrifice, the cutting of the hair or nails or even a toe in dedication to the god. The very widespread practice of circumcision is a partial sacrifice.

In the long sweep of time the ceremony of sacrifice became the ceremony of the sacrament, where symbolic blood and flesh stand for the bringing together of humans and their invisible enemies. Paul of Tarsus expressed the meaning of the life of Jesus as one of a voluntary self-sacrifice that bought redemption from an angry God. His contemporaries easily understood what he was talking about, even though that was not what happened.

SHAMANS AND PRIESTS

As the relationship between humans and the spirit world became ever more complicated, the need for a specialist became clear to all. So came into being the shamans and priests who still today stand between the ordinary religionist and whatever god is being cajoled and worshiped.

The shaman was the ranking fetish man in the tribe, whose power often exceeded the chief's, an early form of the contest between religion and state. Shamans were often epileptic, or female hysterics, or what we would call schizophrenics. They were spirit-possessed, out of the ordinary, and greatly to be respected and feared. They practiced deception in minor matters, but on the whole believed in their own spirit-possession. Female shamans could go into cataleptic trances to communicate with ghosts in the other world and often were accomplished dancers. Later they became spirit mediums. Shamans, who understood hypnotism to some extent, used drugs to induce psychic states that would astound observers. They wore strange clothes and affected a peculiar manner. Admission to their ranks required many years of study and abnegation.

The "minor tricks" shamans played to maintain their authority, for example ventriloquism and slight of hand, did little good when their behavior failed to bring obvious benefit. If the shaman failed in an important matter—if the rains never came as he promised—he was summarily killed. Thus the cleverest shamans survived best, and power in society was first transferred from the old and the physically strong to the clever, shrewd, and farsighted.

The shamans tended to specialize in such skills as rainmaking, healing the sick, or crime detection. They were not doctors as such (although we may call them witch doctors or medicine men); their purpose was to guide a path through the hazards of life. Early men longed for revelation, for a voice from above that would instruct them, especially in all the things they should not do to offend the spirits. The shaman claimed to receive messages from the good spirits and even the gods. He understood the highly complex rituals and incantations that alone could summon the invisible powers. He understood their will by casting lots. Modern games of chance and such oracles as the casting of yarrow sticks before consulting the *I Ching* descend from such practice. So do the "counting out" rhymes like "eeny-meeny-miny-mo" and children's games where someone is *it*. In early times the person who was *it* was killed and often eaten.

The shaman could also read signs and omens and interpret the meaning of the flights of birds. The stars bore messages, and a crude astrology could be found among all peoples. Even today intelligent people believe that the stars reveal their personal destiny or the destiny of the world (they do not). Dreams, too, held secret meanings, which the shaman could explain. Women who could fall into a trance could actually converse with the dead or allow them to speak through her.

Only recently has shamanism been abandoned in most human societies, although it continues to prosper in hidden guise. The shamans and their priestly successors took control of important roles in society. They gave children their names and sprinkled them with magical water. They presided over the burial of the dead and announced when the dead had arrived safely in ghostland. They circumcised the males, a ransom for ancient sacrifice. The shamans became rich as recipients of offerings to the ghosts, sometimes controlling all the wealth in a tribe. They dressed in fine clothes and had many wives. They were often of low mind and morals and killed their rivals through charges of witchcraft. They often controlled the chief or king. They were the original aristocracy, living off the fat of the land and the labor of others.

No one liked the shamans, but they feared them and respected them for their knowledge. Most of their knowledge was fake, but the common people did not know that, and some of it was real. Shamans were the first surgeons, opening the skull to allow the headache spirit to escape, and they could set bones and lance boils. At first the sick were left alone because of the dangerous spirit that had seized them (which unintentionally forestalled the contagion of infectious disease), but the shamans went into the sick, followed by the whole tribe, and through hooting, blowing, drumming, dancing, singing, cupping, and sucking blood drove the ghost out of the sick person. Because people believed in the shaman's power, through their faith they were in fact sometimes healed. The shamans also experimented with herbs and discovered the valuable properties of quinine and raw cocoa. They purged the sick with various drugs.

The shamans were becoming priests, serving the higher spirits and taking control of the highly complex rituals that kept away bad luck. Such rituals needed to be performed precisely; any error in their performance would arouse the anger of the ghosts, clearly a job for professionals. The office became hereditary and priestly castes arose. Thus for tens of thousands of years endless pointless and burdensome rituals, under priests' direction, have held civilization back and retarded progress.

THE FIVE EPOCHAL REVELATIONS

Religion never wants to change but always to stay the same. "And if you will make me an altar of stone, you shall not build it of hewn stone, for, if you use your tools in making it, you have polluted it" (Exodus 20:25). Only two things can change the old religion: the advancing social mores (because the mores establish the forms of the religion, not the other way around), and periodic direct revelation of truth from higher nonhuman sources. *The Urantia Book* is one of these revelations. Evolutionary religion in the fear of ghosts laid the foundation for the acceptance of revealed religion, which will eventually destroy the superstitious roots of its own conception.

Revelation comes at the end of an epoch in planetary history, but suited always to the level of the evolutionary religion. Revelation is always limited by the capacity to receive. There have been five epochal revelations on this planet:

(1) The teachings of Dalamatia, about five hundred thousand years ago, when the concept of the First Source and Center was first promulgated by the one hundred material members of the Prince's staff. But the teaching was lost after the Caligastia secession about two hundred thousand years ago.

(2) The teachings of Adam and Eve, who reintroduced the concept of the Father of all. In spite of the failure of the first Garden, the Sethite priesthood carried the Adamic teachings, never wholly lost to the world, although by 2500 BC they had become almost unknown.

(3) The teachings of Machiventa Melchizedek of Salem, who taught that one must *trust* in the omnipotent beneficence of God and that through *faith* one earned God's favor. His teachings blended with various forms of evolutionary religion to create the religions strong in the Near East at the time of Christ.

(4) The teachings of Jesus of Nazareth, who for the fourth time taught of the Universal Father known through *love* and pleased through *service*. This teaching has persisted ever since.

(5) The fifth epochal revelation in *The Urantia Book* is different from earlier revelations because not one, but many different beings contributed. Of course no revelation short of knowledge of the Universal Father can be complete, but must always be restricted by the time and place of its composition.

SEVEN PERIODS IN RELIGION BASED ON REVELATION

There have been since the second Garden seven major periods in the history of religion on Urantia based on revelation:

(1) *The Sethite Period.* The Sethites (followers of Seth, the first son of Adam and Eve born in the second Garden) were strong among the Andites during the postAdamic period. Their teachings were clearest among the Greeks, Sumerians, and Hindus. The Brahmans of India are direct descendants of the ancient Sethite priests.

(2) *The Era of the Melchizedek Missionaries.* These followers of Machiventa Melchizedek, who appeared in human form around 2000 BC in Salem (later Jerusalem), in the so-called Middle Bronze Age, did not found any religion but their teachings were incorporated into later religions.

(3) During the *postMelchizedek Era* lived the great Egyptian leaders Amenemope and Akhenaten (second half of the second millennium BC), but the greatest leader

was the somewhat later Moses, who advocated to a group of Levantine Bedouins a rigorous monotheism and actually put to death those who followed the old ghost religions. Thus he founded the Hebrew religion. No other single man so influenced so many during this period.

(4). The *sixth century before Christ* witnessed one of the greatest religious awakenings ever on Urantia. Gautama appeared in India, whose teachings or later forms of them are widespread in Asia today, where he is revered as the Buddha, "the enlightened one." In Persia appeared Zoroaster; deeply influenced by the notion of the good and bad spirit worlds, he nonetheless affirmed the triumph of the light over darkness. Confucius in China was a great moral teacher, and Lao-Tse in China taught advanced truths in his description of the Tao, "the way" of the world.

(5) In the *first century after Christ* Jesus' refinement of the cult of John the Baptist turned his followers away from fasts and from adherence to the complex Jewish law. The other great teachers of this era were Paul of Tarsus and Philo of Alexandria, whose religious theories have had immense influence.

(6) In the *seventh century after Christ* a merchant from Mecca named Mohammed established the religion of Islam in opposition to the many debased cults of evolutionary origin that surrounded him. In its ferocious monotheism he hoped to return to what he took to be the original belief of Adam, Abraham, Moses, and Jesus.

(7) In the *fifteenth century after Christ* was broken the unity of the inelastic western Christian church. At the same time in Asia, Nanak and his followers put together the advanced religion of Sikhism fashioned from elements in Islam, Hinduism, and Buddhism.

Today the great interracial and international religions are the Hebraic, the Christian, the Buddhist, and Islam. In the Hebraic religion humans advanced from the evolutionary polytheistic religions to monotheism, the only western people to do this. The Christian religion has complex origins in teachings about the life and mission of Jesus of Nazareth as made to accord with Hebrew theology, various Zoroastrian beliefs about the duality of good and evil, and Greek philosophy, especially Plato. The Christian religion emerged from the work of three men: Peter of Galilee, Philo of Alexandria, and Paul of Tarsus, whose personal views were unfortunately taken to be the "word of God." The Christian religion has undergone many transformations, but been so Westernized as to become "the white man's religion." Buddhism is more a philosophy than a religion because whereas it recognizes the oneness of reality, it does not recognize the personality of God. Islam tripped over the teaching of the Trinity—three personalities in one deity was too much for desert nomads to swallow—and returned to a puritanical monotheism derived from Hebraic theology.

CHAPTER 23. ORIGINS OF THE HEBREW RELIGION

Remember that the Melchizedeks are the first of the four orders of descending sonship called the Local Universe Sons of God, brought into being in the early days of the Local Universe by the Creator Son and the Local Universe Mother Spirit in collaboration with a unique being, the Father Melchizedek. The Creator Son and the Universe Mother Spirit made the Father Melchizedek, of whom there is only one. The Melchizedeks number more than ten million in a Local Universe and govern themselves. This extraordinary order functions on emergency and special missions of all kinds within the Local Universe. They can appear on any level of universe reality, including the physical, and no other order except the Life Carriers is capable of doing this.

MACHIVENTA MELCHIZEDEK

The order of Melchizedeks has been active on Urantia, and we have earlier had a good deal to say about them. Twelve came with the original Life Carriers. Twelve others came after the Caligastia secession to serve as planetary receivers, that is, as the actual governors of the planet. These retired on the accession of Adam and Eve, but after the default again returned until Michael, living as Jesus of Nazareth, took title as Planetary Prince of Urantia.

The twelve receivers who held authority after the Lucifer rebellion realized around 3000 BC that few on Urantia any longer knew anything whatever about the one God. It was a planet of unrelieved spiritual darkness. They also knew that Michael had chosen this troubled sphere as the site of his seventh bestowal, but they did not know when this event would take place. After negotiations with authorities on the Constellation capital of Edentia and the Local Universe capital of Salvington, the receivers decided on their own authority to do what had happened only six times before in all the Local Universe of Nebadon: to materialize one of their own in the flesh, to prepare the way for Michael's coming, whenever that might be.

This materialization they accomplished near the village of Salem (Semitic "peace") in Palestine, much later called Jerusalem ("the heritage of Salem"), working with the resident Life Carriers and various other celestial personalities. One of the writers of the Book of Hebrews inherited a genuine tradition about this Melchizedek and wrote,

"This Melchizedek, priest of the Most High, was also king of peace; without father, without mother, without pedigree, having neither beginning of days nor end of life but made like a Son of God, he abides a priest continually" (Genesis 14.18).

In the year 1973 BC the materialized Machiventa Melchizedek walked into the tent of one Amdon near Salem, a shepherd of Sumerian extraction, and proclaimed, "I am Melchizedek, priest of El Elyon, the Most High, the one and only God." When the shepherd recovered from his astonishment, he invited Melchizedek to dinner. For the first time in his universe experience Machiventa Melchizedek tasted the material food that would sustain him for the next ninety-four years of his earthly career.

In time many followers settled near Melchizedek, who became known as the sage or sheik of Salem and the priest of El Elyon, "God the most high." He was six feet tall, had a commanding presence, spoke six languages, and dressed as a Canaanite priest, except he wore as an emblem the Satania symbol for the Trinity consisting of three concentric circles:

The symbol, which was held to be so sacred that his followers stopped using it, quickly died out.

Machiventa looked like any other human and lived as a human, but he did not marry and he could not have fathered children. His body was like the artificial constructions manufactured for the one hundred material members of the Prince's staff, except that Machiventa did not carry the life plasm of any race. The tree of life was not available to him. Had Machiventa stayed for a long time, gradually his body would have broken down; as it was, he terminated his bestowal long before that time.

Machiventa Melchizedek was indwelt by the very Thought Adjuster that later dwelled within the human mind of Jesus of Nazareth, the only Thought Adjuster ever to inhabit two separate minds on this planet. The Adjuster gained valuable experience through its association with Melchizedek, who was after all a divine being.

Machiventa was in touch with the other eleven Melchizedek receivers during the period of his incarnation, but with no other celestial personalities. He organized his followers into schools much after the patterns of the Sethite priesthood. For simplicity's sake, he allowed his followers to identify the One God, the First Source and Principle, with the head of the Most Highs of the constellation Norlatiadek, who reside on Edentia. To the majority of the Salem students Edentia was heaven and the Most High was God.

To the rank and file Machiventa Melchizedek taught only the rulership of El Elyon, God the Most High, but to a select few he gave instruction in the organization of Nebadon and to some he even described the Superuniverse and the Central Universe of Havona. Moses was a direct descendant of the family with whom Melchizedek resided and received traditions of the teachings of Melchizedek from both his father's and his mother's side. Melchizedek wished to create a category in the minds of Urantia religionists for God as Father, in anticipation of the career of a Son of God on this planet. Melchizedek taught of the coming of another Son of God, after himself, and for this reason Jesus was said to be "forever after the order of Melchizedek" (Psalm 110.4).

Melchizedek's teaching that the One God was known through faith alone utterly baffled those around him, accustomed as they were to the doctrines of sacrifice, forfeit, and indebtedness to the greedy spirits. It made no sense to think that you can get something for nothing, even what counts for most.

Melchizedek reintroduced the moral precepts of Dalamatia and the two Gardens: You will not have any other God than the Most High; you will not steal; you will not lie; you will not have sexual relations with another man's wife; you will not kill other humans; you will not show disrespect for your parents and elders. But of course the old ways went on, and even Abraham sacrificed to appease the angry gods.

ABRAHAM'S COVENANT

There are no chosen peoples, but Abraham was a chosen man, chosen by Machiventa out of all who sought his instruction. Three times Abraham visited the camp at Salem. Melchizedek did lay on him the task of keeping alive the truth of the One God against the primitive illusions of the evolutionary religions. It was not that Abraham alone could have accomplished this task, but he lived in Palestine at the crossroads of the world through which passed the great caravans. Here Jesus too would one day live, largely for geographical reasons.

The Melchizedeks had had their eye on Abraham's ancestors, whose genetic makeup was likely one day to produce a person of the right type. The two surviving sons of a man named Terah, born in this line, were in fact appropriate. They were Abram and Nahor. Terah lived in the ancient Sumerian city of Ur in southern Mesopotamia (see Map 1). He was a half-hearted convert to the Salem doctrine, having heard of it from a Phoenician missionary. He decided to go to Salem with his family, but Nahor had little enthusiasm for the project and persuaded the family to remain at Haran. Haran is an ancient center of trade in the northern Euphrates region at the intersection of three roads: to Damascus to the south, to Nineveh in the east, and to the Euphrates ford at Carchemish to the west.

After Terah died, Melchizedek sent to Abram and Nahor a messenger who said, "Come to Salem, where you will hear our teachings of the truth about the eternal creator, and in the enlightened offspring of you two brothers will the whole world be blessed." Nahor was still not persuaded and stayed behind to found near Haran a town named after himself. Abram went on to Salem with his nephew Lot (by another brother), where they fortified against bandit attack a hilly area near the settlement.

Going down into Egypt for supplies after a drought, Abram discovered that a distant relative sat on the Egyptian throne. Pharaoh recognized him and even made the highly able Abram commander over the Egyptian armies. Returning in triumph from a battle, Abram lived with his wife (and half-sister) Sarah at the Egyptian court. When he returned to Palestine, he carried many treasures with him. He returned to Palestine on Pharaoh's own advice, who well knew of Melchizedek and admired his teachings.

Abram returned to Palestine, but he had kingly ambitions. He placed before Lot his plans to unify through force the tribes of Canaan. Lot wished nothing to do with such a life and went to Sodom south of the Dead Sea to become a merchant and herder (see Map 4). Back at Salem, Abram was recognized as the civil ruler of the colony and of some outlying tribes. Melchizedek with difficulty restrained Abram from rounding up other tribes and forcibly converting them to the teaching of Salem.

In friendly disagreement with Melchizedek, Abram went to Hebron in the Judean highlands south of Bethlehem and Salem/Jerusalem to found his military capital (see Map 4). Spoiling for a fight, he got one when tribes near Sodom raided Lot's property there. Leading an army of four thousand, Abram and Lot fell on the enemy on the plain of Siddim south of the Dead Sea and destroyed them utterly.

Insisting that El Elyon, the God of Salem, had given him victory, Abram gave ten percent of the spoils to the settlement at Salem and took the other ninety percent to Hebron. The fame and prestige of his victory led quickly to the accession of other tribes to his confederacy, including the king of Sodom. Abram was well on his way to creating a state in Canaan, which he hoped to unite under his rule.

Unfortunately Melchizedek, whose spiritual cause he claimed to pursue, would not sanction Abram's military designs. Also, Abram had no heir (although he was not so old as in the biblical tale). In conversation with Melchizedek, the Son of God made Abram a promise: If he would give up his military ambitions and devote himself instead to the promulgation of the Salem teaching that the One God is found through faith alone and not through the killing of first-born sons and similar self-destructive practices, then his descendants would become as numerous as the stars in the sky and easily overcome the backward peoples around them. He told Abram how his descendants would one day live in Egypt, then escape to occupy the land of Canaan. This was the covenant between Abram and Melchizedek.

Soon after, Sarah (not so old as in the biblical account) was pregnant and bore Isaac. Deeply gratified, Abram took his agreement with Melchizedek all the more seriously and had it stated in writing (in cuneiform script, on a clay tablet). At this time he took the name *Abraham*, perhaps meaning "father of many" (Abram seems to mean "the father is exalted"), and he submitted to circumcision, a ritual he earlier had strongly opposed. Abram's circumcision was to be the sign of the covenant with Melchizedek, the agreement with him that if Abraham gave up temporal ambition for spiritual power, his descendants would one day occupy the land.

Constantly in the biblical account it is said to be Yahweh or God or "an angel of the Lord" who speaks with Abraham, but in reality it was always Melchizedek. During the Babylonian Captivity (586–538 BC), when much of what we today know as the Hebrew scriptures took form, the captive Jewish scribes in Babylon systematically altered existing records in order to exalt their lineage and prove that they were a special race chosen by God. After solemnly accepting the covenant, three celestial beings did appear to Abraham on the plain near Hebron, but the biblical story connecting this vision with the destruction of Sodom and Gomorrah is fanciful; Sodom and Gomorrah perished by natural means.

THE RULE OF ABRAHAM

Abraham now again took control of the civil operations of the Salem camp and made many improvements in administration, tithing as many as one hundred thousand neighboring peoples, training missionaries, and enhancing dairy production. Abraham was a shrewd businessman and wealthy. He was not especially pious but he believed in the teachings of Machiventa Melchizedek.

Men and women ventured forth from Salem to the farthest reaches of the planet (but not the Americas), bearing the message of Salem. Sometimes and for a while some people accepted the teaching about the One God, but only in Palestine did this teaching survive in relatively good form. Without the technology of printing, and with systems of writing that were difficult and confined to a tiny elite, messages easily are lost. Soon the teaching of Salem was diluted and absorbed into the ancient evolutionary cults of the Great Mother or the sun or something else.

Machiventa Melchizedek realized that his emergency bestowal was coming to an end. Those around him regarded him with ever increasing religious awe, viewing him as nonhuman and a superior being, which in fact he was. One night, ninety-four years after he had appeared in material form, Melchizedek bid goodnight in the usual fashion and retired to his tent. On the next morning, he was gone.

Affairs in the camp soon collapsed. The much disheartened Abraham traveled with Sarah to his possessions at a place called Gerar in Philistia (see Map 4). Fearing that the Philistine king Abimelech ("my father is the king") might kill him to take the beautiful Sarah, he pretended that she was his sister (as in fact she was, because they had the same father). Abimelech took her anyway. But soon Abraham, famous for his long association with Melchizedek, earned converts among the inhabitants of Philistia, and he became a spiritual wise man for all Canaan, again a man of power and influence. He died in Hebron, where he returned in his last years.

PRESENT STATUS OF MACHIVENTA MELCHIZEDEK

The meaning of Melchizedek's emergency sojourn on Urantia quickly faded and took on the flavor of myth. Isaac and his son Jacob attempted to keep the Salem teachings alive, and especially Jacob's son Joseph was devoted to them. Sold into slavery in Egypt by his brothers, Joseph was recognized there by Pharaoh as the great grandson of Abraham and offered a military command; but Joseph preferred a post in the Egyptian bureaucracy where he could work more effectively to further the Salem teachings.

Three days after disappearing from his tent, Machiventa reappeared in morontia form among the eleven fellow Melchizedek Urantia receivers. After Christ Michael took the title of Planetary Prince of Urantia two thousand years later, the twelve Melchizedek receivers withdrew to Jerusem where Machiventa serves on the council of twenty-four concerned with the affairs of Urantia. Michael has appointed Machiventa as vicegerent Planetary Ruler of Urantia in his own stead, although Michael holds the formal title. The situation is extraordinary in every way, in fact unique in this Local Universe, but certainly we are going to see more of Machiventa Melchizedek.

The mission of Machiventa Melchizedek was to foster a religion of one God in order to prepare the way for the eventual bestowal on Urantia of a Son of that one God. Certainly Michael could not come to Urantia until there was a people there who believed in the Universal Father. Although the teachings of the Salem missionaries were soon diluted after their journeys to Asia, India, Iran, and throughout the Levant, their teaching was nonetheless influential and can be traced in the local histories of various religions. Especially in Egypt the teaching took hold under the remarkable Akhenaten, who lived in the fifteenth century BC. Although his religious revolution from above failed through his inattention to practical and military matters, Akhenaten's teachings never died out, deeply influenced the Hebrews (some of the Hebrew psalms are translations of his writings), and through the Greeks influenced all subsequent religious philosophy. The worldwide religious revival of the sixth century BC was not only an effort to revivify the teachings of Melchizedek, but was coordinated by Melchizedek himself, in ways not altogether clear.

ORIGIN OF THE HEBREWS

As a pattern in the evolution of religion, first there are many gods (polytheism), then one of the many is held to be supreme (henotheism), then the unity of God becomes clear. More or less this is what happened in the case of the Hebrews. The *Kenites*, nomads in Palestine who were expert metal workers, were descended from the Sethite priesthood of the second Garden. They preserved with some clarity the teaching of Machiventa Melchizedek and gave this to the Hebrews, also influenced by Egyptian moral teachings, Mesopotamian theological speculation, and Persian notions of good and evil. The Hebrew religion, which is predicated on the covenant between Machiventa Melchizedek and Abraham, transmitted to modern times the rich thought and religious history of the entire ancient Near East.

Mount Sinai in the northern Sinai Peninsula was an active volcano during the Bronze Age. The neighboring Bedouin called the volcano spirit *Yaweh*, but there were thousands of other spirits in the primitive religion of these desert-dwellers. The Semitic-speaking Bedouin envisioned the spirit of the volcano as a golden calf. Yaweh was to become the supreme god of the Hebrews, transformed after a long process from the tribal fetish god of Mount Sinai to the loving and merciful creator father of the universe.

The Semites were in no sense a race but combined in their genetic heritage traces from all nine races; they were the most mixed of all "races." They are really a linguistic group, speaking related dialects that shared some vocabulary and structure. In the second millennium BC from the hills of the Zagros range separating Mesopotamia from the Iranian plateau, horse-riding eastern Semites put constant pressure on the ancient settlements in Babylonia (central Mesopotamia) and Sumer (southern Mesopotamia) and blended with them. In the north, the better organized northern Semites and the Andonite Hittites from Anatolia repeatedly expelled these mounted invaders. On the coast lived the seafaring Semitic-speaking Phoenicians. In the desert lived the nomadic and warlike Semitic Bedouin, a Semitic word for "wanderers."

In an unusually severe drought, large bands of these Bedouin migrated into the northeastern Nile delta and worked the land as contract laborers. They were quickly enslaved. These were to become the Hebrews (perhaps "those who cross over"). By no means were all Hebrews the descendants of Abraham, but a core of them were. Slaves in Egypt, they scarcely had a religion worthy of the name and were without hope in the world. But by far most of those later to be known as the children of Israel, or Hebrews, or Jews, did not have ancestors who had lived in Egypt. They descended from the clans of Bedouin who followed the leadership of Moses as he led his followers through north Arabia. Machiventa's teaching of salvation through faith had by then been long forgotten; these downtrodden and rude Bedouin were mostly adherents of the volcano god Yahweh, served through magic and sacrifice.

MOSES

Moses was the greatest religious leader on Urantia between Machiventa Melchizedek and Jesus of Nazareth. His vision of the nature of God has influenced one half of those who live on the planet. His mother was of the Egyptian royal family (his name is Egyptian for "son"), and his father was a Semitic liaison officer working with the enslaved Bedouin.

Resisting the allure of life in the Egyptian royal court, where he was a highly educated and trusted administrator, the brilliant organizer Moses decided to throw in his lot with his father's people. He would be assisted by an elite corps of gang supervisors, who thanks to Moses' intercession had received training in the Egyptian bureaucracy. Pharaoh agreed to allow Moses to lead his people out of Egypt, as he requested, and to pay them a small sum for their long labors. Then Pharaoh changed his mind and reneged, saying that he feared that they would organize the desert nomads in raids on Egypt. Later that year during a simultaneous attack on Egypt from the south by desert dwellers and from the north by Greeks, Moses led his followers in a daring escape. Pharaoh himself with a small band of warriors pursued them, but in a trap Moses' band killed them and took rich booty from the trappings.

Both Moses' Egyptian mother and his Semitic father had inherited the Melchizedek tradition of the One God, and this common religious bond underlay the highly unusual union between a member of the royal family and the leader of a foreign race. They called this God El Shaddai, a name of uncertain meaning but a religious concept combining the monotheisms of Machiventa Melchizedek (around 2000 BC) and the great Akhenaten (around 1350 BC). From Egyptian religious philosophy, the El Shaddaists had taken the Egyptian notion of divine providence: God will reward you and make you rich if you are good, a teaching that Moses adopted in his own preachments.

Moses' father-in-law Jethro, however, was a Kenite priest who proclaimed El Elyon, the Most High, as supreme God, and under Jethro's influence Moses began to call his one God El Elyon too. Any such distinctions were in any event far over the heads of the ignorant Bedouin rabble, who preferred the great volcano spirit Yahweh. When under Moses' leadership the fleeing Bedouin camped at the foot of Mount Sinai in the Sinai Peninsula, Moses accepted Yaweh as the great god, but presented to the Bedouin an enlarged concept of Yaweh that reflected his own background and education in Egyptian wisdom. Other peoples had other gods, but Yaweh was greater than they, certainly as far as they, the Hebrews, were concerned. Yaweh was the one. Moses virtually created the Hebrews at this time and with this teaching.

While encamped at the foot of Mount Sinai, Moses promulgated his ten commandments (but not in writing, which did not exist in the Hebrew encampment). He offered a refined ceremonial to celebrate their one God Yaweh, but made little headway until in the third week of their stay a chance explosion of the volcano astounded the mob, making a sensational impression. It was pretty clear that the great god Yaweh really was very mighty, just as Moses said, a devouring fire, fearful, and all powerful.

Moses struggled against the fetish religion strong among these early Hebrews and forbade the making of images of any kind. Moses told the newly established Hebrews that Yahweh had picked them out to be a chosen people, and that if they devoted themselves to Yahweh, they would be rewarded with riches and temporal power in the Egyptian style. He could never proclaim a God of love—a merciful creator would have no meaning to these unruly Bedouin—and so told them that Yaweh was quick to anger and vengeful in retribution for a failure to follow his commandments. Yahweh was an angry God, a terrible God, a fiery God, a God of power, a God who destroyed their enemies no matter what god supported them! In this way he exalted a primitive nature spirit into the Lord God of Israel. Not until the Babylonian Captivity of the Jewish elders in the sixth century BC did Yahweh become, in theory and in some quarters, the god of all nations.

When Moses died a quarter century later, he had welded his band of followers into a partly self-regulating nation of pastoral warriors. Never before had anyone done so much with such unpromising material. But after his death his followers quickly reverted to their ancient ways, adoring the fetish god of Sinai, the big boom-boom in the mountain, as a golden calf eager for magic and sacrifice.

After invading Canaan (that is, Palestine, named after the Philistines, Greek refugees from Crete), the Hebrews changed almost overnight from herders to rather sedate farmers, and the teachings of Moses were forgotten in the welter of local cults and superstitions. It began to appear to celestial observers that the Hebrews would be unable after all to preserve the teaching of the One God until a Son of that God would actually live on Urantia. Apparently the mission of Machiventa Melchizedek had failed.

SAUL, DAVID, SOLOMON

The followers of Moses, and then of Joshua, were warrior nomads from the desert. There never were twelve tribes of their followers, but only three or four. They did not destroy the local Canaanites (as reported in the Old Testament), but lived side by side with them and intermarried with them. The southern Judahites (supposedly named after a son of Jacob) or Jews, who later constituted the kingdom of Judah, hated and vilified the northern Ephraimites ("the doubly fruitful"), another tribe of Hebrews who lived in the hill country between Judah and Galilee, the later Samaria. These Ephraimites would later form the kingdom of Israel, rival to the kingdom of Judah.

After being governed in Canaan by local sheikhs, it was clear that the Hebrews were not going to survive as scattered clans without a centralized power. Therefore after the tribal leader *Saul* won a battle on the east side of the river Jordan (called Gilead, modern northern Jordan), the local tribes proclaimed him king. In fighting this battle he had about three thousand men, although the Bible records three hundred thirty thousand!

The first Hebrew prophet *Samuel* flourished at this time and introduced the notion that Yaweh was merciful, at least to the merciful, and unchanging, not always shifting in mood as in the anthropomorphic conception of him. But Samuel had nothing to do with the inauguration of Saul as in the fictional Old Testament account.

David from the south sometimes served with Saul and sometimes led his band of six hundred mostly nonHebrew misfits and criminals in service to the Mycenaean Greek Philistines. After Saul's suicide when defeated by the Philistines on Mount Gilboa, a ridge that runs northeast-southwest bordering a valley in central Palestine, David's army proclaimed him king of Judah; Samuel had nothing to do with this either. David made his headquarters at Hebron, a nonJewish city where Abraham had lived long before (see Map 4). David's mostly Canaanite followers were herders and respected the southern desert-dwellers' attitude toward the ownership of land.

To enhance his prestige, David married many prominent women, including Saul's daughter, as well as Bathsheba, whose Hittite husband he murdered. The Ephraimite elders came down from the north (the later Samaria) and anointed David as king of Israel too. Making a pact with the local people, the Jebusites, David moved his capital around one thousand BC from Hebron to their fortified city at Jebus near the ancient Salem, midway between Judah and Israel, now called Jerusalem. Still, the bulk of his army were nonHebrew Canaanites, followers of the Canaanite storm god Baal, although David would attribute his success in battle to Yahweh.

David got possession of the so-called ark of the covenant, a tribal fetish, brought it to Jerusalem, and made Yahweh the official god there. He exacted heavy tribute from the outlying tribes, speculated in land in the north against the customs of Judah, and took possession of the caravan tariffs formerly controlled by the Philistines. He was a kind of gangster and soon became rich. No wonder his son Absalom, whose mother was a Canaanite, rebelled against him. When David died many contended for the throne, but Bathsheba's son *Solomon* was successful.

Solomon continued the systematic extortions instituted by David, but bankrupted the state through his lavish expenditure on palaces, temples, and walls. He built a great navy with Syrian sailors. He kept one thousand women in his harem. He concentrated the worship of Yahweh at Jerusalem, discrediting the former cult center at Shiloh in the hill country of Ephraim.

PROPHETS AND THE TWO KINGDOMS

In the tenth century BC the Hebrew nation split into two, the northern being called Israel ("the descendants of Jacob") and the south Judah (so-called after the tribe of Judah, to which David belonged). Although the Hebrew people slid back ever more into the idolatrous practices of the Canaanites and their own forebears, important reformers appeared in both north and south. In the north *Elijah*, followed by Elisha, smashed the altars of Baal and attempted to revive the teachings of Samuel. Under Elijah's influence the name Yaweh was gradually replaced by *Elohim*, a Semitic plural form meaning "gods." The term went back to Sumerian groups who taught a Trinitarian God founded on the traditions of the days of Adam and Melchizedek.

The cult of *Baal*, which in Semitic means "lord" or "owner," was strongest among the Canaanites and in the north generally. Baal was the god of the private ownership of land and sponsor of that land's fecundity. Contracts and bills of sale and mortgages and land covenants were certified by his priests. The Baalites had houses, owned land, and lived in cities. Each local Baal had his priesthood, a special place for worship, and his "holy women," ritual prostitutes. Every respectable woman aspired to serve as temple prostitute early in life, to ensure her fertility after marriage. In the rural south, by contrast, the Yahwehites were strong. Still tied to desert customs, they did not acknowledge the private ownership of land or approve of speculation in land. From the time of the Hebrew penetration into Canaan at the end of the second millennium BC until the destruction of the northern kingdom of Israel by the Assyrians in 722 BC, the Yahwehites fought the Baalites.

Other important prophets in the eighth century BC were *Amos* from the south, who proclaimed that the tribal god Yaweh was really a God of all nations, who did not tolerate wrong-doing among his so-called chosen people. His contemporary *Hosea* called Yaweh a god of love, who desired repentance from his devotees, not sacrifice. A little later appeared the great *first Isaiah*, who in beautiful language explained how Yahweh was there even for the most downtrodden wretch. Other leaders attacked the bloody ceremonial of Hebrew ritual.

In the late seventh and early sixth centuries BC *Jeremiah* refined the concept of Yahweh as a universal God, angering his contemporaries by agreeing with earlier prophets that Yahweh was God of all peoples, not just the Hebrews, and that he could therefore not be counted on to give the Hebrews victory in battle. When Jeremiah foretold the destruction of Jerusalem, the priests threw him into a dark dungeon. He wrote of God's love for all his children and, like Isaiah, he celebrated God's perfect justice.

In 586 BC the Babylonians under Nebuchadnezzar II burned Jerusalem to the ground and carried to Babylon the upper echelons of Jerusalem society, the scribes and the priests. So ended the ancient quarrel between the Baalites and the Yahwehites. The prophets of Yahweh as an international God appeared to have been right, because Judah was destroyed and Yahweh did not save them. The spectacle of the burning Jerusalem and the slaughter and captivity of its inhabitants shocked the remnant of the Hebrews into a rigorous monotheism.

There in Babylon between 586 and 538 BC, when the Persians freed them, the Jewish religious officials created the religion of Judaism, explaining how in a world ruled by a now international God they were nonetheless a chosen people destined to fulfill a special purpose. They manufactured the many fables decorating the Old Testament accounts that glorify and honor the history of the Hebrews. Their purpose was to rehabilitate the Hebrew nation, crushed to the ground as it was, by exalting their racial history. Their purpose was simple and by no means did they claim that they were divinely inspired to create a sacred book. Nonetheless these scholars deliberately and with intention changed the secular history of the Hebrews, for which they had good written sources, into the so-called sacred history of the Jews.

In the meanwhile, and opposing their interests during the days of the captivity, the teacher called the *second Isaiah* reaffirmed the loving and merciful nature of God towards all people. The second Isaiah was an agent in the worldwide dispensation of the sixth century BC, and his writings are preserved in the Book of Isaiah as chapters 40–55. His words are the most uplifting ever about the One God. The second Isaiah made clear that the universe is governed by law and order and that God is dependable, not capricious. God is love. God is an all-loving father.

PART IV

JESUS OF NAZARETH, WHO HE WAS AND WHAT HE TAUGHT

CHAPTER 24. PRELUDE TO THE BESTOWAL OF CHRIST MICHAEL ON URANTIA

By living as different orders of creatures, the Creator Sons acquire experience they can find in no other way. The Eternal Son himself set the model when he lived on each of the seven circuits of Havona in primordial times. When a Creator Son goes out from Paradise to organize and maintain a local creation, he takes an oath to the eternal Trinity not to assume full responsibility of that local creation until he has experienced all seven levels of creature existence and the Ancients of Days on the capital of the relevant Superuniverse have certified his ascendancy. The bestowals are a kind of education in justice and mercy; the Sons are just and merciful by nature, but incarnation as a creature of the realm gives unique understanding. Although the bestowals bring advantages in their train to local personalities and beings, the purpose of the bestowal is to educate the Son himself and not to manage universe affairs.

MICHAEL'S FIRST SIX BESTOWALS

Today the Local Universe of Nebadon is ruled by a Creator Son who has completed his seven bestowals. Michael of Nebadon is registered as the 611,121st bestowal of the Eternal Son on the universes of time and space. He began the organization of Nebadon about four hundred billion years ago. His first bestowal adventure took place about the time that Urantia assumed its present form, one billion years ago. His bestowals have taken place around one hundred and fifty million years apart, the last on Urantia two thousand years ago.

Here is a summary of Michael's earlier bestowals:

(1) One billion years ago Michael announced somewhat mysteriously on the capital sphere of Salvington that *Immanuel* would govern while he himself was absent for unstated reasons. Immanuel belongs to the order Union of Days, high beings of Trinity origin. He resides on Salvington as an ambassador of Paradise; Michael too comes from Paradise. Only he, the local Creative Spirit, and Gabriel, a Bright and Morning Star, knew details of the mission.

Michael appeared alone on the dispatching field of Salvington, as often when preparing transport to Uversa, capital of the Superuniverse of Orvonton, or to

Paradise. He disappeared and was not seen again for twenty standard Salvington years. Records on Salvington report that three days after disappearing from Salvington an unidentified being of the *Melchizedek* order appeared on the first of the seventy primary spheres that encircle Salvington, the so-called Melchizedek university through which each ascender must pass. This Melchizedek was assigned to the emergency corps where he served with distinction on twenty-four missions. He disappeared as he had come. How such transactions take place is unknown, that a being of one nature becomes another, but take place they do.

All was well in Nebadon for one hundred and fifty million standard years when a serious rebellion broke out in System 11, called Palonia, of Constellation 37. There the System Sovereign Lutentia, a Lanonandek Son, dissatisfied by a ruling of the Constellation Fathers in a dispute, led the system into rebellion. Michael again made clear his intention to be away from Salvington and that Immanuel would govern in his absence. Three days later appeared a new member of the reserve corps of *Lanonandek Sons,* with orders from the Ancients of Days on Uversa, certified by Immanuel of Salvington, that he be appointed as the new System Sovereign over rebellious Palonia. There this Lanonandek Son deposed Lutentia, who refused all mercy, and served as System Sovereign for seventeen years before turning it over to another Lanonandek Son. Three days later Michael reappeared on Salvington.

The third bestowal, too, took place in troubled circumstances, in the second great rebellion in Nebadon. On the third day after disappearing from Salvington there appeared, unannounced, on the headquarters world of System 87 in Constellation 61 a strange *Material Son* bearing accreditation from the Ancients of Days of days on Uversa and from Immanuel of Salvington. The System Sovereign standing in for the rebellious ruler immediately appointed this Material Son as acting Planetary Prince of world 217. After a generation's work, the acting Prince reclaimed the loyalty of the earlier Prince and restored the planet to ordinary communication with the universe circuits. Soon after his departure another Material Son and a Material Daughter came to the planet.

As a fourth bestowal Michael appeared in the headquarters of the *seraphim,* one of the seventy primary spheres that encircle Salvington. There, although never recognized, he served as a seraphic teacher for forty years of Local Universe time, finally as the seraphic associate of a Trinity Teacher Son on a bestowal mission to world 462 in System 84 of Constellation 3 in the Local Universe of Nebadon. Increasingly, as he descended into the lower orders of universe life, Gabriel served as liaison between the incarnated Michael and Immanuel of Salvington.

The fifth bestowal took place about 300 million years ago of Urantia time when an unnumbered and unknown *spirit mortal,* an ascendant pilgrim of mortal origin, appeared on Uversa, capital of the Superuniverse of Orvonton. The identity of this being was never in doubt because Gabriel of Salvington accompanied him. On Uversa this spirit mortal pursued a career similar to those of spiritized ascenders from different parts of the Superuniverse of Orvonton.

The sixth and seventh bestowals Michael announced in advance: He would live as a *morontia mortal* at the court of the Most High Fathers on the headquarters planet of the fifth Constellation; then, when his experience was complete, he would live on an evolutionary world in the likeness of mortal flesh, as a *human being*.

THE SEVENTH BESTOWAL

Not until after the defection of Adam and Eve on Urantia did Michael announce that he had chosen this planet for his final bestowal. Unlike his earlier experiences, this incarnation was performed with full publicity, his every act on Urantia broadcast on the universe circuits even as it happened. His bestowal was a lesson for the universe and not just Urantia.

Michael's incarnation as a helpless infant was unexpected and unique in the succession of his incarnations. How it was accomplished is completely unknown. He was conceived as any other human, apart from this mystery. Certain men knew of his coming. The seraphim once attached to Adam and Eve, with the help of midway creatures, announced his arrival to a group of Mesopotamian priests, who visited the infant in the manger.

After the seventh bestowal, Michael received full authority over Nebadon, having earned his sovereignty through experience of the beings who actually live in Nebadon. The small, troubled, and obscure planet of Urantia has earned renown through Michael's final bestowal throughout the entire Local Universe of ten million inhabited worlds—Urantia, "the planet of the cross."

As we have seen, a Creator Son may at any time rule his creation in his own right, but if he wishes to represent the Paradise Trinity, he must undergo the seven bestowals. During his seven bestowals, Michael subordinated himself to the wills of the diverse associations of the persons of the Trinity. On the first bestowal he was subject to the combined will of the Father, Son, and Spirit; on the second to the will of the Father and the Son; on the third to the will of the Father and the Spirit; on the fourth to the will of the Son and the Spirit; on the fifth to the will of the Infinite Spirit; on the sixth to the will of the Eternal Son; and on the seventh and final bestowal, on Urantia, to the will of the Universal Father.

Before departing, Immanuel advised Michael about his adventure: He was to reveal the Universal Father not just to Urantians, but to the whole universe of Nebadon; he was also to terminate the Lucifer rebellion, the third in Nebadon; he was to leave no writing on permanent substances; he would probably not marry, because a Bestowal Son of Paradise must leave behind no human progeny; on successful termination of the bestowal, he would be invested with endless power over Nebadon, his own creation, as a direct representative of the Paradise Trinity.

THE TIMES OF MICHAEL'S BESTOWAL

In the Mediterranean world in the first century before Christ took place a great spiritual awakening, stronger than anything on this planet since the Adamic bestowal or since that time. Not by accident did Michael appear at this time and in this place. The Mediterranean and ancient Eastern lands were unified by Greek culture and the revolutionary technology of the Greek alphabet, which the Romans also used: For the first time since the invention of cuneiform writing in the fourth millennium BC, the Greek alphabet allowed one to record by means of symbols anything that one could say. This alphabet (comparable in power to the modern Internet as a facilitator of the exchange of information) was to become the medium for the four Christian gospels, the most influential books ever written on this planet.

The rule of the Romans was tolerant. The Jews, a Levantine people of mixed western and eastern heritage, were ideally placed to disperse new religious teachings widely to East and to West. Paul of Tarsus is typical: A Roman citizen, but a Hebrew of the Hebrews, he spoke and wrote in Greek in his proclamations about a Jewish Messiah.

The Romans brought a stable social order, cleared the sea of pirates, and for the first time on Urantia built good roads that interconnected population centers. Trade and cultural exchange boomed. Never did such conditions exist again on Urantia until the nineteenth century AD. Yet for most inhabitants of the Roman Empire life was short, miserable, and impoverished. A tiny elite possessed all the wealth, the traditional aristocracy and the business class. There was a small middle class, mostly tradesmen. Many of the Jewish Pharisees belonged to this class. All the rest, the great majority, belonged to the free proletariat or were slaves.

About half the Roman Empire were slaves, taken in military conquest. The free proletariat competed with slaves at a disadvantage, and many slaves bought their freedom and went on to hold powerful positions in state and commerce. Early Christianity held great appeal for slaves and the lower classes but flourished too among the tiny middle class. There was no class resentment at this time or feeling that things could be otherwise; in no sense whatever did Christianity begin as an attempt to reform the economic and social orders of the day.

In the days of Jesus, Palestine and Syria were centers for international trade. They enjoyed peace and relative prosperity. The Jews were a Semitic-speaking people, ethnically close to the Phoenicians, Babylonians, and Rome's great enemy, the Carthaginians. They were the most important Semites of the day because Jewish territory was of great strategic and economic importance. From time immemorial caravan traffic went through some part of Palestine as goods moved from Mesopotamia to the few seaports on the eastern Mediterranean and from there to the farthest west.

Half of this caravan traffic passed through the small town of Nazareth in the Galilean hills that surrounds the odd inland Sea of Galilee (Map 4).

Jewish traders were scattered everywhere in the Roman world. At the great Jewish festivals as many as two and one-half million Jews would gather in Jerusalem, towards which all Jews looked no matter where they lived. The temple of Yahweh was there and there would the Messiah appear. "Messiah" means "anointed one" in Hebrew, translated by the Greek "Christos," because at coronation a king would be anointed with a special oil. The Jewish Messiah would sit on the throne of David.

Through a network of over two hundred synagogues (Greek "places of gathering") and well-organized religious communities the new teaching about the kingdom of heaven first took hold, then spread to the rest of the world. Each synagogue tolerated a fringe of gentile hangers-on, of believers in Yahweh. In the temple at Jerusalem there was a "porch of the gentiles" where they gathered. Among this gentile fringe the preacher Paul made most of his early converts, especially in the Greek city of Antioch near the mouth of Orontes estuary in northern Syria, where his followers were first called Christians.

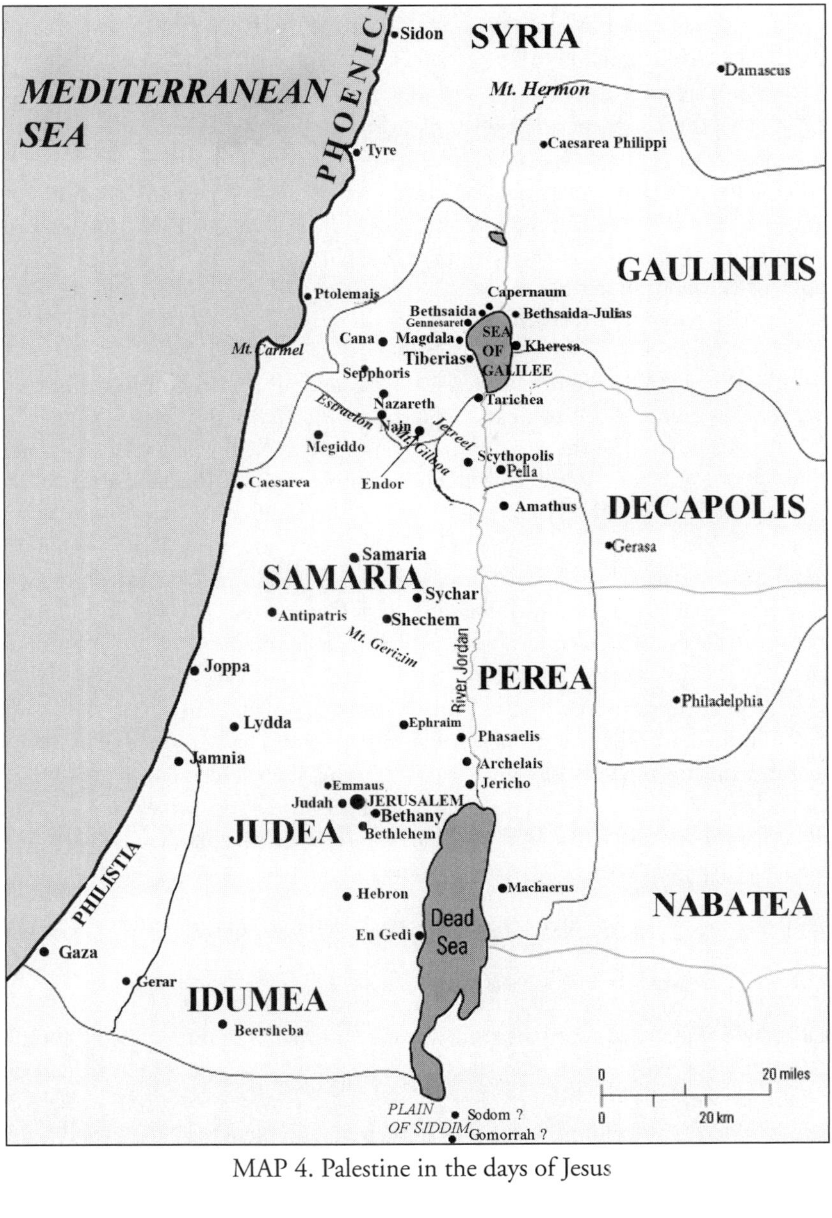

MAP 4. Palestine in the days of Jesus

THE JEWISH LONGING FOR A KING

Although the Jews were under Roman suzerainty, they enjoyed a considerable degree of self-governance. Remembering the fairly recent heroic efforts of the family of the Maccabees against the Macedonian overlord Antiochus IV (167–160 BC), which established the Hasmonean independent kingdom of Judea intact until 63 BC ("Hasmonean" refers to the family of the Maccabees), the Jews expected a Deliverer, a greater Maccabeus, who would drive back the Romans. The Messiah would restore the ancient Davidic kingdom.

The politically naïve Jews did not understand that their relative freedom under the Maccabees was a direct result of Roman foreign policy that established Palestine as a buffer state between Seleucid Syria to the north and Ptolemaic Egypt to the south. These two kingdoms came into being from the conquests of Alexander the Great (356-323 BC). Rather, the Jews attributed the success of the Maccabees to the power of Yaweh, their racial god who they thought had chosen them from all other peoples to rule and master the earth. In a tragic and bitter misunderstanding, whose consequences remain with us today, the Jews never learned that their mission was spiritual, not political.

Not only did the Jews suffer Roman overlordship, but the titular ruler of Palestine in the late first century BC was the scarcely Jewish *Herod the Great*, "king of the Jews," who came from a southern family that had only recently converted to Judaism. Herod the Great won his position through intriguing with the Romans, whose power he represented. Although claiming to be a Jew, he built many temples to strange gods in addition to rebuilding the temple to Yahweh at Jerusalem. In his behavior he showed little respect for traditional Jewish custom. Herod was sympathetic to Greek philosophy and encouraged a blending of it with Hebrew thought. He built the great port of Caesarea on the Mediterranean, which further enhanced Judea's position for trade.

Herod died in 4 BC and passed the rule of part of his kingdom, Galilee and Perea (Greek "the land beyond," across the Jordan River), to his son, *Herod Antipas*, who ruled during the lives of John the Baptist and Jesus and died in 39 AD. The Jerusalemite religious teachers and rabbinic teachers viewed the Galileans in the north with contempt and suspicion. In fact Galilee was more gentile than Jewish when Jesus was born.

RELIGION AND PHILOSOPHY IN THE DAYS OF JESUS

There were many religions and many philosophies in the days of Jesus. The old pagan religions attached to myths and patriotic sentiment continued, while complex philosophies offered the wise man rules for the good life, going quite over the heads of common people. The "mystery cults" were strong because they promised personal salvation whereas the older forms of religion were conducted for the group by the group. The mystery cults strongly influenced early Christianity but were based on myths; their exorbitant ritual could be outrageous or bizarre or repulsive.

Paul's cult of Christianity, too, offered personal salvation, but was based on an historical event, had a higher moral tone, and abolished the blood sacrifice as a form of religious observance. Early religions had little to do with morals, and at this time only the Jewish religion had associated moral behavior closely with religious observance. In this they were superior to the peoples around them.

Ancient Judaism fell into two branches, the western and the eastern or Babylonian. The Judaism of Jerusalem was of the western or Hellenized branch, deeply influenced by the translation of a collection of Hebrew scriptures into Greek in the third century BC in Egypt (the *Septuagint*). Three languages were spoken in Palestine in the days of Jesus. The common people (and Jesus) spoke Aramaic, a Semitic dialect widespread at this time, originally the dialect of Damascus (Aram). The priests and the rabbis spoke Hebrew, the language of Jewish learning. The Jewish upper classes and well-to-do people generally spoke Greek. Paul was deeply influenced by the western Hellenized Judaism, wrote in Greek, and helped to turn Christianity into more of a western religion than a religion for the world.

Clear examples of Greek philosophical influence are found in the Fourth Book of Maccabees, which owes much to the teaching of the Stoics: One must follow the laws of nature in response to the intelligence, or *logos*, that indwells all things. The *Book of Wisdom*, which pretends to be written by Solomon, depends on the idealism of Platonic philosophy and on Stoicism too. The chaos from applying Aristotelian principles of interpretation to Hebrew texts fell into order in the writings of Philo of Alexandria (20 BC-AD 50), a contemporary of Christ whose combination of Greek philosophy and Hebrew theology found broad agreement in Palestine when Jesus lived and taught. Philo was one of the great teachers of the ancient world, and his influence is enormous (though he is little studied in modern universities). Paul read Philo and used his teachings in fashioning his own cult of Christianity. The Christian gospel of John, the last of the four to be written, speaks to Western peoples, and especially the Christians in Alexandria who knew Philo's writings well.

In some respects Philo's teaching's were superior to Paul's. Philo spoke against the notion that forgiveness comes only by the shedding of blood, and he glimpsed the reality of the Thought Adjusters more clearly than did Paul. Paul's theory of original sin—that one is born in guilt and evil but can be redeemed—depended partially on the Persian cult of Mithras, a god of resurrection. The teaching has little in common with Hebrew theology and Philo's philosophy and nothing to do with the teachings of Jesus. Some parts of Paul's teachings about original sin and his theory of the atonement were of his own devising.

The Jews of Jesus' day had arrived at a firm vision of their origin and destiny. Believing that God had chosen them as a special people, they looked with contempt at all nonJews. Their theology was fixed and could not change. They lived by the letter of the law, both the written and even more the unwritten law. They took pride in their intolerance and hatred of outsiders and could not understand Jesus' teachings about tolerance, kindness, and the brotherhood of man. The scribes, the rigorous Pharisees, and the tradition-bound priests held the Jews in an abject bondage wherein every small detail of daily life was regulated. After all, Abraham had set down the unwritten rules for behavior, they said, and the law was from Moses. Whence do you derive your authority?

EARLIER RECORDS OF THE LIFE OF JESUS: THE GOSPELS

The oldest gospel is by John Mark, who was a boy at the time of the crucifixion and witnessed many events first hand. The gospel of Mark is really the gospel of Peter, who thought that the Roman church needed a written narrative in spite of the fact that Jesus had himself carefully refused to write anything down. Mark worked with Peter to outline the narrative until Peter was executed in Rome in AD 67. Mark wrote up the notes in the following year. The existing narrative has been changed, some passages dropped out and new matter inserted at the end where the original final one fifth of the manuscript was lost before it was copied. All other later accounts depend on Mark, along with notes made by the apostles Andrew and Matthew.

The gospel of Matthew was not written by the apostle Matthew, the tax-collector from near Capernaum, but by a certain Isador, a follower of Matthew who attributed authorship to his master according to a common custom. The account is based on Matthew's memories together with notes that Matthew made directly after the crucifixion of the sayings of the master (the so-called document Q of modern biblical scholarship); these notes were in the Aramaic language and script, while Isador wrote in Greek. The gospel is written to appeal to Jewish Christians, emphasizing how Jesus was the fulfillment of ancient prophecy, a descendant of David (he was not in fact), and how he respected the law and the prophets. Isador escaped from Jerusalem, taking a copy of Matthew's notes with him, in AD 70 when the city was invested by the Roman armies of Titus. He went to Pella on the eastern bank of the Jordan, and there in AD 71 he wrote the gospel of Matthew.

Luke was a physician in Antioch in Pisidia in what today is west-central Turkey. He became a follower of Paul in AD 47. He pieced together his notes and wrote his gospel in AD 82 in Achaea in the northwestern Peloponnesus in Greece, showing Jesus to be a "friend of publicans and sinners," a man of the people. He planned three books on Jesus, but died in AD 90 after finishing the second, the Acts of the Apostles. For his account of Jesus' life he depended mostly on Paul, and in this sense the gospel of Luke is the gospel of Paul; but he had numerous other sources, including eye-witness accounts, the record of Isador, the first four fifths of Mark, a written account from AD 78 by one Cedes, a follower, and a mutilated document purporting to be notes by the apostle Andrew.

The gospel of John was not written by John Zebedee, one of the apostles of Jesus, but was inspired by him. John realized that much had been left out of the earlier three gospels, which he possessed. Therefore in AD 101 he commissioned his associate Nathan, a Greek Jew from Caesarea, to compile the account. John did write the document known as First John as a covering letter for Nathan's composition, and he did write the Book of Revelation, although it has been so altered and in parts lost that the surviving version is distant from the description that John wrote of his extraordinary vision.

Imperfect as they are, these documents have changed human life on planet earth. The account in *The Urantia Book* makes use of these documents, in addition to other sources available to a commission of twelve midwayers, supervised by a "Melchizedek of record."

CHAPTER 25. THE BIRTH AND EARLY LIFE OF JESUS OF NAZARETH

MICHAEL HAD EXAMINED THE MELCHIZEDEK REPORTS on all the segregated worlds in the rebellious system before choosing Urantia as the site for his bestowal. A special commission reviewed the various peoples living there and chose the Hebrews in the Levant as the best suited for geographical, spiritual, and intellectual reasons. Three families seemed ideal, and after a personal visit to the planet Gabriel chose Joseph and Mary. He then appeared to Mary and informed her that she would bear the bestowal child.

THE PARENTS OF JESUS

Joseph was of ancient Hebrew stock. His family line went back to Abraham and beyond to the Sumerians, the Nodites, the southern blue man, and to Andon and Fonta. He was not, however, in direct line of descent from Adam or from David and Solomon. The genealogies of Joseph and Mary given by later commentators are mostly fiction, as are the so-called Old Testament prophecies of the life of Jesus.

The family of Joseph were craftsmen—carpenters, masons, smiths, builders. He was himself a carpenter and later a contractor. Six generations earlier, one Zadoc, a descendant of David, had adopted an orphaned ancestor of Joseph and for this reason Joseph was counted as belonging to the "house of David." During the Roman census Joseph therefore enrolled in the city of Bethlehem, the city of David, although he knew that he was not in David's blood line.

Mary was an average woman of her day with an average temperament, but her lineage was distinguished, tracing back even to Eve and to other distinguished women of the past. She was Jewish by culture and outlook, but had little of the Hebrew blood, being a mixture of Egyptian, Hittite, Phoenician, Greek, and Syrian stocks.

Mary and Joseph were ordinary people of their day, and Michael wished to incarnate as someone who could speak to average humans. They were well educated in comparison with their peers, and Joseph had a thorough knowledge of Jewish ceremonial procedures. Joseph, a dark-eyed brunet, was of a sometimes melancholy disposition, but thoughtful and spiritual. Mary, brown-haired, nearly blonde, was always cheerful. She was a skilled and devoted homemaker. Had Joseph lived longer he would probably have accepted his son's mission, but to Mary, whose ancestors were prominent in the Maccabean revolt, Jesus was destined to be a political leader, a savior of the Jews. Mary never understood or accepted her son's spiritual mission.

Joseph and Mary married when Joseph was twenty-one, after a two-year courtship. They lived in a one-room stone house with a flat roof that Joseph had himself built at the foot of the highest hill in Nazareth at the edge of town. An adjoining building housed the animals. The house had a low stone table, earthenware pots, a lamp stand, stools, and mats for sleeping on the stone floor. In the courtyard was an oven and a mill for grinding grain.

THE BIRTH OF JESUS AND JOHN

The mission of Jesus began with the work of *John the Baptist*, the son of Zacharias and Elizabeth. Zacharias was a member of the Jewish priesthood, and Elizabeth was Mary's cousin, belonging to a more prosperous branch of a large family. Although married for years, Zacharias and Elizabeth had no children.

In June of 8 BC Gabriel, the Bright and Morning Star and chief executive of the universe of Nebadon, appeared to Elizabeth just at noon when Zacharias was at the temple in Jerusalem. He declared that the Deliverer so hoped for was about to come, but that the child she would bear would prepare the way for him. She was to name him John. The Deliverer was to be born of her cousin Mary, to whom he would also appear.

The visit terrified Elizabeth and she told no one of it, even her husband. When five months had passed, and it was clear that she was pregnant, she told him what had happened. Zacharias was incredulous, then after a powerful dream accepted that Elizabeth was in fact to bear a child of promise.

In the middle of November Gabriel also appeared to Mary as she was at work in her Nazareth home. On the day before, she had conceived a child. Gabriel informed her that she was to be the mother of a special son who would inaugurate the kingdom of heaven on earth. She was to name him Joshua (that is, Jesus). Gabriel informed Mary of his visit to Elizabeth.

At first Mary kept the experience to herself, then told Joseph. Like Zacharias, Joseph struggled to accept the reality of what had happened, then also like Zacharias had a vivid dream that convinced him that a divine son would be born to them, even the Messiah, although Jewish tradition did not expect the Messiah to be a divine being. In truth Gabriel never referred to Jesus as the "Deliverer of the Jews," because his mission was to all humankind.

When it was clear that Mary was pregnant, she persuaded Joseph to allow her to visit Elizabeth in the small town of Judah just west of Jerusalem where Elizabeth lived. There she remained for three weeks as the two women compared their experiences and prepared for the promised extraordinary events. Gabriel's' appearance to Mary was the only supernatural event connected with Mary's conception, bearing, and birth of the special child.

John was born on March 25, 7 BC in the city of Judah. Eight days later he was presented for circumcision and given the name of John. In March 8 BC, Caesar Augustus had decreed a census for the entire Roman Empire, but because of Herod the Great's domestic difficulties the census was delayed for one year. Only Joseph needed to go to Bethlehem to register, but Mary insisted on coming, hoping that she might again visit with Elizabeth in nearby Judah.

On August 18, 7 BC, the financially strapped couple set out for Bethlehem, the pregnant Mary riding on their only beast of burden. After spending two nights on the road, they reached Jerusalem around noon on August 20, visited the temple, then journeyed on to Bethlehem. No rooms remained to rent in the town, but the innkeeper allowed them to sleep in the grain storage room of the stables carved into the rock beneath the inn, cleaned and prepared for travelers.

Around noon on the next day, August 21, Mary gave birth to a male child. He was wrapped in cloths Mary had brought just in case and laid in a manger. On the eighth day he was circumcised and named Joshua (Jesus).

On the day after the birth, Joseph registered the family. They found lodging in the inn, where they remained for three weeks, then stayed in town with a relative of Joseph. Convinced that the child was to be the Deliverer of the Jews and to sit on the throne of David, Mary persuaded Joseph to remain in Bethlehem, the City of David, for a full year, where he worked at local carpentry jobs.

Mary immediately informed Elizabeth of the birth. Zacharias and Joseph met in Jerusalem and took counsel. Three weeks after the birth, Zacharias sent three priests visiting in Jerusalem from Ur in Mesopotamia to go see the child: A religious teacher in Ur had dreamed that the "light of life" was about to appear as a child among the Jews, and Zacharias had met these men in the city. No star guided them, as in the popular account, a legend that originated from an extraordinary conjunction of Jupiter and Saturn three months earlier in the constellation of Pisces on May 29, 7 BC. The priests of Ur left gifts with Mary and returned to Jerusalem.

According to Jewish custom, Joseph and Mary visited Jerusalem to pay five shekels to the temple to ransom the life of their first-born son, according to the ancient Mosaic code. They sacrificed an animal to purify Mary from the supposed pollution of childbirth. At the temple, a poetess, informed by Zacharias about the prophecy of Jesus' greatness, sang to the child a song of praise, to the future "king of the Jews."

MASSACRE OF THE INNOCENTS

When the jealous Herod heard a rumor about this incident, and about the visit by the Mesopotamian priests, he brought the priests before him and questioned them closely, even giving them a purse to take to the child so that he might follow them. When the priests disappeared instead, after a year of searching in Bethlehem, in October, 6 BC, he ordered that all males in Bethlehem under two years of age be killed. Sixteen children died, hardly an extraordinary event from a household where murder was commonplace. The night before the massacre, Zacharias learned of the plan and warned Joseph, who swiftly left town with Mary and Jesus. They traveled to Alexandria in Egypt. There they stayed for two years with well-to-do relatives of Joseph, until Herod the Great died in 4 BC.

THE FAMILY RETURNS TO PALESTINE

The child grew normally in Egypt, where late descendants of the great Akhenaten (fourteenth century BC) learned that he was a child of promise. They tried to persuade the parents to stay in Alexandria, the center of Jewish learning, but once Herod was dead Mary and Joseph decided to return to Palestine. They left by ship for Joppa, the only port in Palestine, in late August of 4 BC. Alexandrian friends had presented Joseph with a complete copy of the *Septuagint*, the Hebrew scriptures in Greek translation, to give to the child when he grew older.

From Joppa the family went straight to Bethlehem and remained there one month. Mary thought that Jesus should be raised in the city of David, but Joseph was not persuaded. He did not believe that Jesus was to become the Messiah, the temporal leader of the Jews. For one thing, he knew that he was himself not a blood descendant of David. He also feared Archelaus, a son of Herod who now ruled the Roman province of Judea after Herod's death, as one likely to pursue his father's murderous policies. Herod's other son, Herod Antipas, who ruled over Galilee in the north and Perea east of the river Jordan, was likely to be more tolerant. Galilee would be a better place to raise the child, he thought.

In October they departed for Nazareth, avoiding Jerusalem and the dangerous Jordan Valley and western routes by going over Lydda southeast of Joppa, then north through the hills of Samaria to Scythopolis south of the Sea of Galilee, where the valley of Jezreel leads northwest to the plain of Esdraelon (see Map 4). Nazareth lies in the hills north of this plain. For safety five kinsmen of Joseph from Bethlehem accompanied them. Jesus was three years, two months old. Nothing was said to anyone in Nazareth about the child of promise.

THE FAMILY GROWS

In April, 3 BC, Mary bore a second child, James. Joseph built a shop away from the house near the village well, a tarrying spot for caravaneers. His brothers moved about on carpentry jobs, but Joseph remained in the shop manufacturing yokes and other wooden implements as well as working in canvas, leather, and rope. When Jesus was not helping his mother in the house, he spent his time at the shop, listening to the travelers and their tales of faraway places.

On February 11, 2 BC, in the year he would turn five, Jesus made his first moral decision. The same Thought Adjuster that was attached to Machiventa Melchizedek immediately came to dwell within him, although Jesus was no more aware of it than is any normal human today.

Mary gave birth to a third child, Miriam, in July. Jesus greatly enjoyed his little brother and sister. The alert and inquisitive Jesus questioned Joseph about how such things come about, and for the next five years he directed a barrage of questions to his parents about many curious things in the world. The parents could not always answer him.

Jewish family life was superior to that among most gentile peoples, and Jesus fully benefited from its virtues. Educated by his mother until the age of five, he was then turned over to Joseph. Joseph spoke fluent Greek as well as his native Aramaic. He taught Jesus Greek, using as a text the *Septuagint* that they brought from Alexandria. There were only two copies of the *Septuagint* in Nazareth, and Jesus' possession of one made his house a place to visit for those seeking spiritual truths. The Hebrew texts were far too difficult to read, a specialty of the rabbis. Jesus learned to read this book with great facility.

In 1 BC Zacharias, Elizabeth, and the child John visited Nazareth and for the first time the two boys met and played together, building with blocks in the sandbox on the roof of the house. Soon Jesus showed a strong interest in Jewish festivals and ceremonials, which his father explained to him in detail.

Joseph turned over his shop to his brothers and began a career as contractor, working in many of the villages near Nazareth and in Nazareth itself. A fourth child, Joseph, was born in March AD 1. The family income trebled and all prospered in this well-ordered household. The aggressive child Jesus lay awake at night admiring the stars, worked with plants in his mother's garden, and traveled with his father to other towns, always observing and learning about how humans live on this planet.

JESUS IN SCHOOL

At age seven in August of AD 1 Jesus entered the synagogue schools. He already knew how to speak, read, and write Aramaic and Greek and was eager to learn Hebrew, the language of the rabbis. There were no textbooks, but the chazan, the teacher, sat on the ground with the boys in a semicircle in front of him. He would read from the Hebrew text, and the boys would repeat his words in unison. When a written copy was available, the student would repeat the writing aloud over and over until it was memorized.

In the next three years of elementary school Jesus studied the books of the Torah in Hebrew (the first five books of the Old Testament, also called the Pentateuch, "roll of five"). In the following three years of advanced study he read the prophets and psalms and memorized large amounts of the law. No other documents were available for study. His skills were so great that a rabbi from Jerusalem who knew Zachariah, hearing of Jesus, invited him to come to Jerusalem and enter the schools there. But Jesus wished to remain with his parents.

All the while Jesus continued to learn from the Greek-speaking gentiles who passed through Nazareth, a town more gentile than Jewish and known for its liberal interpretation of the law. In Jerusalem was current the snide remark, "What good thing can ever come from Nazareth?" In this way Jesus came to know men of every background and devotedly to love them.

Jesus was a student of nature, taking long walks with his father to the top of the high hill behind Nazareth from which they could see the Mediterranean to the west and Mount Carmel; snow-capped Mount Hermon lay to the north; the Jordan to the east; and the shining Greek cities of the Decapolis (a group of ten Greco-Roman towns) to the south (see Map 4). He learned to milk the family cow, how to weave, make cheese, and to throw pots on the wheel. He got into serious trouble with the chazan when he modeled animals in clay and drew likenesses of them with charcoal, activities strictly forbidden by Jewish law. When caught, he stopped, but objected to the reason in it.

Jesus learned to play the harp, whose beautiful sound he loved. He spent time on his uncle's farm south of Nazareth where he learned to harvest grain. He learned to fish on the Sea of Galilee from an uncle who lived in Magdala on the western bank (see Map 4). He became an expert fisherman and considered pursuing it as a profession, but when his father made him a bench, gave him his own tools, and Jesus became skilled at the making of yokes, he decided to become a carpenter, like his father.

Jesus' third brother Simon was born in April AD 2, and his second sister Martha in September, AD 3. In AD 4, in his tenth year, Jesus got the first inklings of who he really was, but neither parent would tell him what they knew. His fourth brother, Jude, was born in June, AD 5, the seventh child, after which Mary became so ill that at age eleven Jesus assumed the obligations of the first son in managing the household while Joseph worked. The sunny years of childhood were coming to an end.

Jesus' restless, inquisitive mind inspired his teachers and friends, but brought deep consternation with its independence. When on a visit to Scythopolis, greatest of the Greek cities of the Decapolis at the confluence of the valley of Jezreel and the Jordan River, he praised to his father the Greek athletic competitions that he saw there. Joseph was horrified, staggered by the boy's approval, and angered by his compliment: Jews did not do things like that. Jesus learned to keep things to himself, troubled by the conflict between what appeared to be right and what his father and mother said was right.

In January AD 7 Amos was born, the eighth child, with still another to come, Ruth. In his thirteenth year Jesus' voice began to change as his child's body matured into that of a man. The rabbis dismissed him from the synagogue schools, the custom after six years of instruction, and gave him back to his parents. He was now a "son of the commandment," a "child of the most high," a true son of Israel. But he would need to attend the Passover in Jerusalem to confirm his maturity.

FIRST VISIT TO JERUSALEM

The Passover feast of this year fell on Saturday, April 9, AD 7. The dangerous Herod Archelaus, son of Herod the Great, was now dead. Twelve years had passed since Herod the Great's persecution in Bethlehem and there seemed to be no danger.

On the Monday of that week a group of over one hundred from Nazareth set forth down the valley of Jezreel, skirting Mount Gilboa on the right. A shorter route led

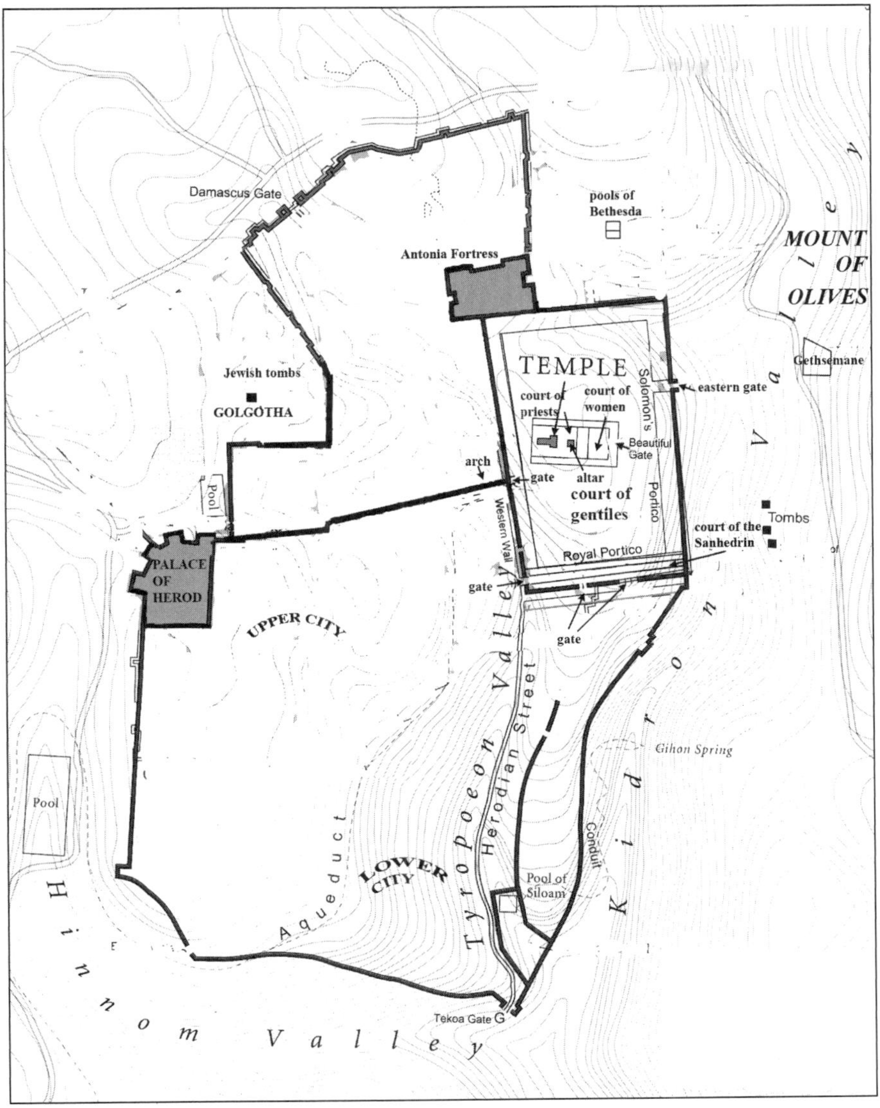

MAP 5. Jerusalem in the days of Jesus

through Samaria over the towns of Sychar (where Jesus later met the widow at the well) and Bethel north of Jerusalem, but the Jews wished to avoid the Samaritans, whom they despised. As they traveled, Joseph pointed out to Jesus where Ahab and Jezebel had ruled in Jezreel, where Saul perished on Mount Gilboa, and other tales from the scriptures.

The group avoided Scythopolis on the right, whose gentile population would pollute them, turned down into the Jordan Valley, and were at once in a tropical climate. They camped by a brook. On the second day they passed where the river

Jabbok enters the Jordan Valley from the east, where Gideon fought the Midianites (see Map 4). They camped near Jericho after whose conqueror Jesus was named.

On the fourth day they climbed up from the Jordan Valley toward Jerusalem, on the "robbers' road." As they came over the crest, for the first time Jesus glimpsed the Mount of Olives, or Olivet, that stood above the city from the east. They rested in a village called Bethany near the ridge overlooking the city and were warmly received by one Simon. Simon of Bethany's children—Martha, Mary, and Lazarus—were about the same age as the children of Mary and Joseph. Thus began a lifelong friendship between Jesus and this family. Many times in his life Jesus and his followers stayed at this house, and the events later to surround Lazarus were some of the most sensational in Jesus' career.

On the next day they mounted the crest of the hill. Jerusalem lay before them. Never again was the human Jesus so thrilled as by his vision of the city, crowned by the great temple to Yahweh rebuilt by Herod the Great. They pressed down the road into the city now thronged with Jews from all over the world in an intense traffic unlike anything Jesus had imagined. They passed the mighty temple, and went to lodge with relatives of Mary (Map 5).

JESUS IN THE TEMPLE

On the next day, Friday, the city was agog in its preparation for the Passover Sabbath. Joseph walked Jesus around the city and showed him the academy where he hoped that Jesus might be enrolled as a student. He took him up into the temple grounds, so vast that the complex could hold two hundred thousand people.

When Jesus entered the court of the gentiles in the southern portion of the temple grounds, he was shocked by the babble of the mob, the bleating of sheep, the shouting of the money changers, and the pressing demands of painted whores who sold themselves in plain view. Spiritual ugliness engraved the faces of the people. The sight of the priests' court beneath the rock ledge in front of the temple, however, where he was taken for a perfunctory ceremony of confirmation, sickened him even more, even to the heart. The priests' hands were drenched in gore as they killed one animal after another on the great altar, washing in between at a bronze fountain. Rivers of blood ran over the pavement. The drone of the dying filled the air, and the stench of the blood filled the nostrils. Jesus loved nature and its creatures and begged his father to remove him from this scene as quickly as possible. Even the vulgarity of the court of the gentiles seemed a relief.

Joseph rejoined Mary, forced to wait in the court of women, to Jesus' perplexity, and he showed Jesus the rest of the temple grounds. The ceremonies were impressive, but Joseph's explanations of them returned repeatedly to the portrait of God as an angry creator, a furious and indignant gigantic power. God, the thirteen-year-old thought, could not be like that. The explanation had to be wrong.

That night they ate the Passover lamb at the house of Simon of Bethany, with unleavened bread and bitter herbs. Dreams of slaughter kept Jesus awake. A messenger from Salvington appeared to him, sent from Immanuel, his first communication from the celestial realm. The messenger said, "The time has come for you to be about your father's business."

And so around age thirteen ended the childhood of Jesus of Nazareth.

JESUS WITH THE DOCTORS

In the following seven days of Passover Jesus attended the discussion groups held in the porticoes that ran along the inner walls of the temple complex. Questions tortured him, but because of his youth he could not participate in public discussions until the Passover was over. He was much in the company of Lazarus, son of Simon, and one night returned with him to Simon's house in Bethany. From this moment Mary and Martha, the sisters of Lazarus, loved Jesus.

One day mother Mary and father Joseph took Jesus back to the rabbinic school where they hoped to enroll him at age fifteen, but he continued to refuse interest in their proposal. His strange behavior and odd indifference to his parents' ambitious projects and his frequent disagreements with Jewish custom and religious observance were beginning to alarm the parents. The Nazareth party had agreed to meet near the temple in the morning of the first day of the week after the Passover. They set off back to Galilee, men walking with men and women with women according to Jewish custom.

Jesus had come down to Jerusalem with his mother, but now that he was a son of the consecration he ought to return with the men. When Joseph did not see Jesus, who had stayed overnight in Bethany, he assumed that he was going back with Mary anyway, as he had come, while Mary assumed that Jesus was with Joseph. Jesus had returned to the temple, however, where he became so absorbed in a conversation about angels that he forgot the time, until the discussion ended at noon. Only when the group from Nazareth reached Jericho did they discover that Jesus was not with them.

On that afternoon, indifferent to the situation, Jesus returned to the discussions in the temple, and that night went up to Bethany. He arrived just as Simon's family was sitting down to dinner. He spent the evening alone in the garden, thinking of all he had seen. On the next day he returned to the temple again and now took part in the discussions.

Because his questions were bright and unassuming, the learned doctors tolerated them, until he questioned the justice of the recent execution of a drunken gentile who had wandered into the sacred precinct surrounding the temple. "How old are you?" asked one doctor. "I'll be thirteen years in four months." "Then you should know that you are too young to speak," the man answered. Jesus explained that he had graduated early from school. Where was that? To Jesus' reply the doctor snorted, "Well, that makes sense. That's the way they do things in Nazareth!"

In the meanwhile Mary and Joseph hurried back from Jericho and spent the day searching frantically for Jesus throughout the city but could find no trace.

On the third day, crowds gathered around this remarkable youth, the child from Galilee who contended in argument with men of the law. Joseph and Mary even came into the temple grounds, but never thought he might be in one of the discussion groups. The child's questions touched on the following: What really exists in the temple's holy of holies, behind the veil? Why should mothers be segregated from the male temple worshipers? If God is a father who loves his children, why do they murder animals to gain God's favor—was this really the teaching of Moses? Because the temple is dedicated to the worship of the Father in heaven, why do they admit into the temple grounds those who engage in barter and trade? Will the expected Messiah become a temporal prince on the throne of David, or will he function as the light of life in the establishment of a spiritual kingdom?

His questions were respectful and good-humored and even those who resented his youth enjoyed the conversation.

In the afternoon, Joseph and Mary, by now beside themselves, decided to go out to Elizabeth and Zacharias in the nearby town of Judah, in case he had gone there. On the way they again looked into the temple grounds in case Zacharias might be there. To their astonishment they heard the voice of their son. Mary stepped into the crowd of observers and doctors and roundly upbraided the boy, asking why he had treated them so rudely, being gone for three days without explanation! What could have brought him to behave in this manner?

Jesus was now a child of the consecration, he had completed the synagogue schools, and he was engaged in enlightening conversation. After consideration he replied, "But mother, why has it taken you so long to find me? Would you not expect to find me in my Father's house? Is it not the time when I should be about my Father's business?"

As he crossed the brow of Olivet that night, Jesus turned, saw the city, and his heart was filled with pity for the Jewish people, enslaved in the spirit by religion and in the flesh by the Romans. Later, back in Nazareth, Jesus assured his parents of his devotion and promised not to embarrass them again. The incident fired Mary's conviction that he was one day to be that great leader who would restore the throne of David and cast off the hated yoke of gentile power, the longed for Messiah, the anointed one.

CHAPTER 26. ADOLESCENCE AND EARLY MANHOOD OF JESUS OF NAZARETH

THE NEXT TWO YEARS WERE THE MOST TRYING in the life of the young Jesus as he struggled to sort out his reality as a human being and his reality as the creator of the Local Universe of Nebadon, in which he now found himself to be a living being made of flesh and blood. He respected the honest Pharisees and the honest scribes, but he was indignant at the hypocrites and thieves who only pretended to understand who God was and what was expected of them.

TWO CRITICAL YEARS

While Jesus worked in the carpenter shop, Joseph set aside rents to pay for Jesus' Jerusalem schooling, supposed to begin in the fall in spite of his lack of interest. On September 25, AD 8, a runner arrived from the town of Sepphoris several miles to the northwest, where Herod Antipas had a palace and Joseph was working. Joseph had been badly injured in the fall of a derrick. Before Mary arrived on the scene, Joseph was dead. They buried him on the next day in Nazareth.

Thus at age fourteen Jesus became the head of a large family, responsible for the wellbeing of his mother and of his seven brothers and sisters. There could be no question of education in Jerusalem now, even had he wished it. Certainly for the moment Jesus could not "be about my Father's business," because he was himself a father to the only human family he would ever know.

The family was overcast with sorrow, but Joseph had saved a good deal of money and for the moment they were not in want. In April AD 9, Ruth was born, the ninth child. Jesus was virtually the child's father.

In his meditations Jesus realized that he could never be the Jewish Messiah whose coming the prophets foretold. He could not lead Jewish armies against the enemy, and he would never sit on the throne of David. Nor was he on earth to bring light to the Jewish people alone. Reading the rather obscure and pseudo-epigraphical Book of Enoch, he was intrigued by the story about the Son of Man who came to earth from heaven to promise salvation to hungry mortals.

He spoke less and less about his inner thoughts and assumed the mask of an ordinary person who had ordinary thoughts. But his strangeness continued to disturb his mother. She comforted herself by remembering that a child of promise could be expected to be strange. She remembered when Gabriel had appeared to her and the promise he had made.

As money ran low, Jesus sold property in Nazareth and reduced the household expenses. The family returned to the poverty of earlier times. Hemmed in by reality, the future looked dim. A carpenter's wages declined steadily. They grew vegetables and acquired more animals whose products they could consume. They struggled to pay the civil taxes and also the synagogue tax and the half-shekel tax to the temple. The local tax collector, impatient with their indigence, threatened to confiscate Jesus' harp. Fearing they would find his *Septuagint*, Jesus gave the precious manuscript to the synagogue library as his maturity offering to Yahweh. A further blow fell when after a long suit Herod Antipas refused to pay the family money owed as compensation for Joseph's death on the job. Their poverty deepened. The life of Jesus is the comfort of every disappointed idealist.

ADOLESCENCE

Girls were not allowed into the synagogue schools but in accordance with his role as surrogate father Jesus taught his young sisters at home. His adolescent body was strong and virile, his personality quiet, engaging. Amidst the always intense sufferings of adolescence, Jesus began to remember his earlier existence and to understand something of his mission as he reached full physical development.

In AD 11, when Jesus was seventeen, a group of Zealots came to Nazareth. They plotted the violent overthrow of the Roman hegemony, but Jesus would not join them. The Pharisees awaited the coming of the Messiah, who would destroy the Romans, but the Zealots wanted to bring armed revolt immediately. Mary begged Jesus to reconsider, but he flatly refused, causing his family turmoil and dividing his friends. Mary's own brother had joined the Galilean brigade of the Zealots. Never again was Jesus in full favor in the town of Nazareth.

In his eighteenth year the family sold all their properties except their home to pay taxes and to pay for incidentals. In this year Jesus took James to the Passover celebration in Jerusalem, this time going through Samaria. They spoke as brothers about every kind of thing. Soon James would be old enough to manage the family, allowing Jesus to begin his life's work. Simon of Bethany—the father of Mary, Martha, and Lazarus—had died, but Jesus and James stayed with the family for three nights, then returned to Nazareth by Jericho and the Jordan Valley.

In September AD 12 Elizabeth and John came to Nazareth to visit; Zacharias was dead. Jesus and John were nineteen. They "talked over some very intimate and personal matters" (*Urantia* paper 127), the only apparent reference in *The Urantia Book* to the sexual natures of John and Jesus. John saw that Jesus must spend years in the raising of his family. Following his model he decided to return to Judea and care for his mother on their small farm. They agreed not to meet again until the time of their public service. In fact they did not meet again until the baptism in the Jordan River.

In December little Amos, Jesus' youngest brother, died after a week-long fever. Jesus consoled Mary as if he were himself the father. Their income continued to decline, but Jesus kept the family optimistic. When the young Jude entered school, a firebrand and a disciplinary problem, Jesus at last sold his harp to pay for school expenses. He had delighted to play it, his last form of recreation, but the tax collector did not take it.

REBECCA

Among the Jews of Galilee wealth or the lack of it did not determine social class. Jesus was of high social standing in the community, admired for his intellectual brilliance, his skill as a craftsman, his loving service as a surrogate father, and as a frequent reader from the scriptures during the Sabbath service. He was also comely and attracted the attention of Rebecca, daughter of a wealthy merchant and trader. Rebecca fell in love with Jesus and went to Miriam, the sister of Jesus, and confessed. Miriam told Mary and in a panic mother Mary and daughter Miriam told Rebecca that Jesus was a man of destiny who could not therefore consider marriage.

Rebecca wanted him the more, thinking how she could be a helpmate to a great prophet and leader. She persuaded her father to support the union and to compensate Jesus' family for the loss of its substitute father. Jesus explained that he could not abandon his family, and in private he told Rebecca that, if he was a man of destiny, he could not tie himself to another person before his destiny became clear.

Rebecca was heartbroken. She persuaded her father to move the family to Sepphoris. She never married but adored Jesus from afar, following his career. She was in the crowd when he came into Jerusalem on the back of an ass, and she stood "among the other women" at the foot of the cross.

MEATLESS PASSOVER

Wishing to visit with Lazarus, Mary, and Martha, whom he loved best after his own family, Jesus again went down to Jerusalem for the Passover. This time he took the Megiddo road, then followed along the coastal route through Antipatris and Lydda, nearly the same route his parents had taken in bringing him out of Egypt (see Map 4, page 217). Lazarus was now head of the household in Bethany, after the mother too had died. Lazarus was the same age as Jesus, Martha one year older, and Mary two years younger.

Not expecting Jesus, they had planned to go down the Jericho road to dine with friends. Jesus proposed that they instead celebrate the Passover without the lamb there in Bethany, because the Universal Father, whose nature is love, was unlikely to be impressed by the death of an innocent animal. And so they held the first Passover ever celebrated by devout Jews without the lamb.

Thus Jesus grew in wisdom and experience, learning as a human what it was to be human, until as a man he could fulfill his destiny to reveal the Universal Father to all humankind.

EARLY MANHOOD

It is worth repeating that the mission of Jesus was twofold: to learn how his lowest form of intelligent life lived so that he might complete his sequence of bestowals and take full authority over Nebadon; and to reveal to humankind the reality and nature of the Universal Father. By age twenty-one, in AD 15, he had united his divine and mortal aspects into a single self, Jesus of Nazareth. He decided to call himself the *Son of Man*, after the phrase that so impressed him in the Book of Enoch.

Jesus availed himself of no supernatural power until after his baptism. Nothing out of the natural course of events had occurred in his entire life, except the appearance of the angelic messenger in Jerusalem who told him to be about his Father's business. He was born of man and woman, lived, experienced, grew wise in the world, suffered, and died as a man. At the same time he was creator of this Local Universe, virtually God to those who dwell within it.

How the union of separate natures is possible remains unknown except in the records of the planet of Sonarington, one of the seven spheres of the Father that orbit the innermost circuit around Paradise, the receiving world of the Eternal Son and Paradise headquarters of the ascending and descending Sons of God.

Gradually in his twenty-first and twenty-second years Jesus weaned his family from his own supervision, living in Sepphoris for six months doing construction work and turning over the family finances and administration to James. He returned on Sabbaths, but allowed James to run things and to train the younger Joseph. With four boys working, and the girls selling dairy products and weaving, living conditions improved. Mary had nearly forgotten the visitation of Gabriel, and scarcely any longer believed that Jesus had a mission to perform.

Jesus took Joseph to the Passover in Jerusalem, and in the following year Simon. During his visit to Jerusalem with Simon he met one *Stephen*, a Greek, with whom he had long conversations. Fifteen years later this same man was first to die for the emerging Christian faith, never guessing that he had spoken with Jesus himself years before. In the crowd watching the execution of Stephen, deeply moved by Stephen's faith, stood Saul of Tarsus, later called Paul, the philosopher and nearly sole founder of the Christian religion.

Jesus lived as a respected citizen of Nazareth, but his fame as a teacher, because he often spoke at the synagogue, attracted the attention of a consortium of businessmen in Damascus who wished to found a school there to rival that of Alexandria in Egypt. Jesus turned down their fine and attractive offer, as he turned down a similar generous offer by a committee of Alexandrian scholars who wished him to become assistant chazan in the chief synagogue in Alexandria. But Jesus kept aloof from formal institutions of teaching and learning.

James married and soon after Miriam married, and Joseph took over management of the household.

JESUS IN THE WORLD

After four years of preparing his family for his departure, on January 21, AD 21, when Jesus was twenty-seven years old, he quietly left to go to Tiberias on the western shore of the Sea of Galilee, a new city built to replace Sepphoris as the capital of Galilee. Herod Antipas would live there, the son of Herod the Great and ruler of the Roman provinces of Galilee and Perea. Jesus never again lived at home.

Finding little of interest in Tiberias, he traveled north along the shore of the Sea of Galilee through Magdala and then Bethsaida and Capernaum, small towns on the northern shore of the lake (see Map 4, page 217). Bethsaida, "house of fishing," was the fishing port of Capernaum (a better-known Bethsaida-Julias lay further east where the Jordan River entered the Sea of Galilee). In Bethsaida Jesus met with Zebedee, a boat builder who had been a good friend of Jesus' father Joseph. Zebedee's fishermen sons were to play a central role in Jesus' life and in the foundation of the Christian religion.

Zebedee hired Jesus, a master carpenter, to build boats. Jesus gladly worked in Zebedee's shop in Bethsaida and made many innovations in the construction of boats, including a more effective way of steaming the boards to fit them. Soon Zebedee's boats were in great demand in this fishing community and in the next five years would gradually replace all the boats on the lake. Jesus worked in the shop over one year, living in the house of the prosperous Zebedee down the lake shore from Capernaum, near Bethsaida.

Zebedee's wife, *Salome*, was a distant cousin of *Annas*, the high priest in Jerusalem until he was deposed eight years before, but he was still powerful in the Sadduccean group of elite Jews. The *Sadducees* (named after a legendary priest under Solomon, Tzadok) were a religious and political group installed under the Maccabees, the secular rulers of the second century BC who maintained a precarious independence from outside powers between 164–63 BC, the so-called Hasmonean dynasty. The Sadducees were opposed politically and doctrinally to the *Pharisees*, whose power derived from their knowledge of the scriptures. The Sadducees were a social aristocracy who controlled the Jewish priesthood and reaped immense wealth from it. They lived in the upper part of Jerusalem (see Map 5, page 229) in splendid houses.

Salome and Zebedee had seven children, including *John* and *James*, who were to become apostles, and *David*, whose messenger corps made Jesus' missions possible. When not building boats, Jesus often fished with the Zebedee boys, for Jesus was an expert fisherman.

A Roman military post, Capernaum lay directly on the caravan route leading from inland Damascus (gateway to Mesopotamia) to the Mediterranean coast. A wealthy Roman officer, friendly to Yahweh, had built the fine synagogue in Capernaum where Jesus, "the carpenter from Nazareth," often spoke. Jesus also held discussion groups for young and old. On the tax records he was listed as "a skilled worker from Capernaum," the only town he ever claimed as residence.

In March AD 22, Jesus left Capernaum and never again remained in any one place for a full year. Jesus sent money to Nazareth every month, a service that John Zebedee continued from money that Jesus left on deposit. Jesus went to Jerusalem carrying a letter to the priest Annas, from Salome, calling Jesus "the same as my own son." Annas politely took Jesus around to the rabbinic schools, but Jesus asked no questions. At Passover the Zebedee parents and their three boys came down to Jerusalem and stayed in the house of Annas.

During the Passover Jesus met a wealthy merchant from India traveling with his son to Italy. Jesus agreed to hire on as his interpreter and tutor to his son. For the next two years the carpenter from Nazareth dropped from sight as he traveled around the Mediterranean, to see how humans lived on this planet. He only told John Zebedee of these travels, to make financial arrangements for his family while he was away.

During his extensive travels Jesus met every kind of human being, and he learned their minds. This was the period of personal ministry, as opposed to his later public mission. As he traveled, he attuned his mortal mind to the presence of the indwelling Thought Adjuster. By the end of his twenty-ninth year he knew for certain who he really was and why he had come here. He remembered life on Paradise before this universe began, and he began to remember his farewell conference with Immanuel before he descended into the life of mortal flesh.

JESUS TOURS THE MEDITERRANEAN

Jesus with his Indian employer and son left Jerusalem in April AD 22, and they did not part company until December AD 23 in a town called Charax on the Persian Gulf. The course of their journey took them to *Joppa* on the coast, up the coast to *Caesarea*, thence by sea to *Alexandria* in Egypt, then west by sea to southern *Crete*, stopping at *Cyrene* on the African north coast, and on to *Carthage* (Map 6).

From Carthage they took a boat to *Naples*, putting in at *Malta*, *Syracuse*, and *Messina*. From Naples they traveled north to Capua, then on the Appian Way to *Rome*. In Rome Jesus actually met the gloomy emperor Tiberius when he served as translator for an Indian embassy.

After Rome they traveled to *Taranto* in the instep of the boot of Italy, then across the Adriatic Sea to *Nicopolis* ("city of victory," where Augustus defeated Marc Antony in a naval battle), *Corinth*, and *Athens*. From Athens they crossed to *Troas* (fifteen miles south of ancient Troy), then to *Ephesus*, where they took a boat to *Cyprus*, stopping at *Rhodes*. They spent a good while on Cyprus, then traveled the seventy-five miles overseas to *Antioch* in Syria; from there south to *Sidon* in Phoenicia, then overland to *Damascus*, then down the Euphrates to Babylon. They visited the site of the ancient Nodite city of Ur (see Map 1, page 123). From Babylon they traveled to Susa in southwestern Persia, and from Susa to the town called Charax on the gulf, where the Indians took ship for home.

A highlight of the trip was the six months spent in Rome, where Jesus met and had long talks with thirty-two of the leaders of the Cynics (who best preserved the

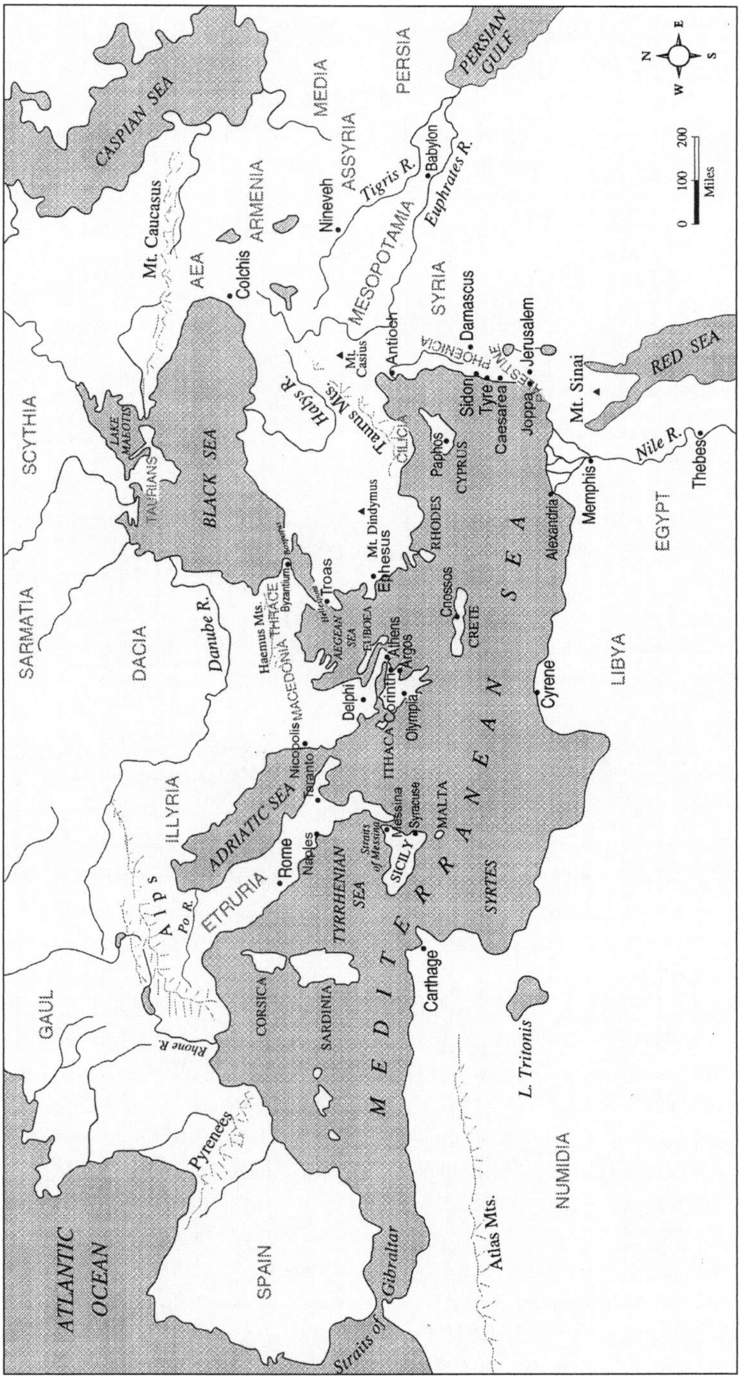

MAP 6. The Mediterranean of Jesus' travels

Melchizedek teachings, after the Jews), the Stoics, and the mystery cults, especially the followers of Mithras. He never discredited what such men believed, but exalted the truths in their belief and refined them in earnest discussion. Thirty of these men later joined the early Christian church in Rome, but none ever connected the "scribe from Damascus," as Jesus called himself, with the Jesus of Nazareth who claimed to be the Son of God. Jesus' individual work in Rome at this time was vital to the later rapid spread of the church in Rome, and then into the empire.

Elsewhere in his extensive travels Jesus spoke with many men and women from all walks of life. When he could he advised them and explained the presence of the Universal Father within them. Many of these people, whose lives his words and presence altered, later learned of the strange carpenter from Nazareth, executed in Jerusalem, whose teachings reminded them of someone they once knew; but none ever made the connection.

JESUS TOURS ASIA AND PALESTINE

After separating from the Indian merchant and his son, Jesus traveled from Charax to Ur, then to Babylon where he joined a caravan traveling to Damascus (see Map 1, page 123). From Damascus he went down to Nazareth, stopping by Capernaum on the way. Jesus' brother James had taken the place of Jesus at Zebedee's boat shop. In Nazareth Jesus visited with his mother and youngest sister Ruth, now fifteen. Simon and Jude were married in a double wedding, with their elder brother's blessing.

Now that all were married except for Ruth, just Mary and Ruth lived in the Nazareth house. Jesus now owned a small house in Capernaum that John Zebedee, together with his father, had bought for him as an investment with the money Jesus left with John Zebedee while he traveled. A caravan just then in Nazareth was looking for a leader, and Jesus signed on for the year's journey. At Jesus' suggestion, his mother and sister moved into his house in Capernaum, while Jesus' brother Joseph moved into the old house in Nazareth with his wife. Mary lived in the house in Capernaum until she died in AD 31.

On April 1, AD 24, Jesus left with the caravan, heading over Damascus northeast to Lake Urmia, then south through Assyria and east to Media and Parthia at the southeastern end of the Caspian Sea (see Map 1, page 123). In this, and in his earlier journey, Jesus made contact with all the surviving races on Urantia, except for the red man. Many were those to whom he brought comfort in his personal ministry.

On the return journey, at the Persian town of Urmia, Jesus gave up the leadership of the caravan and delivered a series of lectures at a unique local university dedicated to spiritual investigation. He then joined another caravan heading to Capernaum and returned there in April, AD 25, where in addition to his mother Mary and his sister Ruth his brother James also now lived. But when in Capernaum, Jesus always stayed with the Zebedees near Bethsaida.

During the following year Jesus wandered throughout Palestine and Syria. From Nazareth he went to Tyre, then up the coast to Sidon and eventually to Antioch, where he stayed for two months, working for three weeks as a tentmaker. From Antioch he returned south, staying briefly at Caesarea, then to Joppa, then inland to the ancient Philistine settlement of Gaza, from where he traveled inland to Beersheba (see Map 4, page 217). From Beersheba he traveled north the whole length of Palestine as far as the ancient city of Dan, now Caesarea Philippi, going up the center over Hebron, Bethlehem, Jerusalem, Sychar, Samaria, Magdala, and Capernaum.

Jesus withdrew up the slope of Mount Hermon above Caesarea Philippi. He wished to fine-tune his adjustment to his indwelling Thought Adjuster. He lived alone for six weeks high on the slopes of Mount Hermon, being provisioned by a boy he retained for the service who placed food in a stone cairn twice a week for Jesus to retrieve.

JESUS ON THE MOUNTAIN

During the last week of his stay on the mountain, Jesus met his enemies in Satania not as the Creator Son, but as Joshua ben Joseph, Jesus the son of Joseph. Both Satan (who represented Lucifer) and Caligastia were fully visible to him. To the many proposals of Lucifer's emissaries, Jesus replied, "May the will of my Paradise Father prevail and you, my rebellious son, may the Ancients of Days pass judgment upon. I am your Creator-father. I can hardly judge you justly, and my mercy you have spurned. Therefore I commit you to the adjudication of the judges of a greater universe." And with such words he effectively ended the Lucifer rebellion.

The so-called "Temptation of Christ," known through the folktale of the devil offering Jesus the profits of the world and set in the forty days he spent in the wilderness east of the Jordan, did not take place after his baptism in the Jordan, but before on Mount Hermon. At the end of the six weeks on Hermon, Michael of Nebadon, incarnate as Jesus, received the full sovereignty of his universe. His seventh bestowal was complete, although not formalized until the personalization of his Adjuster at the time of Jesus' baptism in the Jordan. The Lucifer rebellion and the Caligastia secession were in any event settled on the mountain and the rebels soon to be placed under arrest. The remainder of Jesus' earth career was for him to live as he thought fit.

Jesus returned to Capernaum and with John Zebedee traveled down to Jerusalem, going by the east of the Jordan through Gerasa. John saw a profound change in his manner. He was serious and dark, he spoke little. They stayed three weeks with Mary, Martha, and Lazarus in Bethany. Jesus enjoyed the merriment of the weeklong Feast of the Tabernacles, but left John in the middle to spend time alone in the hills of Bethany. He returned to Capernaum after stopping two days on Mount Gilboa bordering the valley of Jezreel, near where, in ancient times, Saul had committed suicide.

For the next few months Jesus worked beside his brother James in the boat shop, waiting "until my time has come." Rumors circulated about one John who was moving up the Jordan River, baptizing and proclaiming: "The kingdom is at hand! Repent and be baptized!"

CHAPTER 27. JOHN THE BAPTIST

As an only child John received the close attention of his father Zacharias, a priest, and his highly educated mother Elizabeth. They took him at age fourteen to Engedi (Semitic "kid spring") near the Dead Sea and enrolled him in the order of the Nazarites (Hebrew "consecrated ones"). Members of this ascetic order took a solemn vow to abstain from all intoxicating drinks, to let the hair grow, and not to touch the dead (Samson and the prophet Samuel were both Nazarites). Nazarites were the object of veneration among the Jews; only the high priest and a Nazarite of long experience could enter the holy of holies in the temple at Jerusalem.

A VOICE IN THE WILDERNESS

Impressed by tales about the prophet Elijah who worked on Mount Carmel, John began to wear only a shirt covered with hair and a rough belt. At sixteen he was six feet tall, an impressive sight as he tended his father's sheep outside the town of Judah. Zacharias died in AD 12 when John was eighteen. When John was twenty, as the family grew ever more impoverished, they relocated to Hebron. John tended his sheep near a tributary of a brook that entered the Dead Sea near Engedi (the "wilderness in Judea") and the colony of Nazarites. One *Abner* was then at its head. John became fast friends with Abner, who would play an important role in the formation of the early Christian church.

John's mother, whom he visited regularly in Hebron, kept him abreast of world affairs. He was a careful student of the Book of Daniel and believed in its prophecy of the coming kingdom of God: "And the kingdom and dominion and the greatness of the kingdom under the whole heaven shall be given to the people of the saints of the Most High, whose kingdom is an everlasting kingdom, and all dominions shall serve and obey him" (Daniel 7:27). His mother had told him that Jesus was to be the Messiah and that he was to prepare the way for him. John believed this fervently. John came to believe that the age of man was ending and that the age of God was about to begin, that he would himself be the last of the olden prophets and the first of the new. In his heart burned a passion to announce to the people their need to prepare for the new world that was even then dawning.

After the death of his mother Elizabeth in AD 22, John fasted and meditated in the wilderness for over two years. In March of AD 25, wondering if he were Elijah returned as prophesied in the Book of Malachi, this clear-thinking, highly literate but uncultivated man of nature, a fine orator and a ferocious critic, left his flocks and went forth to proclaim the beginning of a new world. He was no example for others, but he terrified his enemies. His public career was brilliant, but also very short.

THE KINGDOM OF HEAVEN IS UPON US

John skirted around the western side of the Dead Sea to the Jordan River where he took up position at the ford near Jericho. In his proclamations of the imminence of the kingdom of heaven, he only echoed what every Jew then believed, that after a hundred years of oppression by the Romans at last the Messiah, a king on the throne of David, would overthrow gentile power and reward the long-suffering Jews for their obedience to Mosaic law.

The event was certainly imminent, just around the corner, already upon them. Only their earlier disobedience could explain the lowly condition into which the Jews had fallen, so it would be necessary to live strictly in accordance with the law. As for the nature of the coming kingdom, there was broad disagreement, some thinking it a temporal kingdom, and others seeing it as a kingdom ruled directly by God through his divine agent, the Messiah. In any event a great war would ensue in which the Messiah would not only obliterate gentile power, but all the unrighteous would be punished and destroyed. The righteous would be elevated to positions of power under the Son of Man (a phrase that Daniel also uses) who would rule the nations in everlasting righteousness. The righteous (some thought) would even never die, but live in eternal prosperity and influence.

The ford at Jericho was the most heavily used on the entire Jordan. Joshua and the children of Israel had crossed here in ancient times to enter the promised land. Nearly a steady stream of travelers crossed here. Before them John proclaimed his message. From his thunderous orations and his strange appearance and his authority that answered to their deepest longings John immediately made a strong impression. Soon throughout Palestine people spoke of him and of his message: Repent for the kingdom of heaven is at hand!

A new feature was John's baptism of the repentant. Baptism already existed in Judaic practice to cleanse gentiles who wished to enter the court of the gentiles, but baptism for the remission of sin was never a Jewish practice. Although John only preached for fifteen months before Herod Antipas placed him under arrest, he baptized more than one hundred thousand hungry penitents.

John was a powerful but utterly tactless speaker. He called Pharisees and Sadducees who came for baptism "a nest of vipers" and "no sons of Abraham." The priests asked if he were the Messiah, but he only answered with scripture, "make ready the way of the Lord, make straight a highway for our God. Every valley will be filled, and every mountain and hill be brought low. The uneven ground will become a plain, while the rough places shall become a smooth valley. And all flesh shall see the salvation of God" (Isaiah 40.3–5).

After four months John moved north along the Jordan. Certain that the kingdom was to come, he was confused about its nature. He could not be sure that Jesus would really be the Messiah within the new kingdom, or whether it would be a temporal kingdom or a spiritual one. But he was sure that the kingdom was at hand. Many came from near and far to hear him speak, and many remained near him. When asked if he were the Messiah, he began to answer that another one would come after him, one far greater than himself.

By December AD 25 John had reached Pella on the east side of the Jordan, near Galilee. His fame had spread throughout all Palestine, and he was a topic of conversation everywhere in Galilee. Jesus, still working in Zebedee's boat shop, spoke favorably of his work. John and James Zebedee went down to Pella and were baptized. On Sunday, January 14, AD 26 Jesus and his brothers James and Jude (who also had moved to Capernaum), headed down to Pella to submit to baptism. "My hour has come," Jesus remarked.

BAPTISM AT THE JORDAN

On the next day at noon the three brothers—James, Jude, and Jesus—presented themselves to John for baptism, standing in the line of penitents. Busy with his baptism, John did not see Jesus until he stood before him. When he recognized him, he said with surprise that he should not baptize Jesus, but Jesus should baptize him. "Bear with me for a moment," Jesus said to John, "so that they may see that my hour has come."

After baptizing the three brothers, John dismissed the crowd, and just the four men stood in the river. They heard a strange sound. An apparition appeared above the head of Jesus, and all four men heard the voice, "This is my beloved Son in whom I am well pleased." Something changed in the face of Jesus. Without speaking, he left the river and alone headed east into the hills above Pella. He was not seen again for forty days.

When Jesus had gone, John told a group of his followers about Gabriel's' visit to his mother and Jesus' mother and what Gabriel had said to them. Soon word went out that the Messiah had appeared at the Pella ford. Large crowds quickly gathered, but no one knew where Jesus had gone.

On the fortieth day John looked up from where he and his company sat eating and saw Jesus coming toward them from the north. John climbed to the top of a large rock and in his booming voice announced, "Behold the Son of God, the Deliverer of the world! This is he of whom I have said, 'After me there will come one who is preferred before me because he was before me.' For this cause I came out of the wilderness to preach repentance and to baptize with water, proclaiming that the kingdom of heaven is at hand. And now comes one who shall baptize you with the Holy Spirit. For I beheld the divine spirit descending upon this man, and I heard the voice of God declare, 'This is my beloved Son in whom I am well pleased.'"

The next morning Jesus headed toward Galilee, where James and Jude had already gone. What was John to do now? What was to be his role? His teaching? "My Father will guide you," Jesus said to him before he left, his last words to John the Baptist.

THE END OF JOHN

John returned south, continuing to preach and to baptize, but the fire had gone out of him. He remained for several weeks in one small town, and in a speech viciously attacked Herod Antipas, the ruler of Galilee and Perea (which included the Jordan Valley), because Herod had unlawfully taken another man's wife. He railed against the corrupt political and religious leaders. John headed further south again to the ford near Jericho, where he had begun preaching one year earlier. Fearing a rebellion and resenting John's vicious criticisms, Herod Antipas arrested him at the ford in June AD 26. His followers scattered into Galilee, looking for Jesus.

In the dark, cold, and hateful prison, John, a man of the outdoors, languished and despaired. He could not understand why Jesus did not help him or even send word. He doubted the divinity of Jesus and his own mission. He doubted God. John's followers brought him news of Jesus, but still no word from Jesus until days before his execution, when Jesus sent word of his love and assured John of the glory to come. The message restored to John his wavering faith.

Because John was arrested in southern Perea, he was imprisoned in the fortress at Machaerus fifteen miles southeast from the mouth of the Jordan, where Herod Antipas maintained a residence. Herod did not know what to do with John. If he released him, it would only lead to trouble, but if he executed him the many thousands of his followers might rise up. Therefore Herod let him lie alone and wretched in the prison of the fortress for a year and half.

Herodias, Herod's unlawful wife, hated John for his public denunciations of her. At the birthday party of Herod Antipas, Herodias' daughter danced so prettily that Herod drunkenly offered her anything she wanted. On her mother's advice she asked for John's head on a platter, which was soon brought to her in the banquet hall. The followers of John claimed the body and sent word to Jesus about what had happened.

WHAT HAPPENED AT THE JORDAN

Some Jews thought of the coming Messiah as a powerful mortal, a combination prophet, priest, and king; others thought he would be a supermortal, the Son of Man; still others thought he would be a spiritual being, not a man at all but a Son of God. All such views could be supported by scripture. But none envisioned the truth about the nature of Jesus, who in an impenetrable mystery was both man and God at the same time. He was not a dual being, but a single being with a dual nature.

Ordinarily when a mortal personality has achieved perfect attunement with its indwelling Adjuster, after preliminary phenomena accompanying the personality's spiritual elevation, the mature soul unites with the indwelling Adjuster. A mortal witness would behold a great flash and the person would disappear. Because Jesus was a man, had lived as a man, had found his Adjuster as a man, the many celestial personalities that observed Jesus as he waited in line at the Jordan expected just this to happen. Jesus had completed his experience as a mortal of the realm already on Mount Hermon and the next step should be his union with the indwelling Adjuster, the same Adjuster that had lived within Machiventa Melchizedek during his life in the flesh, the only Adjuster ever to serve twice on this planet.

Instead, something unprecedented happened. As John placed his hands on Jesus, his advanced Adjuster departed from him, returned to Divinington, the Adjuster headquarters in the first ring that encircuits Paradise, and a moment later returned as a Personalized Adjuster. A Personalized Adjuster is an Adjuster that has received a personality not through the ordinary process of union with the mature soul, but by special dispensation as a direct gift of the Father, always in extraordinary conditions. The Personalized Adjuster spoke the words, "This is my beloved Son …" but only John, James, Jude, and Jesus heard the voice; the others of John's followers had retreated up the bank. Jesus alone saw the apparition.

Looking toward the apparition, Jesus prayed to his Father and "the heavens were opened." The Personalized Adjuster presented to him a vision of his own being before incarnating in mortal flesh, and how he would be once he left the flesh behind. Only Jesus saw this vision.

Jesus was almost thirty-one and one half years old. In his forty days in the Perean hills after the baptism Jesus adjusted himself to the new reality of having a Personalized Adjuster with whom he was in constant contact. He spent the forty days in an ancient rock cavern in the side of the hills near a village called Beit Adis and drank from the small spring that came from the side of the hill near this rock shelter.

Gabriel, the Bright and Morning Star of Nebadon, appeared to him and confirmed that he possessed complete sovereignty over Nebadon and that his mission was complete. As he spoke with Gabriel, the Constellation Father of Edentia (one of the three Most Highs) appeared to him as well and confirmed his registry as Michael No. 611,121, sovereign of the Local Universe of Nebadon. He had no further obligations as a mortal creature, except those he should choose on his own authority for whatever reason.

FORTY DAYS

During the forty days of his absence Jesus did not fast, and he was not tempted. Rather, he carefully weighed how he might live out his earthly life. He was acting not just for Urantia, but for his whole universe, for which his experience would be exemplary. He carefully pondered the advice he had received from Immanuel before setting out, always to submit his actions to the Father's will, always to subordinate his personal and sovereign will to the will of the Father, and to show to the world the Father's love. These then would be his guiding principles in what remained of his ministry.

Jesus had a vision of the vast, numberless celestial hosts who now served him. He decided he would use none of them, unless it was the Father's will. He was not always aware of their presence in the remainder of his earth career, but they always attended him and were in constant communication with his Personalized Adjuster. His personal seraphic guardian left him at the time of the baptism.

Michael was now fully self-conscious of his creative powers. How would he use them? He decided not to use any power that did not depend on natural law, such as even then in his need for food to nourish his human body. There were to be no miracles to prove that he was divine. He would not transcend natural law as the Jews expected of their Messiah, although the unboundedness in time of his Personalized Adjuster might hasten the execution of such laws if a matter were placed in the Adjuster's hands.

Certainly he never walked on the water. Although he would take ordinary precautions to protect his body, he would not defend himself against physical aggression. He would not employ supernatural means to delay the cessation of mortal life, although his countrymen expected a Messiah who would stand above natural law. Nor would he use supernatural power to attract attention. Although he occasionally allowed the compression of time in the ministration of natural law, he always told the recipients of such procedures "to tell no man." Never did he respond to taunts to "show us a sign." Such performances would not show to humankind the love of the Universal Father or satisfy their personal hopes to survive mortal existence.

In rejecting such behavior, he knew that the Jews would reject him as the expected Messiah, and for long he discouraged his followers from ever using the term. Throughout his ministry he had constantly to deal with the clamor to be fed; demands for a miracle; and the desire to make him a king, but he remained close by the decisions he made while in the hills of Perea.

CHAPTER 28.
THE TWELVE APOSTLES

JESUS HAD MINGLED WITH THE FOLLOWERS of John at the encampment near Pella. *Andrew,* an ardent follower of John, approached him and offered to follow him as teacher. Jesus accepted Andrew as the first apostle, then took on his brother Simon too, one of John's foremost supporters. To Simon, Jesus said, "You are enthusiastic, which might bring harm to the kingdom. Therefore I change your name to *Peter.*" Peter means "rock" in Greek, a firm foundation.

Shortly thereafter *John* and *James Zebedee* returned from the hills where they had gone in search of Jesus. At first they were annoyed that Jesus appeared to have preferred Andrew and Peter before them, who were his old companions; Jesus rebuked the Zebedee brothers for seeking what was never lost. He assured them of his love. It was the next morning, Sunday, February 24, AD 26, that they bid goodbye to John the Baptist and headed up to Galilee. Jesus never again spoke to John in the flesh.

The five men—Jesus, Andrew, Simon "Peter," James Zebedee, John Zebedee—traveled north along the east side of the Jordan. As they were about to cross the river they met *Philip,* who had known Jesus in Capernaum, traveling with a stranger from Cana named *Nathaniel.* The two men were heading to the ford at Pella to hear John preach about the coming kingdom. When Philip and Nathaniel learned from Andrew that John had proclaimed Jesus to be the Messiah, they immediately asked if they could join them. Jesus now had six of the twelve apostles who would follow him. The term apostle (Greek "one sent forth as a messenger") was reserved for this closed group. Five of the first six apostles he already knew.

The group traveled first to Jesus' boyhood home in Nazareth and stayed with Jesus' brother Joseph and his family. While there Jesus made a strong impression on his followers when he went through his childhood home and systematically destroyed youthful writings that he had posted, including a version of the ten commandments. He never wrote again, except on sand.

THE WEDDING AT CANA

Two days later the group traveled to the nearby town of Cana north of Nazareth to attend a large wedding between a prosperous bride and groom. His mother Mary and his brother James came up from Capernaum too, and his brother Jude came from

Magdala where he now lived and worked as a fisherman. Everyone had heard rumors about what happened at the Jordan. Mary once again believed that Jesus would be the king of Israel, just as she had always hoped. She thought that he might announce his coronation at this very wedding and reveal his majesty and power in some magnificent display.

Two hundred and fifty guests were invited, but over one thousand appeared, eager to see the man whom John the Baptist had proclaimed "the Deliverer." All hoped to witness some wondrous deed of power. When nothing happened in the course of the festivities, Mary went to Jesus and begged him for a display of his power. He sharply upbraided her for placing him in such a position, and she left sobered and depressed.

Later, as the party wore on, the bride's mother realized that the wine would soon run out. Mary went again to Jesus and complained, in spite of his earlier reproof, "My son, my son, they have no wine." Jesus answered, "Good woman, what has that got to do with me?" "I believe your hour has come," Mary said. "Can you not help us?" Jesus replied, "Again I tell you that I have not come to do things in this way. Why do you trouble me again with such matters?"

Mary broke down in tears. Mary begged, "But, my son, I promised them you would help us. Won't you please do something for me?" "Woman, what have you to do with making such promises?" Jesus replied curtly. "Please do not do it again. We must in all things wait on the will of the Father in heaven."

Mary wept still more copiously and Jesus was touched. He said, "Now, now, mother Mary, gladly would I do what you ask if it were part of the Father's will—" He hesitated, realizing that he had said—or thought—too much. Thrilled at what she was sure would be a miracle, Mary ran back to the celebrants. Jesus soon noticed a hubbub around six large jars of water holding twenty gallons each, kept for a final purification. Servants were drawing forth wine from the jars by the pitcher full.

Jesus was the most surprised of all at what had happened. Of course as a Creator personality standing outside of time, his simple desire had actuated the relevant celestial powers to assemble the correct chemicals and substances to fulfill his desire. The transformation had taken place in accordance with natural law but independent of time. His Personalized Adjuster had determined that such a transformation was not against the Father's Will, the principle by which Jesus had chosen to live. Therefore his desire to provide more wine became fact.

"What a party this is," said the best man, the so-called ruler of the feast. "Usually the host serves the finest wine first and reserves the poorer wine for when our tastes are dulled. But surely this is the finest wine of all!"

Jesus must be very cautious in allowing his human mind to desire things, or to feel sympathy or pity. But on many occasions he was not cautious enough.

TO BE BIG YOU MUST BE SMALL

Jesus and the six apostles left Cana early the next morning without saying goodbye, mystifying all. The apostles were walking on air, anticipating the new kingdom soon to come and their own privileged position within it. They returned to the house of Zebedee in Bethsaida and Jesus sat down with his six apostles.

He told them not to discuss the incident with the wine at Cana. He explained to them the real purpose of his life. He explained that the kingdom of heaven, which he was inaugurating, was a spiritual experience whereby God is enthroned in the hearts of men. They had no idea what he was talking about. They did understand that he was claiming not to be the king of Israel, destroyer of the Roman legions, bringer of boundless prosperity and everlasting supremacy of the Jewish people over all other peoples of the earth. What, then, was he?

Jesus told the six men to go to their nets for the time being. They would lie low until the career of John the Baptist was complete. That night Jesus sat by the seashore and did not sleep. He realized that he could never dislodge from the Hebrew mind their fantasies about the Deliverer to come. He decided therefore to accept the role as "Messiah" constantly thrust upon him but to remake this role as he went. After all, the more spiritual of the prophetic announcements in the Hebrew scriptures did in fact refer to him.

On March 2, AD 26, he spoke in the synagogue at Capernaum, packed to overflow with an expectant crowd, waiting for another miracle. He read from Isaiah, and the crowd dispersed, disappointed. Later that day he went out in a boat with the six apostles and his brothers James and Jude and explained the hardships that lay ahead and the difficulties they would face in revealing God's love for his human children. In the next four months Jesus every evening taught his disciples as much truth as they could absorb. During the day they fished and built boats. Mary fell from ecstasy into despair about her son's intentions, repeating constantly to herself, "I just can't understand him."

John the Baptist was arrested in June. On June 18, AD 26, Jesus put up his apron for the final time. In the synagogue that Sabbath he did his first public teaching. After reading from the scriptures, he announced that he had come to establish his Father's kingdom. His message was for Jew and gentile, for God is no respecter of persons. God's love and mercy is spread out for all. His own kingdom was not of this world. He would have nothing to do with armies and the enslavement of captives. As the Son of Man he would one day be known as the Prince of Peace, the revelation of the Father. Those who enter the kingdom would have all they desire—joy, righteousness, and abundant life, and faith alone was all that the Father required for entrance into this kingdom of heaven. Those who would be great in the kingdom must serve their fellows. The kingdom was like a seed that needed time to grow. Whatever the personal cost on entering the kingdom, the reward in the life to come would be great. Dwellers in the kingdom would ascend to be with the Father himself. Nor did he speak to the righteous alone, but to the unrighteous who in their hearts longed for freedom and the truth.

The audience was astounded at his words. One third vaguely accepted what he said, although they understood it poorly. One third decided to reject any such outlandish message, whatever it was, and one third thought that Jesus of Nazareth had gone completely insane.

SIX MORE APOSTLES

Jesus called together the six apostles and instructed them. For the next two weeks they were to split into three groups, one to visit the village of Kheresa on the mid-eastern shore of the Sea of Galilee, one to visit the town of Tarichea (Greek "place of dried fish") on the west bank of the Jordan where it flows south out of the Sea of Galilee (another town on the northwest coast of the sea, near Magdala, was also called Tarichea), and the third to go to Capernaum. They were to do no public speaking but in private conversation to encourage entrance into the kingdom.

He did not invite his brothers Jude or James to this conference. They were deeply hurt and offended, but the Master had decided not to involve his family directly in his ministry. So began a rift between Jesus and his family that widened ever more up to the breaking point, until after the resurrection when they again believed in him.

The six apostles expected something flashier. Jesus then told them each to find another man to join the group, because he wanted twelve apostles, so great was the labor that faced them. At first the six felt that their rights of priority were threatened, then under Andrew's leadership they saw the wisdom of expansion and the trust that Jesus had showed them in asking each to pick a member of the inner circle.

Jesus thought much about John the Baptist, but in this matter as in all others he had decided "to do the Father's will." The apostles returned after two weeks, having learned a great deal from their experience: how people were hungry for spiritual knowledge and how to deal with people on a one-to-one basis. Each presented his candidate for apostleship, and after a unanimous vote of the six, they were all appointed. The six new apostles were:

Matthew Levi, a customs agent who lived east of Capernaum near the border with the territory of Gaulinitis, part of the province that belonged to the "tetrarchy" of Philip, brother of Herod Antipas. Herod Antipas ruled the tetrarchy of Galilee/Perea. A tetrarchy was one of the political divisions into which the realm of their father Herod the Great, who died in 4 BC, had been divided; in fact there were three tetrarchies in Palestine—Judea; Galilee/Perea; Gaulinitis—in spite of the name, which means "rule of the fourth part."

Andrew chose Matthew Levi.

Thomas Didymus (both names mean "twin," the first in Hebrew and the second in Greek; Thomas lost his beloved twin sister when he was nine). He was a fisherman in the Tarichea at the end of the Sea of Galilee. He had earlier worked as a carpenter and stone mason in the Greek Decapolis.

Philip selected Thomas Didymus, who was then still in Tarichea.

John Zebedee chose one of the twin fisherman sons of a man named Alpheus in the village of Kheresa on the east coast of the Sea of Galilee, *Judas Alpheus*, and James Zebedee chose Judas Alpheus' brother, *James Alpheus*. These men were also not present but remained still in their village.

Hence there were three men named "James" in close proximity to Jesus: his brother; James Zebedee; and James Alpheus. There were two men named Judas: Judas son of Alpheus and Judas Iscariot. Judas and James, sons of Alpheus, were also called Thaddeus and Lebbeus.

Peter chose *Simon Zelotes*, that is, Simon the Zealot, who had been a member of the nationalist party determined to force change through political revolt. Before joining the Zealots Simon was a merchant.

Nathaniel chose the only Judaean in the group, *Judas Iscariot*, the only son of a prosperous family of Sadducees in Jericho who disowned him when he became a follower of John the Baptist. He was experienced in finance. He was then staying in Tarichea.

ALL MEN ARE MY BROTHERS

That night Matthew held a banquet in his house near Capernaum to celebrate his decision to follow Jesus. Jesus was the guest of honor. Matthew's friends came, who were mostly involved in business and in tax-gathering, an old family occupation. They were an irreligious and pleasure-seeking lot.

On such occasions, when the guest was eminent, people would gather in the banquet room to observe the banqueters and listen to the conversation. In this case, all the Pharisees from Capernaum were present. When they saw how the wine ran freely and everyone was having a good time, one of them went up to Peter and complained that Jesus claimed to be righteous, but he dined with "publicans and sinners." Certainly Matthew was a publican, a Latin word for "tax-gatherer," a hated class in Judea because they served the Roman overlords by extorting taxes from fellow Jews. Peter passed the remark to Jesus.

For the concluding toast Jesus declared, "I am happy that we are enjoying ourselves, but we rejoice even more when we think how many here will enter the coming kingdom of the sprit, where you will abundantly enjoy the good things of the kingdom of heaven. As for those of you who criticize me because I make merry with these friends, let me say that I have come to proclaim joy and liberty of the spirit to the downtrodden. Those who are whole do not need a doctor, but those who are sick. I have come not to call the righteous, but to call those who have gone astray."

He made similar remarks the next day when they took a boat to Kheresa to gather in the Alpheus twins, to whom Jesus said, "All men are my brothers. The kingdom of heaven is open to all men and women. We will sit at meat with all who desire to hear of the kingdom. As our Father in heaven looks down on men, they are all alike. The door of the kingdom is open for all who desire to know the truth and to find God."

On the following day the group, now numbering eleven, crossed the sea to Tarichea. The fisherman Thomas Didymus and the wanderer Judas Iscariot met them at the dock. As in every case, Jesus looked into their eyes and said, "Follow me." Later he began to instruct them in the nature and function of the Holy Spirit, and in the nature and origin of so-called unclean spirits, and about the ministry of seraphim. But they understood little of what he was saying. They understood not at all that he was bringing to humankind an entirely new way of finding God, a new revelation of God the Father.

Back at Capernaum on the Sabbath, staying at the house of Zebedee, Jesus cautioned the twelve apostles about entering into conflict with the civil authorities, both Jewish and Roman. They were not to excite their suspicion and fear. They must leave to him any criticism of such authorities.

ORGANIZING THE TWELVE

For an intensive week he trained them further and established the rule that Wednesdays were to be a free day in which the group would pursue their recreation. Simon Zelotes was placed in charge of group recreation and the other apostles were assigned other roles within the group. Judas Iscariot would carry the purse, but Matthew would organize the finances. Andrew was to be head apostle. John and James Zebedee and Peter were to attend to him personally, accompany him and see to his wants. The Alpheus twins would keep order in large gatherings.

Jesus cautioned the apostles about allowing teachings *about* himself to take the place of *what* he taught. He planned to send them into Galilee to spread the good news in a quiet five-month campaign. When Simon Peter, Judas Iscariot, and James Zebedee asked him whether he was to proclaim the kingdom in Capernaum or Jerusalem, and what position would they themselves hold in the coming kingdom, he strongly rebuked them for their failure to understand that the kingdom of which he spoke was not of this world. They simply did not understand him, but as a person they adored Jesus and believed in him. All twelve declared their loyalty, wherever he should lead.

Because they had money for only two weeks sustenance, Jesus sent the group again to go fishing and in this way build up the purse. "And then you will go forth to be fishers of men." For the next five months the group alternated between two weeks of fishing and two weeks of personal work on behalf of the kingdom, an exhausting trial.

The apostles went out two by two to towns in Galilee, with Jesus circulating among the six groups. In this way Jesus became a close personal friend of the apostles. It was their friendship, not the Master's teachings or his famous deeds that held the apostles together after the crucifixion (except for Judas).

They taught the forgiveness of sin through faith in God, not through penance or sacrifice. John the Baptist had taught the need for "repentance—to flee the wrath that is to come," but they taught that God loves all humans equally without distinction. The apostles were shocked by Jesus' treatment of women as spiritual equals to men, but soon got used to it. Jesus strictly enjoined the apostles from discussing the imprisoned John the Baptist, the voice heard at the river, and the incident at Cana.

The apostles made a strong impression in their journeys. They were men of lowly status, although all (except the Alpheus twins) had graduated from the synagogue schools. Most were in their late twenties and early thirties, like Jesus himself (John Zebedee was the youngest at twenty-four). They had read some scriptures, but knew nothing of the refined interpretations in which the rabbis specialized, which the rabbis thought to be essential for the truly righteous and pious man, for the highest sons of Abraham. Rather, these were men of the world with practical experience and practical knowledge about human life on Urantia.

Most of the apostles were married, and many had children. They now made arrangements for the permanent support of their families, sometimes accepting subsidies from what became the apostolic funds. Nathaniel was in charge of seeing that the needs of all families were met. Nathaniel was from Cana and the only one of the six first apostles who had not known Jesus. He and Judas Iscariot were the best educated of the apostles.

ORDINATION OF THE TWELVE

On Sunday, January 12, AD 27, Jesus called the apostles together for their ordination as public preachers of the gospel of the kingdom. They went up into the highlands north of Capernaum. There he sat down before them as they gathered around him, and he instructed them in the talk called the sermon on the mount, reported in the gospel of Matthew:

They were to be ambassadors of the Father's kingdom. This kingdom would never die, and it was open to all human beings, Jews and gentiles. This kingdom was not won by force of arms, but by entering into the spirit. One need only seek the kingdom and all else will follow and you will never die.

Those greatest in the kingdom will be the servants of all, but through their humility will be glorified as the sons of God. Not what you say, but who you are will spread the kingdom over the face of the whole earth.

They were to represent him on earth, even as he himself represented the Father in heaven. Therefore they were to learn to show a fatherly love toward those around them, not just the brotherly love contained in the golden rule. A father's love for unlovely humankind can attain levels of devotion far beyond a brother's love. A father's love can return good for evil. They were to love other humans even as Jesus loved them.

He then stood up as they kneeled in a circle around him. He placed his hands on the head of each man and prayed that the Father give him wisdom to carry on once he himself was gone—the creator of a universe placing on commonplace humans the responsibility for establishing a divine brotherhood!

He then delivered the ordination sermon:

They were no longer like ordinary humans. They had dedicated themselves to a special task. Henceforth they were to bring comfort to the downtrodden by declaring the exact opposite of what most believed: Those who are humble before God will know God; hunger for righteousness will be satisfied; the meek, not the violent, will come to rule the world; those free of suspicion and revenge, the pure of heart, will know God; those who mourn will be happy; the man of peace will be a man of God; those who are persecuted will know God and be happy.

They would be like salt, enriching the flavor of life. They will be a candle, casting light. They were not to resist aggression, but if struck on the cheek should turn the other cheek to be struck as well. They were to behave as himself, always showing kindness and mercy and praying for those who hated them bitterly. One day they would be perfect, even as their Father is perfect.

They were not to judge others but to save them. In their promulgation of the Father's kingdom, they were to be as wise as serpents, never casting the pearls of truth before the swine of the unspiritual. When the false prophets appear, they would be known by their fruits. The tree that bears no fruit will be cut down and cast into the fire.

The "sermon on the mount" contains much useful instruction, but it is not the gospel of Jesus. The ordination was meant for the twelve apostles alone, who led special, unrepeatable lives in the scheme of things. The apostles were stunned by it, and still expecting the Son of Man to occupy the throne of David, they understood little. A silent group, they descended from the highlands around dusk.

FURTHER ADMONITIONS

Later that night, in the garden of Zebedee, at the request of Andrew, Jesus further expounded his instructions:

They were having a hard time understanding him because they clung to their old religious beliefs.

They must be reborn and as small children enter God's kingdom.

Their ideas about the Son of Man were wrong. Rather than seeking the Father's favor through almsgiving, fasting, and prayer, they must seek his favor through mercy, truth, and love—the desire to do the will of the Father. Humans judge other human by their deeds, but the Father judges them by their intent. The law says not to kill and not to commit adultery, but to be angry and desire vengeance, or to lust in one's heart for another man's wife is the same as to kill or actually to commit adultery.

James asked about divorce, made easy by Mosaic law, but Jesus replied that it was not his mission to pronounce on social, economic, or political realities. He had come to reveal the Father's love for all humans and to reveal to humans that they were all the children of God. He was concerned solely with the inner spiritual life of man. His mission was to bring a fresh beginning for the whole human race, not to offer solutions to social problems.

He went on to emphasize that "an eye for an eye" was not the law of his kingdom: They must return good for evil. Retaliation for wrongdoing was a legitimate activity of the state, but not of the individual. They were to perform their good deeds in secret, not letting the left hand know what the right was doing. There were to be no ostentatious displays of piety.

They were not to seek worldly riches, but riches in heaven. They were not to concern themselves with food or clothing. There was nothing wrong with property and wealth, but spiritual values must come first. The Father knew their needs in advance, even before they asked. They were not to be anxious for the next day when the problems of this day were sufficient to occupy their attention.

The group decided to remain in Bethsaida for a full week, discussing what Jesus was teaching. Many of the curious and spiritually hungry had by now gathered in Bethsaida, and Jesus turned them over to the twelve, who did their best to answer questions.

During this week Jesus took Peter, James, and John out in a boat where he attempted to instruct them further, so that they might clarify his teachings to the others. On the following Sabbath they returned to the highlands where he went over some of what he had said the week earlier. He warned them: They were about to go as sheep among wolves. They should be as harmless as doves. They would be persecuted or killed, but they should not forget that his spirit and that of the Father was always with them.

One by one he consecrated the apostles to the work ahead. But they never understood that their Master lived his life of spiritual inspiration for every person of every age on every world of a far-flung universe. Jesus lived his earth life not as a personal example of mortal living for the men and women of Urantia, but to create an inspirational ideal for all mortal beings on all worlds.

CHAPTER 29. WORK BEFORE THE DEATH OF JOHN THE BAPTIST

On January 19, AD 27, the group gathered around noon to begin their journey toward Jerusalem. The families and friends came to bid them God's speed. When Jesus did not appear, Andrew searched and found him at the seashore, sitting in a boat and weeping. He had never seen Jesus weep and he asked why. Jesus replied that he wept because not one member of his family had come to see them off. Only Ruth remained loyal to her brother, and she was at that time at Joseph's house in Nazareth.

GOING DOWN THE JORDAN VALLEY

Jesus wished to leave the province of Galilee and go into Judea because he knew that Herod Antipas, who governed Galilee/Perea and had imprisoned John the Baptist, had already taken notice of his work. On the first night they went along the west coast of the Sea of Galilee to Tarichea, and on the next day down the Jezreel Valley to the Jordan River. They rested at Pella for two weeks in the very camp that John had used one year before. A sizable crowd assembled, coming in some cases for a considerable distance from Syria and Phoenicia. About three quarters of them were once followers of John. Already tensions emerged between John' followers and those of Jesus. Many in John's following could not understand why Jesus did not do something about John's imprisonment, if he really was who he said he was. Jesus refused to speak about this matter or in any way advise his followers how to deal with the followers of John.

While the apostles spoke to the crowd, Jesus spoke at night to the apostles, going over what he had earlier taught and emphasizing that when the Father's will became *their* will, then were they in the kingdom, then were they the free children of God, liberated sons of the kingdom. But the apostles understood little of what he said.

They journeyed south to Amathus, a fortress town east of the Jordan and twenty miles south of Pella. They remained for three weeks. Peter, James Zebedee, and John Zebedee did most of the preaching, while the others worked on a personal level or managed aspects of the encampment. Jesus spoke each Sabbath.

During their stay Jesus explained that they were in spirit as one, a brotherhood, because of the fragment of the Father that lived within them, but that in expression everyone must create one's own unique vision and experience. As sons of God they were in harmony, but as individual personalities they were diverse. As sons of God, they lived within the kingdom, but there were to be no creeds or traditions based on his own life and his teachings that compelled men to do this or that. Anyone could through faith enter into the kingdom as a child of God and thereby receive the assurance of eternal life and eternal growth in the divine plan. There were no creeds to control this fact.

On February 26 the apostles and a large group of followers continued down along the Jordan to the ford near Jericho, where John had begun his preaching. Here they remained for four weeks. Andrew, head of the apostolic corps, sent apostles two by two up to Jericho to minister to those who were sick, according to a plan that Jesus advanced. There were three kinds of sickness, Jesus said: of the body, of the mind, and possession by "unclean spirits." Really "unclean spirits" were errant midwayers isolated since the Caligastia rebellion, although this he could not explain to his followers. In fact the glad news of the kingdom brought solace and healing to many of the sick and ailing.

On the last day of March, as the Passover approached, the apostles went up to Bethany above Jerusalem, leaving the crowds behind, and stayed at the house of Lazarus, where Mary and Martha were thrilled to serve them. On April 6 and most days for the rest of April they descended into the city.

JESUS AND THE APOSTLES IN JERUSALEM

During the week of Passover both Jesus and the apostles taught in the temple, or to the large crowds gathered outside the temple. Their message was: The kingdom is at hand, but the kingdom is of the spirit and through faith alone one enters it. By doing the will of the Father and realizing in one's own life the fruits of the spirit, one obeys the only law of the kingdom. About eternal life the Jewish religion held confused and contradictory notions, but as members of the kingdom they were the sons of God destined for eternal life and a career of eternal progression. God is your father and loves you as a human father loves his children. In the meanwhile, in this life, love your brother and sister as yourself.

In private instruction Jesus explained how man's relationship to God is not really that of a kingdom at all, that he used the metaphor "kingdom" only because people expected it and because John had spoken of "the kingdom of heaven" to come. Really, the metaphor of the family is more accurate, Jesus taught, with God as loving father and other humans as beloved brothers and sisters. As a father loves his children individually, so does God love all his children individually and not just the group. Many were present at the Passover from all over the Roman Empire and far beyond it to the east, and during this Passover began the worldwide dissemination of the teachings of Jesus.

On his first day in Jerusalem Jesus visited Annas, the relative of Salome wife of Zebedee, and former high priest, whom he had met as a youth, who had showed him the scribal schools. But Annas, knowing of Jesus' teachings, greeted him rudely.

Jesus turned and left his house and did not see Annas again until his trial before the Sanhedrin, when Annas was instrumental in bringing about his execution.

Jesus attempted to clarify the apostles' confusion about the nature of God, whom as good Jews they saw as a terrible God, an avenging God, an angry God. The Jewish God had a complex history, he explained, reflected in God's several names in scripture. The Hebrews' God was always changing, but moving ever closer to a universal being. *Yaweh* was the primitive god of the Sinai tribes, whom Moses exalted into the Lord God of Israel. *El Elyon* was the Most High God of Melchizedek. *El Shaddai* was built on Egyptian notions of the god of heaven. *Elohim*, "the gods," was a plural concept going back to teachings about the Trinity in the days of Adam. The Supreme Yahweh was Isaiah's conception of a universal god, all merciful and all powerful. And now he presented to them God as the *Father in heaven*, a loving being.

Sometimes Jesus stayed overnight with a Greek Jew named Flavius. Many prominent Jews came to visit him in secret at this house. Flavius was a "proselyte of the gate," that is, he accepted the power of Yahweh but he was not circumcised. His house was filled with beautiful statuary and other art. When Jesus admired it, Flavius was surprised because of Jewish censure of representational art. Jesus explained how Moses understandably attempted to stop primitive idolatry by his proscription against making idols, but such prohibitions made little sense in a day when the Father in heaven was revealed as the universal spirit ruler over all. The Father in heaven took delight in beauty and in those who cultivated beauty.

One night *Nicodemus*, a powerful member of the Sanhedrin, came secretly to the house of Flavius to visit with Jesus. The Great Sanhedrin (from Greek "council") was the Jewish ruling council in Jerusalem, consisting of seventy-one members. Nicodemus had heard Jesus speak in the temple. Already the Sanhedrin and the priests were deeply concerned about Jesus and his teachings. In a respectful conversation Jesus told Nicodemus that only if he were born again in the spirit could he enter the kingdom of God. But the sincere and eminent scholar Nicodemus could not understand him.

By the end of April the hostility of the Sadducees and the Pharisees had grown so strong that Jesus and his apostles traveled south to Hebron and Bethlehem, the site of his birth, where in private conversation they worked throughout May. Jesus visited with Abner, friend of John the Baptist, in the Nazarite camp near Engedi. Some Nazarites followed Jesus, but most Nazarites could not accept him as a teacher because he opposed fasting and other forms of self-denial.

In June, when the fuss had died down, the group returned to Jerusalem. They camped in a garden called *Gethsemane* on the western slope of the Mount of Olives near the brook Kidron that flows from north to south along the eastern edge of the high-walled city (see Map 5). The Jewish elders, learning he was in the garden, were determined to arrest him, but because he did no public speaking, nor did his followers, they decided to leave him alone for the moment. However, when one of the Sanhedrin declared himself a believer in Jesus, the rulers realized they must soon act. Jesus and his group quickly departed into Samaria, north of Judea.

IN SAMARIA AND THE DECAPOLIS

For the apostles the stay in Samaria was a severe trial. All Jews learned to hate and despise the Samaritans. The apostles struggled mightily to love the Samaritans in spite of their upbringing, but Judas could not bring himself to love them.

The origin of the hostility between Jew and Samaritan went back 700 years to when the Assyrian king Sargon (reigned 722–705 BC) moved against a revolt in the northern Jewish kingdom of Israel (the southern Jewish kingdom was called Judea). To punish the rebels he removed 25,000 of the northern Jews to locations in Assyria, where they disappeared, and replaced them with various other peoples imported from different parts of the vast Assyrian empire. Later, the Assyrian king Assurbanipal (685–627 BC) established other colonies in the north-central portion of Palestine. These peoples of mixed but nonJewish origins were the *Samaritans*.

Relationships between the imported peoples and the Jews still in Palestine were from the beginning poor, but when those Jews whom the Babylonian king Nebuchadnezzar had deported to Babylon in 586 BC returned to Palestine in 536 BC, the Samaritans, who adopted much of the Jewish religion, opposed the rebuilding of the temple to Yahweh at Jerusalem. Later, the Samaritans befriended Alexander the Great, who allowed them to build their own temple to Yahweh and various tribal gods on the top of Mount Gerizim in central Samaria. There they offered blood sacrifice much in the manner of the rite at Jerusalem until during the reign of the independent Jewish Maccabees (164 BC-63 BC) Judas Maccabeus destroyed this temple on Mount Gerizim, further inflaming the hatred between Jews and Samaritans. On the ruins of this temple, the apostle Philip would preach to the Samaritans after the execution of Jesus.

After spending two weeks in Samaria the group traveled east to the Jordan and the Greek cities of Phasaelis and Archelais (see Map 4, page 217). There they preached to Greeks, Romans, and Syrians; there were few Jews in these towns. The gentiles thought little of the gospel. This is a wonderful teaching, they said, but if one were to live with love and forgiveness and the brotherhood of man one was not going to last in the cutthroat real world: obviously a religion for weaklings that did not ennoble the spirits of courage and endurance encouraged by their own pagan religions.

When the apostles brought such complaints to Jesus, he answered that he had come to reveal the love of the Father for all his creatures, but that a father's love does not preclude harsh discipline to train the character of the child. Yet the child does not recognize the loving behavior of the father behind the discipline. It did not matter what anyone thought in this age or the next, because the Father does rule his creation through love. Truth is a liberating revelation, and the supreme bonds of the world are love. Human progress takes place through the recognition of the love of the Father and the brotherhood of man, and one day this gospel would prevail in the world, whatever the gentiles or the Jews wanted to believe.

Furthermore, do the followers of this teaching seem weaklings? Is John the Baptist a weakling? Jesus may speak to the poor and downtrodden, but God is no respecter of persons. All alike are his children, rich, poor, woman, man, adult, child, Jew, and nonJew. Do not suppose that the service of the kingdom is easy or without danger. Many would be put to death for their teaching and their belief. It is easy to die in battle beside your peer, but quite another matter to die for what you know to be true in the spirit. Their own heroism and that of their followers, hardly weaklings, will astound the world. He spoke with passion and at length and the apostles were greatly cheered by his words.

In another discourse he spoke of the need for self-mastery. He who conquers himself is greater than he who conquers a city. Self-denial and fasting belong to the old religion, although John accepted such practices, because those who enter the kingdom through belief already live in joy and communion with God. All things old are new. Because the spirit of God lives within them, they are free of the bondage of the old rules for pious behavior. Fear is gone. It is not a duty, but a privilege to cleanse oneself of evil thoughts and behavior in seeking perfection in the love of God. Even faith itself is a gift from God.

At this time the nervous irritations of living in close quarters not only with each other, but with the often troublesome followers of John, the hated Samaritans, and the unclean gentiles created a near crisis in the group. Giving Andrew instructions to inform each of the apostles individually of the need for relaxation, they went up into the nearby hills and spent three days there. Jesus spoke of the need for rest and how relaxation diminishes trouble. But when they were refreshed, "We go again into Samaria," he announced.

THE WOMAN AT THE WELL

They came to outside Sychar in central Samaria to a place called Jacob's well after the Hebrew patriarch. The apostles went into the town to find supplies while Jesus waited alone by the well, whose water was prized by the inhabitants of Sychar because of its low mineral content. While Jesus waited, a Samaritan woman came to the well with a jar for water. Jesus, being thirsty, said, "Give me water."

The woman was astonished because from his dress Jesus was clearly a Jew, from his accent a Galilean. Self-respecting males did not speak to women in public, and Jews did not speak to Samaritans under any conditions. "How is it that you, a Jew, ask me, a Samaritan woman, for water?" she replied. Jesus said, "Yes, I have asked you for a drink, but if you understood, it is you who would ask me for a drink of the living water." "But you have nothing with which to draw water, and the well is deep," she said, "unless you are somehow greater than our father Jacob who drank at this well, and his cattle too."

"Everyone who drinks of this water will thirst again," said Jesus, "but whoever drinks of the water of the living spirit will never thirst again. This well of refreshment will become a living water that leads to eternal life." "Then give me this water, so that I will not thirst and need to come again to this well. For anything that a Samaritan girl might receive from a fine Jew such as yourself would be a pleasure."

The woman, whose name was Nalda, had misunderstood Jesus, thinking that he was making a pass at her. Because Jesus was handsome and she a woman of lax morals, she was flattered and she was interested.

But Jesus looked at her intently and said, "Woman go bring your husband here."

Nalda was startled and somewhat afraid. "I have no husband," she said. "Yes, you once had a husband, but you are now living with a man to whom you are not married," Jesus said. "Do not trifle with my words but seek the living water of life that I am offering to you."

The troubled life of the comely Nalda had not been all her fault. Her husband had callously and without justice cast her aside. She took up with a sequence of men. At the present she lived with a Greek as his wife, but she was not married to him.

Nalda realized that she had misunderstood the situation. "Please forgive my words," she apologized, "for I see that you are a holy man or a prophet. But as a Jew, tell me, our fathers worshiped on Mount Gerizim, but you say that God can be found only in Jerusalem?"

As is common, when faced with the question of her own eternal life, the woman preferred to discuss a theological issue. Jesus saw that she nonetheless hungered for truth, and he replied with patience. "I tell you that the day will soon be here, is even here at this moment, when you will not seek the Father on Mount Gerizim or in Jerusalem, but in your own heart. God is spirit and those who worship him must do so in spirit and truth." "Yes, I have heard that John has said that soon he will come, the Converter, the Deliverer, who will tell us the truth about all things." "I am that very man, the Son of the living God," said Jesus. Such was the first direct pronouncement that Jesus ever made of his divine nature and his sonship with God, made not only to a woman, but a Samaritan of questionable moral character! By his behavior Jesus made clear time and again that he had no respect for popular intolerance over issues of sexual morality, that he looked past such matters and past the person's gender to the spirit that dwelled in men and women alike.

Just then the apostles arrived with the supplies and were astounded to see Jesus conversing with the woman. They stood aside, not daring to interrupt, when Jesus said to her, "Go now, woman. You have found the water of eternal life and a new joy will spring up within you as you become a daughter of the Most High."

Nalda ran into the village proclaiming, "Go to Jacob's Well! There is the Comforter that John promised!" On the next day she told the whole story to the apostle John, who kept it to himself until many years later. She told John that Jesus "told me everything about my life," but really he never mentioned the five husbands—it only seemed that way to her.

FURTHER TEACHING IN CENTRAL PALESTINE

The apostles preached for two days in Sychar, then camped on Mount Gerizim, where many Samaritan believers came to request baptism. But Jesus' followers did not baptize at this time.

At night Jesus further taught his followers:

True religion is of the individual, between God and the individual; organized religion is a socialization of individual experience; worship must alternate with service, life in the real world; prayer does not increase knowledge, but enhances insight; worship is the yardstick by which to measure one's distance from the material world and intimacy with the spiritual world; worship anticipates the better life ahead, then reflects this reality back upon the enriched life we are living now; worship is effortless attention, a form of restful spiritual exertion; worship is the part identifying with the whole. As prayer may recharge the spiritual batteries of the soul, so in worship the soul tunes in to the universe broadcasts of the infinite spirit of the Universal Father.

The apostles understood little of what he said, although beings on other worlds did understand him.

In September and October the group went north to the slopes of Mount Gilboa on the border between Samaria and the Decapolis. The messengers of David Zebedee informed them that the leaders in Jerusalem to the south were becoming ever more aggressive in their hostility toward Jesus and his teachings, and Herod Antipas to the north in Galilee, who still held John in prison, suspected that John and Jesus were working together.

On Gilboa Jesus added to his instruction and after persistent requests gave them what is known as the Lord's Prayer, in more or less similar form to the well-known version (but it did not contain the line "For yours is the kingdom and the power and the glory, forever "). Jesus disapproved of such preset prayers, although John had given them out, but the apostles complained that interested followers constantly asked them for a prayer to say.

At the request of Abner, John the Baptist's principal supporter, John had from his prison cell appointed twelve apostles from his own followers, after the model of Jesus. These men and other followers of John gathered at the Gilboa encampment and worked for three weeks to sort out the differences between the two groups. Both groups accepted the prayer Jesus had given them to give out to new believers. Jesus refused to offer advice in their efforts to come to an agreement otherwise and went away by himself for two of the three weeks that they encamped there.

In the end the two groups did achieve a general understanding and a method of rapprochement whereby a follower of John would accompany a follower of Jesus. When someone declared readiness to "enter the kingdom," John's follower would then baptize that person as a symbol of the baptism of the spirit. In this way baptism became Christian practice, but John's message of "Repent and be baptized!" was replaced by the evolving Christian message of "Believe and be baptized!"

During this time Jesus recounted to his disciples some of the things that had happened on Mount Hermon when he was a youth and what had happened during the forty days in the hills of Perea after his baptism. Memorable remarks made by Jesus during the Gilboa encampment included his admonition to ask, because it shall be given; to seek, then you will find; to knock, so the door will open. For every one who asks, receives; he who seeks, finds; and to him who knocks, the door of salvation will open. The gifts of God come with parental wisdom: If the child asks for a stone when he needs bread, what will the father give? If a child asks for a water serpent in the net when he needs a fish, what will the father give?

On November 2, AD 28, the Gilboa camp broke up and followers of both Jesus and John went into several cities in the Decapolis where they quietly preached. Abner became a devoted follower of Jesus, who later appointed him as head of a commission of seventy committed followers dedicated to spreading the gospel.

Toward the end of December they camped near Pella. On January 10 Herod Antipas beheaded John in his prison cell. Jesus called together the twenty-four apostles of himself and John and announced that with John's death a new phase had come. They were now to proclaim the gospel openly and with power. "Tomorrow we go to Galilee," he announced. On January 13 Jesus and the apostles journeyed to Capernaum, where they stayed at the house of Zebedee.

CHAPTER 30: THE PUBLIC MINISTRY OPENS

THE NEWS OF JESUS' RETURN TO GALILEE spread swiftly. To avoid him who had so disappointed her hopes, his mother Mary left Capernaum to visit Joseph in Nazareth. Of his large family, only the youngest sister Ruth visited with Jesus secretly, sitting in a boat with him near the shore.

On the next day Jesus preached on the shore to the gathered people, so many that he was backed up to the water and, signaling to one of the Zebedee sons, he stepped into a boat from which to continue his talk. He had made this boat with his own hands.

JESUS THE HEALER

After concluding his talk, he offered to take the Zebedee brothers out fishing, because he had interrupted their labors. Their earlier efforts were unsuccessful. "Cast your nets there," Jesus said, knowing from his own experience of the habits of fish in these waters that they congregated at that spot at this time of the day; for Jesus was a superb fisherman. The catch was so great that it nearly swamped the boat. Immediately Jesus' followers declared the catch a miracle, but as often in Jesus' career, it was not.

On the Sabbath Jesus preached in the synagogue on the theme that God is the God of the individual, not the group, as Yahweh was regarded by the Jews. After the sermon, a man stood up and fell into an epileptic fit. Jesus took his hand as the fit passed and the man, who mistakenly thought his malady caused by an unclean spirit, shouted that Jesus had driven the spirit from his body.

After Jesus returned to Zebedee's house, once the sun had set on this Sabbath day and Jewish law allowed one to seek relief for pain, a large group of the ailing, learning of the so-called miracle at the service, assembled and begged Jesus to heal them. Their plight touched him in his human heart. He spoke, saying that he had come into the world to reveal the Father, but if it were not against the Father's will that their suffering be made whole—in that instant an army of celestial beings of various kinds, under the direction of Jesus' Personalized Adjuster, descended on the 683 sick and afflicted in the crowd and healed them. As at Cana, Jesus had allowed his human sympathy in combination with divine prerogative to perform what was seen as a miracle. There are no miracles, but all things in the universe progress according to law. If the Son desires it, and the Father wills it, it *is*: That is the law. Very few of those healed experienced benefits of a spiritual nature.

Disturbed by the incident, which threatened to undermine his revelation of the Father's presence within the individual, Jesus went off alone. When Peter, James, and John found him after a long search, he refused to return to Zebedee's house, where the grateful crowd still gathered. He instructed them to prepare the group to leave on the following day. They would proclaim the gospel in other towns of Galilee. The apostles, raised to a feverish enthusiasm by the mass healing, the largest in Jesus' career, were once again downcast and confused. Why, why, did Jesus turn his back on the great crowds who now thronged Capernaum, seeking to be made well? Why did he not proclaim that he was king and these his subjects?

In the afternoon on Sunday, Jesus' brothers James and Jude came to visit their brother. But Jesus had left town.

TOURING GALILEE

During the next two months Jesus and his twelve apostles and the twelve apostles of John visited numerous mostly small towns in Galilee, but avoided Nazareth. They kept a low profile when near Capernaum and Tiberias, the new capital of Herod Antipas on the western shore of the Sea of Galilee. At night he further instructed the disciples, and by day the disciples taught the people, with varying success.

In his instructions Jesus emphasized how deliberately acting with iniquity over time shuts off the channels of communication with the spirit. Prayer keeps such channels open, the breath of the soul, but only prayer launched in correspondence with universal law will be answered. Foolish prayer or prayer for one's personal advantage was not real prayer and could not meet success. Prayer must be in secret and without fanciness or display or self-torture.

Prayer did not persuade the creator; it opened the one who prayed to the creator's power. One prayed to learn the Father's will. It was important to pray for those who hated one most. After prayer, remain in silent receptivity, to allow the spirit to speak to your soul. He spoke of how the degree of love for one's fellow humans measures the spirit's domination of the personality. Such personalities shall never die.

During this tour in a small town in northern Galilee Jesus healed a leper (according to Jewish custom lepers were excluded from the religious community), who against Jesus' explicit request shouted to the rooftops what was Jesus' first deliberate healing. Every such incident drove the group to another town to avoid the flocks of those seeking signs and wonders.

Again in Cana, a prominent man of Capernaum named Titus traveled to Cana to beg Jesus to heal his dying son. When Jesus said that he would live, another miracle was credited to him; but Jesus only used his prescience to see that the child would recover by natural means. Titus later became important in the earliest Christian church and died in the persecution at Rome in AD 69.

At the village of Nain southwest of Nazareth above the plain of Jezreel a widow's only child was being carried to burial, but when Jesus awoke the child, who was in a deep sleep, the crowds thought that he had brought the dead back to life. To avoid the clamoring multitudes that pursued them, longing for healing, for a miracle, they now went south to the ancient village of Endor. Resting there, Jesus explained to his followers how in the celebrated story of King Saul and the Witch of Endor a rebellious midwayer had impersonated the spirit of the dead, but that such cases of spirit possession of the simple-minded would end once he ascended to the Father. From that moment on there would be no more spirit possession on this planet.

They arrived in Capernaum on March 17, AD 28, and spent two weeks at the Bethsaida headquarters. In a famous incident a Roman centurion named Mangus, who admired the Jews and had even financed the synagogue in Capernaum where Jesus often spoke, went to the elders of the synagogue and asked that they ask Jesus to heal his ailing orderly. Jesus made to go to the man's house, but Mangus sent out his friends to say that as an officer Mangus could from a distance order a man to do this, and another man to do that, and that he himself received orders from afar. Therefore Jesus could heal the man from where he was. He would not have to enter the unworthy lodging of a gentile. Jesus acknowledged the man's surpassing faith, and the orderly got better, although whether through Jesus' intervention or the course of nature is not clear from the records.

THE WHORE AT THE FEAST

When this year's Passover approached, the party again traveled down the Jordan, at first stayed at Bethany, then set up camp in the park at Gethsemane at the bottom of the Mount of Olives. There they could better accommodate the crowds of those seeking healing or salvation.

At one point John Zebedee, the youngest of the apostles, took Jesus aside and led him to the pool of Bethesda, a hot mineral spring just north of the temple grounds where the sick and afflicted gathered in hopes of a cure (see Map 5). Periodic bubbles emerged in the pool and many believed that the first to enter the water after such a bubbling, whose cause was thought supernatural, would be healed. "Master," John said, "if you heal these afflicted, will not then the whole city know that your kingdom has come in power?" Jesus reproved him, saying this was not his method, as he had many, many times made clear. Nonetheless he spoke to those gathered around the pool, and several believed they were healed. One man, an imaginary cripple, actually took up his bed and walked, although it was the Sabbath.

During the nearly three weeks the group spent in Jerusalem the now very famous Jesus attended numerous private and semiprivate gatherings and banquets. It was the custom at fancy banquets in the city to leave the doors open to the streets so that those interested and even beggars could enter the dining hall and stand against the wall, in case a diner might toss them some food. Jesus was dining as honored guest at such a banquet at the house of Simon the Pharisee, an important religious leader (although not a member of the Sanhedrin) and a half-believer in Jesus' gospel. Into the hall came a former prostitute, well known for the high-class brothel she once supervised near the court of the gentiles, below the temple mount. Having accepted Jesus' teaching and abandoned her profession, she persuaded most of her girls to accept their personal salvation through faith and to change their lives. Nonetheless she was condemned to wear her hair long, sign of the harlot.

The former prostitute brought with her a vial of precious perfumed oil. She went up to Jesus and rubbed oil on his feet, then wet his feet with her tears. She dried the tears with her hair, all the while kissing his feet, over and over. The impatient and powerful Simon thought, well, this man could be no prophet who could not tell a decent woman from a whore!

Jesus knew his thoughts and said, "Simon, what if a moneylender lent out 500 coins to one man, and 50 to another, but neither could pay. Yet the moneylender forgave them both. Which of the borrowers do you think would be most grateful to the man?" "Why, I suppose he who received the 500." "Even so, Simon. When I came into the room, you did not kiss me, nor did you give me water for my feet. You did not anoint my hair with oil. But this woman has not ceased kissing me, and she has rubbed my feet with precious oil. Who do you think loves me most? Those who are deepest in sin are most filled with love when their sins are forgiven. Even so this woman's sins are forgiven."

He raised the woman to her feet and told her to go, assured that she had entered the kingdom and could never be harmed by the ill opinion of others. Then he and Peter, James, and John, who accompanied him, arose and left the hall. All who heard of the incident were shocked to hear that Jesus implied that he himself forgave the woman's sins.

BREAKING THE SABBATH

Jesus' renewed presence in Jerusalem further alarmed the Jewish religious leaders and in secret meetings they debated what to do. They hoped that Herod Antipas might arrest and kill him, as he had John the Baptist, but Herod was not personally unfriendly to Jesus. Members of his household were believers and assured Herod that the kingdom of Jesus was a spiritual kingdom, that Jesus did not threaten Herod's power. He believed them. He regretted John's blood on his hands and was happy to let religious fanatics go their own way.

At last the religious leaders decided to send six spies to pretend they were followers and to catch Jesus in a purely religious offense; for the Sanhedrin retained the power to try Jews on such charges, while civil charges were restricted to the Roman courts. The six spies caught up with the apostolic group in Jericho, where they had gone after the Passover. John's followers remained in Jerusalem. Jesus and a group of about thirty started out north from Jericho on the Sabbath, heading for Galilee, but at the last moment, seeing the presence of the spies, Jesus ordered his group to go no further than one thousand yards, the legally allotted limit for travel on the Sabbath.

In this short distance some of the followers broke grain from the fields at the side of the road, as was customary for travelers, rubbed their hands to free the wheat from the chaff, and ate the grain. At once the spies loudly criticized them for breaking the

Sabbath. Taking their complaint to Jesus, he replied that surely the Sabbath was made for man, not man for the Sabbath. Furthermore, he, as the Son of Man, was lord of the Sabbath. In the new law that he was revealing, meaningless and mindless ritual had no place, the very ritual that leaders like themselves so zealously practiced.

To later complaints at Capernaum that his followers did not fast he replied that such practices belonged to the past. If the old was good, it should be kept, but for the new wine they would need new bottles. Why stitch new cloth to old?

BACK IN GALILEE

During the next five months the apostolic corps maintained at Bethsaida an enormous encampment from between five hundred and fifteen hundred persons, ever fluctuating, made up of truth-seekers, health-seekers, and the idly curious. There were representatives from all over the Roman Empire and from beyond the Euphrates. These visitors were soon to carry the teachings of Jesus to far parts of the world.

A portion of the camp was a hospital supervised by a Syrian doctor, a believer, who used all available medical resources in addition to prayer and the healing power of faith. Jesus many times visited the hospital but also spent much time alone in the hills where he was "about my Father's business," consulting with various celestial beings about the administration of Nebadon. Two apostles always accompanied him. They reported that when in the hills the features of his face would undergo rapid changes, although no other beings were obviously present.

This was the longest time that Jesus and his corps ever remained in one place. The apostolic corps trained one hundred and more new evangelists. On the second Sabbath before departing on October 3, AD 28, for the second evangelizing tour of Galilee, the Jerusalem spies sent a man with a withered hand up to Jesus as he spoke to the people, tempting him to break the Sabbath in a clear-cut fashion by healing the man. "Is it lawful to save a sheep on the Sabbath, who has fallen into a ditch?" he asked. They agreed that it was. "Then it is lawful to heal the sick on that day," he replied and the man's hand was made whole, the first time that Jesus healed someone in answer to a challenge.

On Friday, October 1, as he spoke in the enlarged front room of the Zebedee home in Bethsaida, occurred one of the strangest events in his career. A paralytic, unable to enter the crowded room through the doors, was lowered down through a hole opened in the roof so that he came to rest directly in front of Jesus as he spoke. When the man begged to be healed, although his illness was a direct result of his dissolute life, Jesus said, "Your sins are forgiven and your faith has made you whole." The Jerusalem spies were outraged. Who *was* this man to forgive sins? Only God could forgive sins. The cripple arose, took up his bed, and left the room.

At this time the spies were recalled to Jerusalem, but three of them had been so moved by Jesus' personality, presence, and teaching, that they themselves were baptized and remained with the group. The three remaining spies returned to the city.

The second preaching tour lasted almost three months, from October 3, AD 28, to December 30. The one hundred seventeen newly commissioned members of the *evangelical corps*, supervised by the twelve apostles, broke into groups and visited many towns, small and large, in the Roman provinces of Galilee, Syria, and the Decapolis. The purpose of the tour was to train the evangelists. Jesus circulated among the various groups, accompanied always by James and John Zebedee. The stalwart David Zebedee, who had forty or fifty runners at his disposal, coordinated the entire effort from the headquarters in Bethsaida. The work of David Zebedee was critical to the coordination and termination of this schedule of preaching, and in fact to Jesus' entire ministry, which would not have been possible without him.

Of the one hundred seventeen evangelists who went out on the second preaching tour, seventy-five returned after a two-week recess to remain in the newly formed evangelical corps.

MISTAKES OF THE CHRISTIAN CHURCH

By now Jesus' fame throughout the Levant as a healer brought a constant stream of the afflicted; many went away cured or improved, apparently through the power of their faith in conjunction with the actual presence of a being who, after all, himself created the universe we are living in. So charming and prepossessing was Jesus' personality that those who knew him, unless blinded by religious prejudice and political ambition, loved him and admired him. The power of his personality was the beginning of one of the two great errors made in the formation of the Christian religion, which became a religion about Jesus and not a religion that promulgated man's sonship with God and the eternal life that comes from believing that this is true.

The second great error was Paul's doctrine of the atonement, which proposed that an angry Yahweh required blood sacrifice of an especially exalted kind ever to be at peace with his children. Paul drafted his teaching to appeal to traditional Jews with traditional views, few of whom were persuaded, but along the way he made the doctrine a foundation of the Christian religion. The modern Christian church cannot reject this doctrine, because for nearly two thousand years they have taken Paul's instructions and admonitions to early followers, and his misunderstanding of the nature of Christ's mission, as truth revealed to man by God. The dilemma of the modern Christian church is that it cannot be reformed from within and it cannot be reformed from without.

The church, as a social outgrowth of the kingdom, was natural and even desirable. Jesus foresaw that a social organization would follow the progress of the true spiritual kingdom, and for this reason he never opposed the apostles' practice of John's rite of baptism. The evil of the church was not that it existed, but that it supplanted Jesus' concept of the kingdom of heaven, really the spiritual brotherhood of those who follow the will of the Father. Paul's institutionalized church became a substitute for the kingdom of heaven that Jesus proclaimed, which exists in the heart of the believer.

In the Christian church the kingdom of Jesus' teaching became submerged in the mystic conception of Jesus as the Redeemer/Creator, the spiritual head of a socialized religious community. Although the church was an inevitable *social* result of Jesus' life and teachings, it was a tragic failure that this social response to the teachings of the kingdom displaced the spiritual concept of the kingdom. Once the church substituted its own organization for Jesus' promulgation of the spiritual brotherhood of man in a family ruled by God the Father, it then proceeded to set the kingdom off into the future, something that would come at the culmination of the Christian age, at the second coming of Christ.

Although Jesus' ideal concept largely failed, Paul still built on the foundation of the Master's personal life and teachings, supplemented by Greek and Persian concepts of eternal life and Philo's doctrine of the temporal contrasted with the spiritual, to establish one of the most progressive human societies ever to exist on Urantia. The teaching of Jesus is still alive in the advanced religions of the world, and sooner or later must come a revival, a restatement that will undo the work of his early followers who built their system on the *fact* of Michael's sojourn on earth, with little respect for what Michael taught and exemplified in his life.

The teachings of Jesus have an eternal nature that bears fruit in the hearts of reflective humans. The kingdom as Jesus conceived it has to a large extent failed on earth, but the outward church that took its place is only the larval stage of the thwarted spiritual kingdom, the cocoon in which the kingdom of Jesus' concept now slumbers, from which the butterfly will one day emerge.

THE WOMEN'S CORPS

Nothing about Jesus' behavior was more shocking or revolutionary in social terms, than his attitude toward women. He many times announced that women were the spiritual equals of males, at a time when a man would not salute even his wife in public, when according to Jewish prejudice it was "better that the words of the law should be burned than delivered to women." Women were not allowed on the floor of the synagogue, and according to some interpretations were not even Jews: Jews were the *sons* of Abraham and Isaac. On January 16, AD 29, Jesus announced that he was forming a women's corps of ten women, chosen from those who had worked in the camp hospital and performed other administrative services.

Shortly thereafter the group launched the third and last tour of Galilee. The evangelical corps of around seventy-five went out in groups of five into the towns of central Galilee. The apostles mostly traveled together with Jesus and the women's corps, soon to be increased to twelve members, although the apostles later broke off in groups of two to preach and baptize. John's twelve apostles, now based in Hebron, came to Galilee to join them for three weeks of the tour.

The women's corps was immediately effective because they could speak freely to other women about the spirit of God that lived within them, and could enter into bordellos and speak to the women there. In just this way was Mary Magdalene recruited in the city of Tiberias, a woman fallen into prostitution through a series of misfortunes. Mary Magdalene was to become the most effective recruiter of all the women's corps.

Mostly Jesus stayed with the women's corps, or instructed the still inexperienced members of the evangelical corps. Although all the apostles except one were to flee at the time of Jesus' execution, all twelve of the women's corps remained with him to the end; not one betrayed him or turned back. Paul's hostility toward women, based on traditional attitudes, was to undo much of Jesus' example, but Jesus' example remained as an ideal.

INCIDENT AT NAZARETH

According to prior agreement, the scattered emissaries all gathered at Nazareth on March 4, AD 29. They camped on the hill north of town. Nazareth had changed for the worse since Jesus left town, and resentment toward him was strong: Had he not ignored them in his earlier ministries? He had performed miracles at Cana and Capernaum but none in Nazareth. Did he not neglect his brother Joseph who still lived there and his sisters? Had he no loyalty to family and fellow townsmen?

That Saturday Jesus spoke at the synagogue in Nazareth, where he had often spoken as a youth, reading from what had once been his own copy of the scriptures, the Greek *Septuagint* from Alexandria, kept in a special chest. After the sermon he mingled with the crowd, but hooligans hired by local enemies taunted him and jostled him so that some of the apostles and evangelicals in the crowd, outraged, resisted.

The hooligans seized Jesus and swept him from the synagogue and hurried him to a nearby hill. They intended to push him to his death. Suddenly Jesus turned around and folded his arms. He walked slowly through the crowd away from the cliff. No one stopped him.

On March 10, AD 29, a much disheartened group reassembled at Bethsaida and awaited the time of this year's Passover at Jerusalem.

CHAPTER 31. OPEN BREAK WITH THE JEWISH ELDERS

JESUS SAT IN A BOAT belonging to Zebedee at the edge of the shore. A great multitude gathered on the shore until one thousand people waited for him to speak. Realizing that many in the crowds were not sincere seekers of the truth, some even hostile to it, and that their ability to absorb truth varied widely, Jesus for the first time taught by means of a parable, the parable of the sower. After this time most of his teaching took this form.

TEACHING IN PARABLES

"A sower went forth to sow, and as he sowed, some seed fell by the wayside to be trampled underfoot and eaten by the birds of heaven. Other seed fell on rocky places where there was little earth, and immediately it sprang up because there was no depth to the soil. But as soon as the sun shone, the plant withered because it had no root with which to find moisture. Other seed fell among the thorns, and as the thorns grew up, the plant was choked and yielded no grain. Still other seed fell on good ground and, growing, yielded, some thirty fold, some sixty fold, and some a hundredfold. He who has ears to hear, let him hear."

The apostles were startled by this new form of teaching. Although there were a few parables in the Hebrew scriptures, Jesus' use of them was his own invention. At first in discussion among themselves the apostles attempted to explain the story as an *allegory* in which every detail could be translated into some worldly fact or event, but the apostle Thomas correctly pointed out that a parable has one meaning, teaches a single truth, in this case that they were going to meet with varied success in their ministry because of circumstances beyond their control. Jesus agreed that, yes, the parable described his experience so far and the experience they were going to have once he was gone.

Later Jesus added one further item to the parable of the sower. "The kingdom of heaven is also like a man who cast good seed upon the earth, and while he slept by night and went about his business by day, the seed sprang up and grew. Although he knew not how it happened, the plant came to fruit. First there grew the blade, then the ear, then the full grain in the ear. When the grain was ripe, he took up the sickle and finished the harvest. He who has an ear to hear, let him hear."

Soon after he told other parables:

"The kingdom of heaven is like a man who sowed good seed in his field, but while he slept his enemy came and sowed weeds among the wheat, then hurried away. When the young blades sprang up and later were about to bear fruit, there appeared also the weeds. Then the servants of this landholder came and said, 'Sir, didn't you sow good seed in your field? Where have these weeds come from?' And the landholder replied to his servants, 'An enemy has done this.' The servants then asked their master, 'Well, do you want us to go out and pluck up these weeds?' But he answered, 'No, because while you are gathering up the weeds, you might uproot the wheat too. Let them both grow together until harvest time, when I will say to the reapers, 'Gather up first the weeds and bind them in bundles to burn and then gather up the wheat to be stored in my barn.'"

Some achieve eternal life by accepting their sonship with God, others do not. At mortal death some will live, and some will perish.

"The kingdom of heaven is like a grain of mustard seed that a man sowed in his field. A mustard seed is the smallest of seeds, but when it is full grown, it becomes the greatest of all herbs and is like a tree so that the birds of heaven can come and rest in its branches."

Once one has turned towards the spirit, growth in the spirit one day becomes mighty.

"The kingdom of heaven is like leaven which a woman took and hid in three measures of meal, and in this way it came about that all of the meal was leavened."

Rebirth in the spirit changes first a part, then all of one's life.

"The kingdom of heaven is like a treasure hidden in a field that a man discovered. In his joy he went forth and sold all he had that he might have the money to buy the field."

When the possibility of eternal life becomes clear, everything else comes second.

"The kingdom of heaven is like a merchant seeking fine pearls. Having found one pearl of great price, he went out and sold everything he had that he might buy the extraordinary pearl."

When eternal life through sonship with the Father is found, nothing can stand in the way of pursuing it.

"The kingdom of heaven is like a sweep net that was cast into the sea, and it gathered every kind of fish. Now, when the net was filled, the fishermen drew it up on the beach, where they sat down and sorted out the fish, gathering the good into vessels while the bad they threw away."

At death those who have found eternal life through faith go to another world, but those who have rejected eternal life cease to exist.

THE LUNATIC OF KHERESA

Wearying of the constant commotion, Jesus wished to cross the Sea of Galilee to the eastern shore and escape the multitudes. They set forth. He was so exhausted that while he slept in an overhang at the stern of the boat, one of those sudden maelstroms that rise on the Sea of Galilee tossed the boat about as if it were a toy, threatening to capsize it. In a panic Peter woke up Jesus and begged him to save them from drowning. Jesus stepped out from the overhang. "Why are all of you so afraid?" he asked. "Where is your faith? Peace, be quiet."

At just this moment, by coincidence, the storm quit as suddenly as it had come. Jesus explained that he was only encouraging them, not controlling the weather, but Peter and the other rowers were certain that by his words he had commanded the storm to stop—easy to believe in an age when all meteorological phenomena were considered the work of spirits and gods.

On the eastern shore they encountered a madman who lived in the cemetery of the small village of Kheresa (the hometown of the Alpheus twins). During a lucid interval this man had heard Jesus in Bethsaida, and when he saw Jesus, the man ran up, groveled like a dog, and begged not to be harmed. "You know that you are a son of God," Jesus said. "You are not possessed by spirits. I command you to come out of this spell."

At Jesus' admonition the man did regain normal control over his turbulent emotions, but by another coincidence unruly dogs at just this moment drove thirty swine over the steep cliff at the edge of the sea so that they fell to their deaths on the rocks below. The madman, the apostles, and the gentile swineherds believed that Jesus had driven out the many spirits from the madman into the swine, although there had been no spirits in the man and the death of the swine was coincidental. Once cured, the man went throughout the Decapolis celebrating the glorious power of the god-man Jesus of Nazareth.

The very unhappy swine-herders asked the party to please leave, and so they embarked again for Bethsaida.

LOAVES AND FISHES

When they stepped ashore in Bethsaida, the boat was mobbed. In their midst stood a ruler of the synagogue. He begged Jesus to come to his home where his daughter lay dying. On the way to the house, in the midst of the crowd, Jesus said, "Someone has touched me. The living energy has gone forth from me." A woman suffering from a chronic hemorrhage had touched the hem of his robe, a superstitious act that disguised that she was cured through her living faith in the close presence of creative power.

When they approached the house a servant came forward to say that the girl was dead. Jesus entered the house. All was chaos as mourners shrieked and musicians played (to drive away the ghost). Jesus looked at the girl. He said that she was not dead and that the mourners should leave. They ridiculed him, but he took her hand and said, "Awake, arise," and the girl, much famished, got up and walked across the room. Jesus explained that the girl had been in a coma after a long fever and that he had not raised her from the dead, but witnesses were sure he had done just that.

Day by day more people poured into Capernaum until on March 27 Jesus led the apostolic corps to take ship and cross the upper end of the sea to a lovely park south of Bethsaida-Julias at the northern tip of the Sea of Galilee, where the river Jordan comes in. But the crowds followed them in boats or walked around the shore until over five thousand had gathered in the park. Abroad in the crowd was the desire to see Jesus crowned king of the Jews. There was breathless expectation that at long last the Messiah had come.

After three days camped in the park, the food ran low. Jesus, in the only deliberate nature "miracle" of his career, took the remaining five loaves of barley bread and two dried fishes and broke them again and again until the entire congregation was satisfied, with many baskets of food left over. Up rose the cry, "Make him king! King of the Jews!"

Jesus stood on a high rock and addressed them. He proclaimed that his kingdom was not of this world and that the kingdom he represented was the presence of the Father in the heart of man. The crowd was devastated. Later, Jesus sent away his apostles while he went alone into the hills.

That night the group returned by boat to Bethsaida. Another sudden storm arose. The exhausted and bitterly disappointed Simon Peter—a ringleader in the plot to make Jesus king—slept on the cushion at the back of the boat. In a vivid, hallucinogenic dream he dreamed that Jesus, who seemed to have abandoned them, approached the boat, walking on the water. When Peter attempted to walk to him, he sank, but through his faith was raised again. Peter woke up, actually in the water, because while dreaming he had stood up and stepped over the side of the boat. Peter himself always believed his dream to be real, but Luke, in his gospel, rightly understood that it was not real and left it out of his account.

On April 3, AD 29, avoiding recognition, the apostles traveled down the west coast of the Sea of Galilee, crossed the Jordan, and headed south through Gerasa and Philadelphia (modern Amman, Jordan, in the Decapolis) to Jerusalem for the Passover. There they did no preaching. The feeding of the multitude in the park had aroused the terror both of civil and religious authorities. The hatred against Jesus in the city was palpable. One month before fifty thousand in Galilee alone professed to follow his gospel, but most decidedly turned away after he refused to be made an earthly king, to make manna rain from heaven, to feed those who had not worked, to destroy the Romans and set up the Jews over other peoples. Only five hundred souls now accepted him and his mission.

The miracle of the loaves and the fishes, although dependent on Jesus' sympathy for the hungry crowd united with his creator prerogatives, was designed to show to his disciples that public sympathy is fickle and that few actually want the truth. Most took his point. Still, some, including Peter, held out hope that somehow their anointed Jesus would sit on the throne of David, the long promised Messiah, king of the Jews.

CRISIS IN CAPERNAUM

Jesus of Nazareth, the incarnated Creator Son Michael of Nebadon, had one more year to live. He spoke little on the way back from Jerusalem after the Passover. They arrived at Bethsaida on Friday. Jesus was scheduled to speak the next day on the Sabbath.

On this Sabbath day Jesus began his sensational final twelve months with a militant speech guaranteed to create an open break with the authorities. Fifty-three fresh agents of the Jerusalem Sanhedrin, Pharisees and Sadducees, were in the audience in the Capernaum synagogue. The Jewish religious rulers were determined to destroy him and waiting for a fatal slip. Thirty leaders from neighboring towns were also there. Agents of Herod Antipas sat in the seats of honor, disturbed at reports that a large crowd had tried to proclaim this man king in the jurisdiction of Herod's brother Philip east of the Jordan River (the so-called tetrarchy of Philip).

The room was packed to overflow as Jesus prepared to speak. He had begun his public challenge of the Jewish notion of the Messiah by refusing to be made king after performing the miracle of the loaves and fishes, and now he would challenge the Jewish notion of who and what the Deliverer would be. After this speech his public support would collapse nearly completely. This man was not the one they awaited. For the first time Judas Iscariot, in his deep disappointment at the turn of events, considered getting out of the group.

Jesus began by reading from the law. Those who do not receive the word of God will not survive. They shall be scattered to the far kingdoms of the earth. Their failure will be a proverb, an astonishment, a wonder. The enemy will come, tear down their walls, starve them into eating their own flesh, then enslave them with shackles.

Then he read from Jeremiah. Jeremiah warned the Jews that unless they believed the prophets of God, they would be destroyed. For his words the priests wanted to kill Jeremiah. They let him down into a cesspool with ropes around his arms. What will the priests and the teachers do to the prophets of today? What do they want as a sign of his divine nature and mission and of the truth of his teaching?

They have had their material wonders and the blessings of a new teaching. He has not come to threaten them. On this day they stand at a parting of the ways. Many want the Messiah who brings bread and fishes so that no one need labor to live, but his own message brings spiritual liberty, truth, and faith.

In the story they say that Moses brought down manna from heaven, but better is the bread of life than the bread you eat. If you ask where to find that bread, the Son of Man is himself that bread, sent down by the Father, and whoever eats that bread will never hunger or thirst again. He had not come to do his own will, but that of the Father who sent him. Those who believed his words would never die. Would they eat of this bread?

One of the Jerusalem Sanhedrin arose and asked, "Do you say that Moses did not send down bread from heaven but that you are yourself the bread of heaven?" "Yes,"

Jesus agreed. "But are you not Jesus, son of Joseph the carpenter, whose mother and brothers and sisters many of us know perfectly well? How can you have come down from heaven?"

Jesus replied that he was all that they said and more. He and the Father were one. The Son does only that which the Father teaches him, while all those given to the Son by the Father, the Son will receive to himself. Everyone who yields to the teaching of the Father's indwelling spirit will eventually come to the Father. Not that any human has seen the Father, but the Father's spirit does live within.

And the Son who came down from heaven has surely seen the Father. Those who truly believe this Son already have eternal life. He is this bread of life. Those who ate manna in the wilderness are dead, but those who eat the bread of life shall never die.

And this bread of life that he gives to all who will receive is his own living and combined nature, the Father in the Son and the Son one with the Father—that is his life-giving revelation to the world and his saving gift to all nations.

After the talk, questions continued for three hours in informal discussion. A Pharisee, mounting a lamp stand, shouted, "You tell us that you are the bread of life. How can you give us your flesh to eat or your blood to drink? What is the use of a teaching like that?"

Jesus said, "I did not teach you that my flesh is the bread of life nor that my blood is the water of life. I did say that my life in the flesh is a bestowal of the bread of heaven. The word of God bestowed in the flesh and the phenomenon of the Son of Man subject to the will of God are equivalent to the divine sustenance.

"You cannot eat my flesh nor drink my blood but you can become one in spirit with me as I am one in spirit with the Father. You can be nourished by the word of God, which is the bread of life, which has been bestowed in the likeness of mortal flesh.

"The thirst of your soul can be quenched by the divine spirit, truly the water of life. The Father has sent me into the world to show how he desires to indwell and direct all men. I have lived this life in the flesh to inspire all men likewise always to seek to know and do the will of the indwelling heavenly Father."

Another Pharisee complained that Jesus and his followers did not properly wash their hands before eating, an egregious violation of Jewish custom: Jews regarded eating with unwashed hands the same as intercourse with a harlot. Jesus replied that the commandment said that everyone should honor his mother and father, but through priestly legal legerdemain pious Jews cheated their parents of all sustenance, all according to traditional law.

"Not what enters the mouth, but what comes from it, from the heart, pollutes and defiles," he said.

At this point one of the Pharisees brought in a man who actually was possessed by an errant midwayer and dared Jesus to cure him. "You know who I am," Jesus spoke, "come out of him." The midwayer did leave and the man was healed, the first time that Jesus had cured someone by the removal of a mischievous spirit. "You see!" the Pharisees announced. "This man is an agent of the devil, in partnership with Satan!

Beelzebub has performed his so-called miracles. The spirit recognized him, he is one of them!"

"How can Satan cast out Satan?" Jesus answered. "A kingdom divided against itself cannot stand. If a house is divided against itself, it is soon brought to ruin. Can a city withstand a siege if it is not united? If Satan casts out Satan, he is divided against himself. How then shall his kingdom stand? But no one can enter into the house of a strong man and carry off his goods unless he first overpower and bind that strong man. If I by the power of Beelzebub cast out devils, by whom do your sons cast them out? But if I cast out devils by the spirit of God, then has the kingdom of God come upon you. If you were not blinded by prejudice and misled by fear and pride, you would see that one who is greater than devils stands in your midst. He who is not with me is against me. Let me utter a solemn warning to you who would with your open eyes and with open and premeditated malice knowingly ascribe the works of God to the doings of devils. Your sins shall be forgiven, even your blasphemies, but whoever blasphemes against God with deliberation and wicked intention shall never obtain forgiveness."

Another Pharisee demanded a sign to prove that Jesus was who he said he was. "No sign will be given," Jesus replied.

The apostles and evangelical corps were deeply depressed by what had happened in the synagogue, and by the almost instantaneous loss of public support. Jesus warned them that such a winnowing and a testing was essential, that they could not be concerned by the departure of those unable to enter the kingdom. He warned the apostles, too, that one of them was in danger.

"I HAVE NO MOTHER"

That very night, Saturday, April 30, representatives of the Sanhedrin met with Herod Antipas in his palace at Tiberias, requesting permission to arrest Jesus within his jurisdiction and transport him to Jerusalem for trial on religious charges. Herod Antipas would not cooperate. He did not believe that Jesus was a threat—he was either a prophet or a harmless fanatic. An apostate Jew, Herod believed nothing and feared everything. When the Sanhedrin threatened to report him to Caesar's representatives as harboring an insurrectionary, he ordered them from his chamber.

For the next week Jesus prepared his disciples for what must come. His followers tried and true were reduced to about one hundred. These he instructed in three sessions daily. He gave his last public seaside talk in Bethsaida on the Sabbath, May 7. On the next day in Jerusalem the Sanhedrin passed an order that all synagogues in Palestine be closed to Jesus, although they held jurisdiction over only the synagogues in Jerusalem. Every synagogue was regulated independently by a local board. Five members of the Sanhedrin summarily resigned at this unprecedented and illegal action. Messengers were sent out and all synagogues obeyed except the synagogue at Hebron, not from friendliness toward Jesus, but from resentment at the Sanhedrin's highhanded behavior. The synagogue was shortly thereafter burned to the ground.

During the next week Jesus dismissed his followers to rest and visit their families. On May 18, AD 29, Herod Antipas, under relentless pressure from many sides, in another meeting with the Sanhedrin at last agreed that they could arrest Jesus within his jurisdiction and transport him to Jerusalem for trial on religious charges. But the Roman authorities in Jerusalem must agree to the plan, Herod insisted. He received their agreement on May 21 and signed the decree allowing Jesus' arrest on religious charges within his jurisdiction.

On May 22 David Zebedee's messenger came in from Tiberias with news that Herod had signed the decree and that Jesus' arrest was imminent. Jesus' estranged family heard of the order and immediately set out for Zebedee's house at Capernaum. The Pharisees had long worked on Mary's sympathy, telling her that her son was out of his mind and needed to be taken home before he did further damage to his community and to his family. Jesus' other siblings, except Ruth, did not really accept his divinity. Mary thought that if she could only bring her son home to live with her, she could be influential and get him to back off from his strange behavior and wild pronouncements. Yet in the back of her mind she had not forgotten the visit from Gabriel. Mary wavered between fear and love, not certain of anything, uncertain of everything.

At 7:30 in the morning Jesus began his farewell address in the front room of the Zebedee home. The apostles and twelve evangelists would remain with him, he said, but the women's corps would stay in Capernaum. At 8:00 AM Jesus' family arrived but, being unable to penetrate the crowd, gathered on the back porch. They sent a message into the front room that his mother and family had come to see him. Simon Peter interrupted Jesus to whisper the message.

"Say to my mother," Jesus said aloud, "and my brothers that they should not be afraid for me. The Father who sent me into the world will not forsake me, and no harm shall come to my family. Tell them to be of good courage and to put their trust in the Father of the kingdom. But, after all, who is my mother and who are my brothers?" Stretching his hands toward his disciples in the room, "I have no mother," he said, "I have no brothers. Behold my mother and behold my brothers! For whoever does the will of my Father who is in heaven, the same is my mother, my brother, and my sister."

When Mary on the back porch heard him say "I have no mother," she collapsed in the arms of Jude, Jesus' brother, who carried her into the garden. Jesus finished his talk and was about to go to his family when a messenger announced to the apostle Andrew that agents of the Sanhedrin would be there in minutes. Andrew interrupted Jesus and spoke to him when a woman stood up in the room to shout, "Blessed is the womb that bore you and blessed are the breasts that nursed you." Jesus turned aside a moment from Andrew. "No, rather is the one blessed who hears the word of God and dares to obey it."

Just then a group of David Zebedee's men arrived at the back of the house. Fearing it was the Sanhedrin, the party of twenty-five hastened out the front door. Jesus gave instructions to apologize to his family, then boarded a ready boat that set off for the eastern shore. They landed in Kheresa, then proceeded northwards through the tetrarchy of Philip to Bethsaida-Julias, then continued north to Caesarea-Philippi (ancient Dan), then west to the coast of Phoenicia. The Sanhedrin arrived shortly after the boat's departure and spent nearly a week searching for Jesus in the environs of Capernaum.

ON THE RUN

Jesus and the twenty-four spent nearly seven weeks in Phoenicia, traveling first to Tyre, then north to Sidon and outlying towns. Many of the Greek-speaking Syrophoenicians, who generations earlier were Canaanites, were eager to hear the message of the gospel. The tour was a success. They then returned south through Tyre, south to Ptolemais north of Mount Carmel (modern Akko), then inland to northern Galilee, then southeast to Gennesaret on the west of the Sea of Galilee south of Capernaum, where David Zebedee awaited them (see Map 4).

They arrived on the first of August, AD 29 and that night took a boat across the lake again to Kheresa where the lunatic was healed. Again they journeyed north and pitched camp in the park where the miracle of the bread and the loaves had taken place. They had been gone seven weeks, and Jesus' enemies hoped that he had been frightened away for good, that he would not return. Philip, Herod's brother, was a half-believer and learning that Jesus was in the park sent word that he could live unmolested in his tetrarchy if he wanted. Herod Antipas himself said that he had agreed to the arrest of Jesus only in Galilee. If he wished to live in Perea, also under Herod's control, that would be fine with him. A popular resentment at the imperious closing of the synagogues throughout Palestine turned many against the Pharisees. Synagogues were once more opened to Abner and his followers, the excuse being given that they followed John, not Jesus.

CHAPTER 32. "WHO DO YOU SAY I AM?"

Through the messengers of David Zebedee Jesus arranged for a secret meeting with his family at Zebedee's boat shop in Capernaum, but the Pharisees were watching Mary's house and the meeting did not go forward. A temple-tax collector spotted Jesus, Andrew, and Peter at the boat shop and wondered if Jesus and his followers were still loyal Jews. Had they paid the tax? None had money because Judas, who carried the purse, was still across the lake in the tetrarchy of Philip. Jesus proposed that they go out and catch a load of fish to pay the tax, but a messenger of David Zebedee overheard the conversation and sold several baskets of his own catch to cover the tax.

On Monday, August 8, Jesus held a conference in Magadan Park south of Bethsaida-Julias, and several Pharisees came over from Capernaum to heckle him and ask for a sign. Odd that they foretold the weather from the color of the sky, he said to them, but could not see the signs of the times!

At the conference the group agreed on a tour of the Decapolis, but first Jesus wanted to go back to Caesarea-Philippi in the north, built by the tetrarch Philip as his capital. Caesarea-Philippi lay in a beautiful valley watered by the spring that became the upper Jordan River. To the north loomed snow-capped Mount Hermon, site of Jesus' earlier encounter with Lucifer and Satan, and from the hills to the south one could see the upper Jordan and the Sea of Galilee in the distance. Jesus wished to begin the final phase of his earth career where he had begun an earlier phase.

THE SON OF MAN IS THE SON OF GOD

On the way to Caesarea-Philippi they stopped for lunch, after which Jesus stood up and asked, "Who do you say I am?" After some discussion, Peter leapt up and said, "You are the Deliverer, the Son of the living God!" "This has been revealed to you by my Father," Jesus said. "The hour has come when you should know the truth about me." Two days later, in Caesarea-Philippi, he returned to the topic of his divinity and declared that he was then and there creating the brotherhood of the kingdom of heaven on the basis of their understanding that he was both the Son of Man, the perfection of humanity, and the Son of God, a divine being. When he was gone from this world, he said, they would carry on with the work unafraid of anything the world might do to them, even if it killed them. Such was the meaning of the many tribulations they had known since becoming his chosen ambassadors.

This brotherhood would never die but would grow until it was present in all the earth. This living temple was built on the eternal foundation of his divine sonship, and those who became self-conscious sons of God would be the human stones that made this living temple to the glory and honor of the wisdom and love of the Universal Father.

In openly declaring his divinity Jesus inaugurated the fourth phase of his earth career. In the first phase, he was a child, only dimly aware of his nature. In the second phase, he became increasingly aware of his destiny. This phase ended with his baptism in the Jordan. In the third phase he attempted to reveal the kingdom of the Father as the Son of Man, but made little progress; those who heard him could not separate their expectations of the Jewish Messiah from his very different nature. In the fourth phase Jesus ceased to oppose the claims that he was a Messiah, because most of his followers continued to believe that he was the Messiah in spite of what he said. Through the open declaration of his divinity he was now known to his apostles as the Son of God. As the Son of Man he was a teacher and a healer, but as the Son of God he was a direct revelation of the eternal Father.

"Although I stand before you in this physical presence," he said, "I came forth from God the Father. I did come forth from the Father into this world as you have known me, and I declare to you that I must presently leave this world and return to the work of my Father. The Father and I are one. He who has seen me has seen the Father. My Father is working with me in all these things, and he will never leave me alone in my mission, even as I will never forsake you when you soon go forth to proclaim this gospel throughout the world."

But trouble stirred in the heart of Judas, as Jesus well knew, and even Andrew feared for his fellow. Judas was resentful of many slights, disappointed when Jesus did not become king and when he would not show a sign, and as the only Judean in the group of Galileans he felt isolated and the object of discrimination.

On Friday, August 12, they set off from Caesarea-Philippi for the slopes of Mount Hermon.

TRANSFIGURATION

The group arrived at the base of Mount Hermon on August 12, AD 29. From this very spot Jesus had earlier climbed the mountain to settle the spiritual status of Urantia and technically to end the Lucifer rebellion. They remained for two days, but in Jesus' opinion none of the apostles was able so to spiritualize himself as to be prepared for what was about to happen. Therefore he set forth on August 15 accompanied only by Peter, John, and James, whose assignment in the group was to accompany the Master and look out for his needs. It was six days since Peter had blurted out the truth about who the Master was. The group was in gentile lands on a gentile mountain.

Halfway up the mountain they stopped for lunch and Jesus told the three men something of what had happened earlier on this mountain, and more about his experiences during his wandering in the mountains of Perea after the baptism. Around three o'clock Jesus separated from the three men, telling them to wait. For three hours he consulted in private with Gabriel and the Father Melchizedek, then returned around six o'clock. He cautioned them against fear and said that he had decided to live out his full life as one of them, and among them.

That night around a campfire Jesus told the three men various details about the Lucifer rebellion until, heavy with fatigue, they lay down to rest. About a half hour later a crackling sound awakened them. They saw Jesus standing near them in conversation with two celestial beings clothed in light. His face and form were illuminated by a heavenly light. They spoke in a strange language, but Peter thought he heard the names Moses and Elijah, whom he took the beings to be. Really they were Gabriel and the Father Melchizedek, whom the Physical Controllers had made visible to the three apostles through the Master's specific request.

"We will make three tents, one for you, one for Elijah, and one for Moses," Peter said as suddenly a silvery cloud descended over the four men. The terrified apostles fell to the earth. From the cloud came the same voice heard at the Jordan, that of Jesus' Personalized Adjuster, "This is my beloved Son. Follow him."

"You will see things still more wondrous than these," Jesus later said to the three silent, frightened apostles as they prepared around midnight to descend from the mountain. On the way down, "Tell no man what you have seen," Jesus said, "until after the Son of Man has risen from the dead."

His words astonished them. What could he mean? Was he going to die just as his divine nature was revealed? During the transfiguration Gabriel had again confirmed to Jesus that he had satisfied all that the Eternal Son required. The Father Melchizedek confirmed that the Infinite Spirit, too, was satisfied with his conduct in this final bestowal before assuming absolute authority over the universe of his own creation. The presence of the Father was revealed through the Personalized Adjuster, whose voice the apostles heard. But Jesus chose to do the Father's will in living out his human life to its natural end.

While Jesus was on the mountain, both enemies and followers had tracked him to the camp at the base of the mountain. While the apostles, who could not get rid of their expectation that the Messiah would rule the land, disputed among themselves about who would receive what position in the coming kingdom, a man came up whose son was both epileptic and possessed of an errant midwayer, an "unclean spirit." The apostles attempted without success to expel the midwayer by calling on the name of Jesus. The Pharisees in the crowd hooted at this ignominious display just as Jesus arrived. The father knelt at Jesus' feet and begged him, saying: "If you can cure him, I beg you to have compassion on us and deliver us from this affliction." Jesus said, "Question not my Father's power of love, only the sincerity and reach of your faith. All things are possible to him who really believes." The father, a minor official in the court of Herod Antipas, gave the famous reply, "Lord, I believe, but I pray you help my unbelief."

The group traveled back to Caesarea-Philippi. While resting there, Jesus upbraided the apostles for arguing about worldly position in his otherworldly kingdom. Great trials lay just ahead when they would be tested to the utmost. His enemies would take him and put him to death, but he would rise again, he said. The astounded apostles could barely understand his words. What could he mean, that he would be "put to death"?

FURTHER TEACHING IN THE DECAPOLIS

On the journey back to Magadan Park south of Bethsaida-Julias the group took a route through Galilee to confuse those who pursued them, knowing he was wanted in Galilee. On the way he repeated that he would be put to death. "Lord, this shall never happen!" said Peter, running to Jesus and taking his hand. Jesus rebuked him stingingly because Peter was appealing to his human sympathy, to persuade him to go against the Father's will. "Get behind me," he said, "you smell of the spirit of the adversary, the tempter. When you talk like this, you are not on my side. You are on the side of our enemy. In this way you make your love for me a stumbling block to my doing the Father's will. Mind not the ways of men but the will of God."

Jesus went on to say, "Whoever would save his life selfishly, will lose it, but whoever loses his life for my sake and the gospel's, he will save it. What good is it to gain the whole world and lose your own soul? What would a man give in exchange for eternal life? Do not be ashamed of me and my words in this sinful and hypocritical generation, even as I will not be ashamed to acknowledge you when in glory I appear before my Father in the presence of the celestial hosts. Nevertheless, many of you now standing before me shall not taste death till you see this kingdom of God come with power." Many of his followers were later to interpret this remark as meaning that Jesus would reappear on earth very soon, even during the lifetimes of the apostles.

They returned to Magadan Park. Over one hundred evangelists and the women's corps awaited them. From the park they set forth on a four-week tour of the Decapolis, visiting all the major cities and many small towns. One apostle accompanied one member of the corps of evangelists, while Jesus circulated from couple to couple. The women's corps stayed with Jesus. There were no "miraculous cures" during this tour.

Memorable talks by Jesus during the tour included the parable of the wealthy man who forgave his debtor a large sum, then discovered that the debtor had cast into prison a man for owing a small sum. Will not the wealthy man cast into prison the man who, although himself forgiven, did not forgive another? Forgive that you be forgiven.

He also spoke of the shepherd more concerned for one lost sheep out of a hundred than the ninety-nine who did not stray. Even so does God go forth to find those who are lost in evil.

As an individual, you must forgive those who harm you. The group has the right and responsibility to manage, punish, or expel wrongdoers, but do not exact your own revenge.

Do not take punishment lying down, but offer positive good to the evil-doer; for good will always overcome evil. If a man strikes you on the cheek, don't just endure it like a dumb animal, but present your other cheek for a second blow. If a man forces you to walk a mile, walk two miles.

Do not wound the self-respect of those you would lead into the kingdom, but respect each person, never using force or fear.

Entering the kingdom does not protect one from the accidents of time; in misfortune knowledge of the spirit will carry one through.

As for the sacred scriptures so dear to the rabbis, much in them has value, although they are the writings of men who shared the prejudices and ignorance of their own day. Certainly God never told the Jews to go out and kill every last man, woman, and child of an enemy people, as reported in the scriptures. Nor are the scriptures very old, but were assembled in fairly recent times from scattered documents. In no sense are the scriptures the word of God. Pious adherence to a stultifying tradition has no place in the new gospel of the kingdom. In his own career he has rigorously avoided promulgation of his teachings in writing, words that followers will have a chance to wrestle over and contest. He has taught the new truth through living the truth.

THE WOMAN TAKEN IN ADULTERY

On Friday, September 16, AD 29, the entire corps of followers reassembled at Magadan Park by prearrangement. They made ready for a week's rest prior to the final tour of Jesus' earth life, in Perea, east of the Jordan. But on September 25 Jesus surprised them by declaring that he was going down to the Feast of Tabernacles in Jerusalem in company with ten of the disciples; Nathaniel and Thomas would remain at Magadan Park in discussion with a Greek philosopher and believer in Jesus, while the evangelists preached in the towns of Galilee and the women's corps went to Bethsaida. The Feast of the Tabernacles, really a harvest festival, was one of the three principal Jewish festivals when Jews from far and wide traveled to the temple in Jerusalem. It lasted seven days. Participants lived in the open air in leafy temporary huts ("tabernacles") meant to remind the Jews of their days wandering the wilderness.

The group traveled along the eastern shore of the Sea of Galilee, then crossed the Jordan near Scythopolis and arrived in Bethany outside Jerusalem on Wednesday. Nathaniel and Thomas joined them the next day. They remained in the vicinity of Jerusalem for four and one half weeks until the end of October, AD 29. Jesus went into the city several times during the feast and preached openly in the temple, terrifying the apostles, who feared that the Sanhedrin might take him. But the Sanhedrin, some of whom believed in Jesus, were afraid to arrest him in the midst of the multitudes from all parts of the Roman Empire. Many thought that he must have made some arrangement with the Roman authorities to behave so boldly.

In his temple talks Jesus announced publicly the fact of his divinity and the meaning of his divinity to humanity. The listeners' response was by now typical: he was mad, he was possessed of a devil, he was a religious fanatic, probably harmless. But some believed what he said. When an onlooker asked him why the Jewish elders wished to kill him, Jesus replied that his teaching freed all humans from the burdensome obligations of elaborate and stultifying ceremony. If you believed him, you no longer needed the oppressive religion of the scribes and the Pharisees. For this reason they wished to kill him.

His words so outraged the elders in the crowds that they determined to arrest him no matter what and sent a man named Eber to take hold of him. Jesus saw Eber coming and said, "You come to do the bidding of those who sent you. But no man will touch me until my time has come." Soon he would be gone, he said. They would seek him but could not come where he was going. "To Greece?" one man shouted. Eber and his assistants, moved by Jesus' words, turned around. They returned empty-handed to the Sanhedrin. The Sanhedrin were thrown into increasing confusion.

During one visit to the temple, Jesus' enemies brought before him a woman allegedly caught in adultery, in fact taken in collusion with the Sanhedrin and her husband, who for many years had used her as a prostitute to make money. According to the law of Moses, adultery was punishable by stoning, but the Romans had forbidden the Jews to execute anyone without their explicit approval. There was the dilemma: What were they to do with this adulterous woman? If Jesus said, "Let her go," he would violate the law of Moses. If he said, "Follow the law," he would be defying Roman power. Jesus did not say, as reported, "Let him without sin cast the first stone"—the Pharisees were the most moral men of their age and in a legal sense sinless—but went over to the woman's husband in the crowd. He wrote something in the sand before him.

The man quickly departed. Then again he wrote in the sand near the woman, and her accusers departed one by one. Finally, he wrote a third time beside the man taken in adultery with the woman, and he left too.

"Is no man left to stone you?" he asked the woman. "No man," she replied. "Then go your way in peace. I do not condemn you." The woman left her depraved husband and became a disciple of the kingdom. *The Urantia Book* does not report what Jesus wrote in the sand.

FURTHER PROCLAMATIONS AT THE FESTIVAL OF TABERNACLES

On the evening of the second to last day of the festival, when the temple grounds were lit by candelabras and torches, Jesus stood up in the midst of the multitudes and declared that he was the light of the world and that those who followed him would have the light of life. Although he himself bore witness of himself, his witness was everlastingly true. He knew where he came from, who he was, and where he was going. Those who would reject this light were from beneath, while he was from above. He was not of this world, but lived in the eternal light of the Father of lights. He was the light of life, and every one who deliberately and with understanding rejected this saving light would die.

On the last day of the festival, as wine and water were poured on the altar, Jesus went on to say, "If any man thirst, let him come to me and drink. From the Father above I bring to this world the water of life. He who believes me shall be filled with the spirit that this water represents. When the Son of Man has finished his work on earth, there will be poured out on all flesh the living Spirit of Truth, and those who receive this spirit shall never know spiritual thirst."

Later, in the afternoon, he proclaimed that if they believed his words and did the will of the Father, they were his disciples. They would know the truth and the truth would make them free. Their leaders sought to kill him because their souls were sealed by prejudice and blinded by the pride of revenge. Would that they could see that he came from the Father, sent by God, and that he did not do this work of himself. Those who would not hear his words were not of God. If those who believed the gospel kept this word of truth alive in their hearts, they would never taste death.

As Jesus turned to leave, he said, "Before Abraham was, I am." Hearing this, members of the Sanhedrin in the crowd shouted, "How can you be before Abraham? You are not even fifty years of age. You are the agent of the devil!" They looked for stones to cast, but Jesus escaped through the crowds and down narrow corridors. He made his way up to Bethany and the home of Mary, Martha, and Lazarus.

During the following week Jesus spent much time in Bethlehem, where Abner and the other eleven apostles of John were now stationed. The agents of the Sanhedrin searched for him but could not find him. Jesus agreed with Abner that John's apostles and his own should soon meet together at the park at Magadan.

AT MAGADAN PARK

Around four hundred followers soon assembled at Magadan Park, including the women's corps, Abner and fifty followers of John, the apostles, and about one hundred fifty dedicated disciples. A committee of three formalized membership in the evangelical corps, now about seventy in number. Family and possessions must, for this elect group, be left behind.

One rich man, a Pharisee and member of the Sanhedrin, wished to become an official member of the kingdom. He deeply believed Jesus' teaching, but because in his heart he loved his wealth, the traditional sign of piety in Jewish society, Jesus told him he must sell everything he had if he wished to become an official messenger. No other thing must stand between the messenger and the gospel to which he devoted himself. No man, having put his hand to the plow and turned back, was suitable for this elect group. Not wealth, but the love of wealth, stood between a believer and entrance into the kingdom. The candidate went away saddened because he could not bring himself to give up his love of wealth. After Jesus' execution, he sold all that he had, too late to enjoy a personal experience with the Master.

Jesus told the apostles the parable of the master who hired workmen at dawn, at 9:00 in the morning, at noon, and at 3:00 and 5:00 in the afternoon, then paid them all the same wage. Was that not his privilege, to spend his money as he pleased? And no one received less than he expected. The Father was like that, receiving any one at any time with the equal promise of eternal life, once they came to believe the message of the gospel.

On November 19, AD 29, Jesus ordained the seventy of the new evangelical corps as they sat in a circle around him. He placed his hands on Abner, the head of the corps, then one by one on the head of each person, then explained how they must act.

They must speak to gentile and Jew alike and be ready for a hostile reception. They must be as lambs among wolves. They must carry nothing with them as they traveled in Galilee, Judea, and Samaria in couples, but depend on the Father for lodging and sustenance. The time had come for an open break with the religious leaders in Jerusalem. Instead of the 613 rules of living expounded by the Pharisees, they were to teach that man's whole duty is summed up in the commandment to love God with all one's mind and soul and to love your neighbor as yourself. They would travel about, preaching the gospel and tending to the sick, then reassemble at Pella on December 30, AD 29.

THE GOOD SAMARITAN

After two months the camp was moved from Magadan near Bethsaida-Julias to Pella on the upper Jordan. There the seventy assembled, as agreed, on December 30, to report successes and failures. Jesus knew that he could now leave this planet, having trained a corps of dedicated believers who would carry on the work into the future.

Shortly after establishing the camp at Pella, Jesus, Nathaniel, and Thomas ventured south along the Jordan, but not until they reached the ford near Jericho did Nathaniel and Thomas realize that he intended again to go up to Jerusalem, now for the Feast of the Dedication, founded in 165 BC by Judas Maccabeus to commemorate the rededication of the altar of burned offerings after its desecration by the Macedonian king Antiochus Epiphanes. They were terrified that the Sanhedrin would arrest him, but Jesus was determined to offer his enemies one last chance to accept his revelation of the Father's love for his children and the possibility of attaining eternal life through faith.

That night they stayed at Jericho. A group gathered around him to ask questions, including a Pharisee who hoped to catch him in a violation of Jewish law. He asked how one attains eternal life. "What is written in the law and the prophets?" Jesus asked. The scholar, knowing the teachings of both Jesus and the Pharisees, answered, "To love the Lord God with all your heart, soul, mind, and strength, and your neighbor as yourself." Jesus replied, "You have answered rightly. This, if you really do it, will lead to life everlasting."

But what really is a "neighbor," the scholar went on, because in Jewish sacred law only another Jew could be a neighbor—nonJews were "gentile dogs." The man knew that the teaching of Jesus was open even to gentiles.

Jesus told the story of the man beaten by brigands on the road from Jerusalem, left to die by a passing priest, then by a Levite (a member of the Jewish priestly caste), but at last bound up and cared for by a passing (and of course despised) Samaritan. Who was the good neighbor? "He who showed mercy," the scholar could only reply, unable to bring himself even to say the word "Samaritan." "Go and do likewise," Jesus said.

THE MAN BLIND FROM BIRTH

On that Sabbath, Jesus, Nathaniel, and Thomas were walking outside the temple where a man blind from birth, named Josiah, begged for alms. He did not do so on the Sabbath, which was against the law, but nonetheless occupied his usual place. The two apostles asked Jesus why it was that this man suffered so. According to traditional explanations, such afflictions came from the sins of the father or mother, or even the child itself before it was born, or when the person lived before in another body and had behaved badly. Although such people were punished for sin, it was nonetheless considered virtuous to give alms to them.

Jesus rejected utterly the popular view. He explained that such conditions were the result of entirely natural events. The soul had no prior existence for which it might now be paying the price. Wishing to challenge the Sanhedrin openly, Jesus now healed the blind man, not from the man's faith, or even from his desire to be healed, but because Jesus wished to make clear his own power to heal whom he chose and when he chose.

Rubbing his spittle with clay, he rubbed the clay in the man's eyes and told him to wash it off in the pool of Siloam, which lay somewhat south of the temple grounds. It was a popular superstition that the spittle of a holy man had healing power. The man did as instructed and to his astonishment he could see. Having nothing else to do, he reoccupied his place outside the temple. When the many who knew him saw that he could see, a great commotion arose that swiftly brought the incident to the attention of the Sanhedrin. Jesus was again in Jerusalem.

Themselves violating the Sabbath, about fifty Sanhedrin gathered in their room in the temple and called Josiah before them. Obviously he was guilty of violating the Sabbath! He had washed the clay from his eyes, hadn't he? And the man who placed it there was guilty of violating the Sabbath. Therefore that man could never be a man of God. Such a man would never violate the sacred rules governing the Sabbath. Only a prince of devils could have done this.

The simple Josiah replied that God's work can only be done by a man of God. "Whether this man is a sinner, I do not know," he said. "But one thing I do know—at first I was blind, and now I see." Did they, too, wish to follow this prophet who had healed him?

Furious at Josiah's insolence, the Sanhedrin sent immediately to expel him from his synagogue—to deprive him of every benefit and contact with the Jewish community forevermore, except the right to buy sustenance.

At the same time as these events took place, Jesus was standing nearby in the porch of Solomon (a colonnade on the inside of the eastern wall of the great temple, see Map 5, page 229), and spoke and answered questions for two hours. When he stated that as the Father was in him, so would he be within those who believed in his gospel, many Jews picked up stones to kill him for blasphemy. But other Jews restrained them, believing what Jesus said. In the meanwhile the Sanhedrin feared to arrest him in the midst of the crowds and bring him before them. Their illegal meeting, which took place nearby, broke up in confusion.

On the next day Jesus and the two apostles returned to Pella in company with Josiah, who became a fervent advocate of the kingdom.

CHAPTER 33. "THE GOOD SHEPHERD" AND OTHER TEACHINGS

On January 3, AD 30, Abner, former head of the Nazarite brotherhood at Engedi near the Dead Sea and head of the twelve apostles of John the Baptist, now head of the evangelical corps of seventy, called the evangelical corps together and gave final instructions for the tour of Perea. The tour would last three months and go into even the smallest villages in this most beautiful portion of all Palestine, the most thoroughly canvassed of any area during Jesus' ministry. The camp at Pella would serve as headquarters. The population was an equal mixture of Jew and gentile. In Perea the Master's message appealed more to the better classes than elsewhere.

The women's corps, now numbering sixty-two, performed most of the ministry of healing on the tour of Perea. Jesus journeyed out with the twelve to visit the evangelists in the field. Having shaken free to a large extent from the wonder-seekers who were first attracted to him, Jesus performed no "miracles" on this tour, except for a single incident outside the city of Amathus where he healed a group of lepers. Abner instructed the seventy to baptize all believers, although Jesus had said nothing about this. In this way the rite of baptism entered the Christian church, the only remnant of anything that John taught or practiced. Abner knew Perea well, where John had done most of his work. He and the seventy evangelists were never to return to the camp at Pella.

THE GOOD SHEPHERD

By mid-January, around twelve hundred people had gathered at Pella to hear Jesus speak in the mornings. In the afternoon he spoke to the apostles, and then to an intimate group of fifty every night, when they were in camp. By mid-March four thousand had gathered at the camp. Jesus was reaching the peak of his popularity in the nonwonder-working phase of his career. But a quarter of those who gathered were hostile Pharisees and other doubters and cavilers.

During his educational sessions Jesus gave some of the speeches for which he is best remembered, including the parable of the good shepherd:

The good shepherd goes in to the sheepfold only by the door. Anyone who enters otherwise is a thief and a robber. The porter opens the door and the shepherd calls to his sheep. They recognize his voice and follow after him as he leads them forth. They will not follow a stranger whose voice they do not recognize. Even so among the multitudes gathered at Pella many—those who hunger for the truth—hear the good shepherd's voice and they follow him. Others do not hear his voice because they are not of his flock, and they do not follow him.

Jesus is this true shepherd, but he is also the door to the sheepfold. Every shepherd who tries to enter the sheepfold without him will fail, because the sheep will not hear his voice. Like a good shepherd, he so loves his flock that he is willing to lay down his life for it. The thief breaks in to steal and kill but Jesus, this Son of God, has come to bring abundant life. His sheep know him and he knows them. When the wolf comes, the true shepherd will not run away and will even lay down his life to protect his flock.

But he is the shepherd of other folds, on other worlds, and under his leadership all will be brought into a single fold, who all shall acknowledge the fatherhood of God. If he lays down his life to protect his fold, he will take it up again. No being can take from him his life; he lays it down and takes it up as he wills. He received this power from the Father before this world came into being.

Obviously Jesus of Nazareth had lost his mind, thought many Pharisees. His words threw the disciples into confusion. Some Pharisees believed what Jesus said, and they remained in his company.

OTHER TEACHING DURING THE TOUR OF PEREA

Jesus told the story about the rich man who owned so much that he enlarged all his barns to store his wealth. That very night robbers broke into his house, killed him, took everything from the barns and burned the rest, and what they missed his heirs squabbled to obtain. The foolish man had filled his barns but had not stored treasure in heaven to feed the hunger of the soul. What will it profit a man to gain the whole world but lose his soul?

In private conversation with his disciples, who had given up everything in service to the kingdom, he cautioned them against doubting the certainty of their support: "Consider the ravens, who neither sow nor reap, and have no storehouses or barns, and yet the Father provides food for every one of them that seeks it. Of how much more value are you than many birds! Besides, all of your anxiety or doubts can do nothing to supply your material needs. Which of you by anxiety added a hair's breadth to your stature or a day to your life? Because such matters are not in your hands, you should not give anxious thought to any of these problems.

"Consider the lilies, how they grow. They do not work, nor spin, yet even Solomon in all his glory was not arrayed like one of them. If God so clothes the grass of the field, which is alive today and tomorrow is cut down and cast into the fire, how much more will he clothe you, the ambassadors of the heavenly kingdom. Seek the greater thing and the lesser will be found within. Ask for the heavenly, and the earthly will follow, for the shadow is certain to follow the substance.

"You must be like servants who watch for their master to return from the marriage feast so that, when he comes and knocks, you may quickly open the door. The master blesses such watchful servants. Blessed are those who are faithful at such a moment. Then will the master make his servants sit down while he himself serves them. But if the servant is lazy and begins to say in his heart, 'My master delays his coming,' and begins to mistreat his fellow servants and to eat and drink with the drunkards, then the lord of that servant will come just when he is not looking. Finding the servant unfaithful, the master will cast him out in disgrace. A defining crisis is just ahead, a trial that will test your soul, and you must watch, be ready, and not be afraid.

"I have not come into the world to bring peace, but division between those who understand the truth and join the brotherhood of man under the fathership of God and those who reject these truths. In the end the truth will prevail, but before that brother will be set against brother, and father against son."

The Perean tour lasted three months. Abner, who took up residence in Philadelphia where the local synagogue was never under the control of Jerusalem, preserved an assembly based more closely on the teachings of Jesus than the familiar cult built around his personality and miraculous deeds. Abner is unknown to the existing records because he was at odds with James (the brother of Jesus), head of the early church in Jerusalem, whom Peter supported on various administrative matters. Then Abner fell out with Paul, who distorted the mission of Jesus by making philosophical and theological claims appealing both to Jews and to gentiles. Abner called Paul the "clever corrupter of the life teachings of Jesus of Nazareth, the Son of the living God."

Philadelphia was really the headquarters of the early church in the south and east as Antioch was in the north and west. Abner died at eighty-nine years of age, November 21, AD 74, in Philadelphia, to the very end a faithful believer in, and teacher of, the gospel of the heavenly kingdom.

PARABLE OF THE WEDDING FEAST

During the Perean tour, on Wednesday, February 26, AD 30, Jesus arrived in Philadelphia followed by a crowd of six hundred. His gospel had been well received in Philadelphia. On the next Sabbath a wealthy and influential Pharisee, who believed what Abner had said of the kingdom, invited Jesus for breakfast at his home. About forty assembled for the meal, all of them leading and learned men, including Pharisees from Jerusalem, who heard that Jesus planned to be in Philadelphia.

While Jesus stood near the doorway talking with Abner, a Pharisee entered, a member of the Sanhedrin so powerful that he went straight for the seat to the left of the host, the seat of honor. The host informed him, to his intense chagrin, that the seat was reserved for Jesus, and the one to his right for Abner, but that he could be seated in the fourth seat to the left.

The meal was friendly because most of the guests believed in the teachings of Jesus, or were not against them, except for those explicitly hostile, who were careful to observe how Jesus did not wash his hands before sitting down and that Abner washed only once, before the meal.

As they finished eating, a chronically sick man, deep into dropsy, came into the room from the street. "Why do you permit a man like that to enter the room?" the Jerusalem Pharisee demanded to know. Because Jesus smiled to the man, he sat on the ground near his couch. Jesus said, "I have a question to ask all of you. Is it lawful to heal the sick on the Sabbath?" But no one would answer his question.

Jesus arose, went to the man, and took him by the hand. "Get up and go your way. You have not asked to be healed, but I know the desire of your heart and the faith of your soul." Jesus returned to his seat and, before the man had left, now healed of his disease, Jesus addressed those at the table:

The Father performs such works not to tempt anyone into the kingdom, but to reveal himself to those already in the kingdom. Which one, having a favorite animal that fell in the well on the Sabbath, would not draw him up? When bidden to a marriage feast, one should not sit down in the chief seat, because perhaps a man still more honored has been invited. The host will need to ask that you give your place to the other man, the honored guest. Thus shamefully you will have to take a lower place at the table.

When bidden to a feast, better to seek the lowest place and sit there, so that when the host looks over the guests, he might say, "My friend, why do you sit in the seat of the least? Come up higher." For every one who exalts himself will be humbled, while he who humbles himself will be exalted. Therefore, when you entertain at dinner or give a supper, don't just invite your friends and family, or your rich neighbors, so they in return may bid you to their feasts. When you give a banquet, sometimes bid the poor, the maimed, and the blind, so that you might be blessed in your heart. For the lame and the halt cannot repay one for loving ministry.

To break the awkward silence that followed this speech, one Pharisee said, "Blessed is he who eats bread in the kingdom of God," a common saying, prompting Jesus to tell the parable of the wedding feast:

A certain ruler gave a great banquet, and having bidden many guests, he sent out his servants at dinner time to say to those who were invited, "Come, for everything is ready." But all began to make excuses. The first man said, "Well, I have just bought a farm, and I must see what it's really worth. Please, give my excuses." Another invited guest said, "Yes, but I have just bought five yoke of oxen. I must go get them, Please, have me excused." And another invited guest said, "I have just taken a wife, and so I cannot come." The servants went back and reported to their master.

The master was very angry, and turning to his servants, said, "I have made ready this marriage feast! The animals are killed, and all is ready for my guests, but they have spurned my invitation. Every one of them has gone after his lands and his property, and they even show disrespect to my servants who ask them to come to my feast.

"Go out quickly, therefore, into the streets and lanes of the city, and into the highways and the byways, and bring here the poor and the outcast, the blind and the lame, so that the marriage feast may have guests." And the servants did as their lord commanded, but there was still room left for more guests. So the lord said to his servants, "Go now out into the roads and the countryside and constrain those who are there to come in so that my house may be filled. I declare that none of those who were first invited shall taste of my supper." And the servants did as their master ordered, and the house was filled.

After Jesus finished, everyone went home. But one of the sneering Pharisees understood the meaning of the parable. Taking it to heart, he was baptized that day and in public confessed his faith in the gospel of the kingdom.

Later that day Jesus preached in the synagogue for the first time since being banned from the synagogues in the rest of Palestine. After the talk, he encouraged a woman bent over with depression for eighteen years. Through her faith her depression vanished and she stood upright, evidently a miracle (but not really), an act that a local Pharisee loudly condemned because it was done on the Sabbath.

On Monday a messenger arrived from Bethany saying that Lazarus, brother of Martha and Mary and one of Jesus' oldest and closest friends, was mortally ill and that Jesus should come at once. Jesus determined to make use of the situation in a last appeal to the Jerusalem intellectuals to overturn their preformed notions of a Messiah's powers.

According to Jewish custom, the dead were buried immediately, a necessity because of the warm climate. Sometimes those buried were not really dead, but comatose from disease or some other reason, so that from time to time a buried person would "rise from the grave." But never after the third day. Apparently the soul departed after the third day, never to return, popular wisdom held.

For this reason Jesus waited for two days before setting out for Bethany. The apostles were wholly alarmed at his decision to go into Judea, where there were plots to kill him, but he cautioned them that he would remain on earth as long as he chose. They set forth on Wednesday, followed by about fifty supporters and detractors.

THE RAISING OF LAZARUS

Lazarus had been dead for four days when on Thursday a messenger reported that Jesus was coming up the road from Jericho. Of the two sisters, Mary had resigned herself to grief for her beloved brother, but Martha still harbored hope that Jesus might do something. Therefore she ran out to meet him and falling down said, "If you had been here, he would not have died." Jesus urged her to have faith, that her brother would rise again.

Martha answered, "I know he will rise again in the resurrection of the last day, and even now I believe that whatever you shall ask of God, our Father will give you." "Yes, Martha," he said, "for I am the resurrection and the life. Whoever believes in me, although he dies, yet shall he live. Whoever lives and believes in me will never really die. Martha, do you believe this?" Martha answered, "Yes, I have long believed that you are the Deliverer, the Son of the living God, the very one who should come to this world."

Mary came out of the house to greet Jesus and together with about fifty mourners they went into the garden where Lazarus was buried in the family's private tomb. Jesus wept as they walked. Among the mourners were some believers in Jesus, but there were many officials from Jerusalem who hated Jesus.

The family of Mary, Martha, and Lazarus, close friends of Jesus' youth, was the richest and most respected in the small town of Bethany, which overlooked Jerusalem on the Jericho road. Their parents had inherited many vineyards and olive orchards. The parents were now buried in the same tomb where Lazarus lay, a natural cave in a thirty-foot high rock at the back of the garden.

According to Jewish superstition, on the third day the angel of death touched the dead with his sword tipped in gall, and by the fourth day the gall had taken full effect. Therefore on that morning of the fourth day they rolled across the entrance a large stone. Standing before the stone, Jesus commanded that it be rolled back. Invisible to the onlookers were the hosts of celestial beings that now followed Jesus everywhere, and Gabriel himself in their charge. Jesus' Personalized Adjuster, now in command of all Adjusters in the universe of Nebadon, had ordered, fifteen minutes before the death of Lazarus, that his Adjuster be held on this planet indefinitely and that it not return to Divinington.

"Lazarus, come forth," Jesus said. Lazarus was in fact dead. Physical resurrection of the dead is a highly intricate operation requiring a complex interaction of many beings and has occurred only this single time on this planet. Jesus himself was not resurrected in his former physical form. But at the word of the Creator Son, Lazarus within seconds began to stir and sat up in the niche where he lay, wrapped in grave clothes, his face covered by a napkin. He demanded to know why he was dressed in this way and why he was in the garden. In absolute terror all in the garden fled to their homes, except for the apostles and Mary and Martha. At the command of Jesus' Personalized Adjuster, Lazarus' Adjuster then returned to him.

The raising of Lazarus was the greatest wonder of Jesus' career, but it only intensified the Sanhedrin's fear and hatred. On the next day in a long and acrimonious session one Pharisee proposed that they abandon their efforts to entrap Jesus in a clear violation of the law and pass an immediate sentence of death on him without trial, in contravention of all Jewish law and precedent. Obviously this man was in league with Beelzebub, the prince of devils. He dined with filth—publicans and whores—and flouted Jewish law. He healed on the Sabbath. He blasphemed God by speaking of him as his father. He claimed to forgive sins.

At this illegal proposal, fifteen members of the Sanhedrin immediately resigned. They indignantly arose and left the room. Later when a vote was taken, five more were dismissed under suspicion of being friendly to Jesus. "It is better that one should die, than that the whole community should perish," said the high priest Caiaphas, son-in-law of Annas, the distant relative of Salome wife of Zebedee. With the resurrection of Lazarus the whole people might now follow Jesus, ruining their machine for profit in the administration of the temple and the religion and endangering their relationship with the Romans. The man must be stopped at any cost.

After the execution of Jesus, the Sanhedrin decreed the death of Lazarus too, the living exemplar of Jesus' power. Learning of this, Lazarus fled to Philadelphia where he worked for the rest of his life with Abner. Mary and Martha soon sold the property in Bethany and joined him in Philadelphia. Lazarus died at age sixty-seven of the same disease that first carried him off at a much younger age.

"THE PRODIGAL SON" AND OTHER PARABLES

Back in Pella, on Thursday Jesus told the multitude the parable of the woman who had ten pieces of silver made into a necklace, but lost one piece. She lit the lamp and carefully swept the house and kept on searching until she found the lost piece of silver. When she found the lost coin, she called together her friends and neighbors and said, "Rejoice with me, for I have found the piece that was lost."

He told his most famous parable:

A great landowner had two sons, the younger one bighearted, carefree, and irresponsible, the older serious, hard-working, and vain. The younger son was so unhappy that he asked his father for his inheritance. He took it and went forth into the world and squandered his fortune in riotous living, until everything was gone. When a famine fell on the land, he was reduced to feeding swine. He envied the swine the food he fed them, until he remembered and said to himself that even the lowest servant in his father's house had plenty to eat. He would return there and be a servant in his father's house!

When his father saw the younger son coming toward the house—for he loved him and missed him and always looked to see if he would come—he ran out to the boy, embraced him, and clothed him in a fine robe and sandals and gave him a ring. He ordered his servants to kill a fatted calf that they might celebrate the son's return.

When the older son returned from the fields and learned that the party in the house was to celebrate the return of his younger brother, he refused to enter and take part. The father came out to him, but still he refused. "Father," he said, "I have stayed at home and done all that you asked, but for me you have not even killed a kid to make merry over!" "Yes, my son," the father said, who loved his older son too, "but you have never been in want and always had what you needed, and could have had a kid to make merry any time you asked. Now you should rejoice with me, for your brother was lost. Now he is found, as if he returned to life."

Jesus also told the story of the wily steward. Once there was a steward who mistreated his master's servants and stole from the master and wasted his funds. When rumors of this behavior came to the master's ears, he called the steward to him and told him to make an immediate accounting of his affairs and to prepare to turn over the stewardship to one more dependable.

Realizing that he was too weak to dig and too ashamed to beg, the wily steward called in the master's greatest debtors. To one that owed one hundred measures of oil, he said, change the figure to fifty on your wax tablet recording the transaction. To one who owed one hundred measures of wheat, he said, make it eighty. In this way the wily steward stored up credit with powerful men who could help him once he was dismissed from the stewardship.

Even the master, when he heard what happened, thought well of the steward for in this way cleverly providing for a future of want and adversity. Even so should one store up credit in the spiritual realm so that when deprived of the material, you will not want.

WHAT JESUS SAID ABOUT GOD

Jesus spoke of the kingdom of God because that is what John taught and what people expected, but God is not a king and Jesus never called him king. He called him Father, the loving parent of the brotherhood of man. His teaching regarding the Father focused on the declaration that he and the Father were one, that he who has seen the Son has seen the Father, that the Father, like the Son, knows all things, that only the Son really knows the Father, and he to whom the Son will reveal him, that he who knows the Son knows also the Father, and that the Father sent him into the world to reveal their combined natures and to show forth their conjoint work.

Mostly, except when quoting the scriptures, Jesus called God one of two names: *Elohim* ("gods," that is the Trinity, God) and *Father*. He did not use the name Yahweh, the tribal god of the Hebrews. Jesus employed Elohim, God, to designate the *idea* of deity and the word Father to designate the *experience* of knowing God. The word God cannot be defined and so stands for the infinite concept of the Father, while the term Father, being capable of partial definition, may represent the human concept of the divine Father as he is associated with humans during the course of mortal existence. Jesus exalted the concept of a deified racial Father, represented by Yahweh, to the notion of a Father of all human children, a divine Father of the individual believer. He further taught that this God of universes and this Father of all humans were one and the same paradise deity.

Jesus never claimed to be the manifestation of *Elohim* (God) in the flesh. He never declared that he was a revelation of Elohim (God) to the worlds, and he never taught that he who had seen him had seen Elohim (God). But he did proclaim himself as the revelation of the *Father* in the flesh, and he did say that whoever had seen him had seen the Father. As the divine Son he claimed to represent only the Father and to reveal his Father's character, in so far as such a revelation was comprehensible to mortals.

Jesus taught only two things about God: that God in himself is *spirit* (to the woman at Jacob's well in Sychar he said, "God is spirit") and that, in all matters of relationship with his creatures, he is a *Father*: "I have come out from the Father, and I have come into the world. Again, I will leave the world and go to the Father."

Jesus never said, "Whoever has heard me has heard God." He did say, "He who has *seen* me has *seen* the Father." To hear Jesus' teaching is not equivalent to knowing God, but to *see* Jesus is a revelation of the Father to the soul. The God of universes rules the far-flung creation, but the Father in heaven sends forth his spirit to dwell within your minds. You can *know* the Eternal as a Father; you can *worship* him as the God of universes, the infinite Creator of all existences.

CHAPTER 34. "HE COMES AS THE LOWLY ONE, RIDING ON AN ASS"

JESUS ANNOUNCED THAT AFTER COMPLETING their tour of southern Perea the group would go up to Jerusalem for the Passover. In spite of all he had said, again and again, the apostles and other followers still believed that he was at last going to Jerusalem to establish a temporal kingdom as the long-awaited Messiah. Even after his death and the complete destruction of the city by the Romans in AD 70, Jesus' followers were confident that Jesus would soon return in physical form to establish his temporal kingdom of God. At this time Salome, the mother of James and John Zebedee and distant cousin of the high priest Annas, sought a special audience with Jesus in which she requested that her sons sit at his right and left hands in the coming kingdom.

Deeply disappointed at their refusal to understand, Jesus turned to James and John, who stood nearby, and asked if they were really prepared to drink the cup he was soon to drink. They assured him that they were, and Jesus agreed that they would, then, drink this same cup of bitterness, but as for who would sit at his right and left hand—that was not up to him but to the Father who had sent him. The other apostles, when they heard of Salome's request, were incensed at the presumption of James and John, and only after the brothers made profound apologies were they reconciled.

PAYING THE PRICE

On March 13, AD 30, the group, together with the seventy evangelists, left Pella for a two-week tour of southern Perea before heading to Jerusalem. Around one thousand followed him down the Jordan River. When they stopped at the ford near Jericho, Jesus mounted a high rock and delivered his sermon on "counting the cost."

"From now on you who follow me must be willing to pay the price of wholehearted dedication to the doing of my Father's will. If you would be my disciples, you must be willing to forsake father, mother, wife, children, brothers, and sisters. If any one of you would now be my disciple, you must be willing to give up even your life just as the Son of Man is about to offer up his life for the completion of the mission of doing the Father's will on earth and in the flesh.

"If you are not willing to pay the full price, you cannot be my disciple. Before you go further, each of you should sit down and count the cost of being my disciple. Which one of you would build a watchtower on your lands without first counting the cost to see whether you had enough money to finish the job? If you fail thus to calculate the cost, after you have laid the foundation you may discover that you are unable to finish that which you have begun, and all your neighbors will mock you, saying, 'Look this fellow began to build but could not finish his work.' Again, what king, when he prepares to make war on another king, does not first sit down and take counsel whether he will be able with ten thousand men to meet him who comes against him with twenty thousand? If the king cannot afford to meet his enemy because he is unprepared, he sends an embassy to this other king, even when he is a long distance away, and he asks for terms of peace."

He went on to make clear that even so must those behave who would still follow him. They must now count the cost of being his disciple. The time of going along with the crowd and listening to his teaching was over. His disciples will face bitter persecutions and crushing disappointment. If they are unwilling to renounce all that they are and to dedicate all that they possess, they are unworthy to be his disciples. Soon will the chief priests and the Sadducees reject the Son of Man and give him into the hands of mocking unbelievers.

"You must therefore examine your motives. If you seek honor and glory, if you are worldly minded, then you are like salt when it has lost its savor. And when that which is valued for its saltiness has lost its savor, how will it be seasoned? Such a condiment is useless, fit for the trash." He warned them to go back to their homes if they were not willing to drink with him the cup now prepared. "Again and again I have told you that my kingdom is not of this world, but you will not believe me. He who has ears to hear let him hear what I say."

Not one understood what he meant. The apostles, and many others, believed that Jesus could not die: After all, he was the resurrection and the life. He brought Lazarus back from the dead! He was the Messiah. He would sit on the throne of David when he went up to Jerusalem.

On Wednesday, March 29, AD 30 in a small nearby town Simon Zelotes and Simon Peter arranged for the secret delivery of one hundred swords to the group, in preparation for the coming institution of the kingdom in Jerusalem.

Jesus called the apostles together and again he told them he would soon fall into the hands of his enemies, who would kill him. Even so they should not fear, because on the third day he would rise again, he said. They thought he spoke in metaphors, and no one took him literally.

Friendly Pharisees warned Jesus that not only had the Sanhedrin decreed his death, but that Herod Antipas too now wanted him dead, being profoundly disturbed about reports that he had brought a man back from the dead.

The party arrived at Jericho toward evenfall. Seeing a man named Zaccheus, a publican, who had climbed into a tree to get a glimpse of the Master, Jesus called out to him that he would stay that night at his house. On the next day at noon, halfway up the "robbers' road" between Jericho and Bethany, Jesus told his apostles the parable of the pound:

A nobleman called together his twelve servants and gave to each a sum of money, saying he would soon go away but that when he returned he would expert an accounting from each. And when he did return, one servant stepped forward and said, "Lord, I have invested your money and made ten times as much." "Good, said the nobleman, "then you will rule over ten cities." A second had earned five times as much; the nobleman entrusted him with five cities, and so forth until the last servant stepped forward and said, "Master, here is the money you gave me. I have wrapped it and cared for it because it is well known that you are a hard taskmaster and take what you want, and I did not want to risk losing what you gave." The master was furious and ordered his men to take away the money and give it to the man who already had ten times as much. For those who have shall gain more, and those who do not have shall lose even what little they have.

THE POOR ARE ALWAYS WITH US

Jesus and the apostles arrived in Bethany on Friday afternoon, March 31, AD 30. Jesus lodged in the house of one Simon, the most prominent man of this small village since the death of the father of Mary, Martha, and Lazarus. The Sanhedrin learned of his presence in Bethany but feared to arrest him in the midst of his friends.

Six days before the Passover, on the evening after the Sabbath, a great banquet was held at Simon's house in honor of both Jesus and Lazarus. Jesus was in an exceptionally good mood, to the puzzlement of his apostles, who dreaded what the Sanhedrin might do at any moment. Many of the apostles were armed with concealed swords that the two Simons had obtained in Perea.

Everyone enjoyed the meal. At its end Jesus' friend Mary came forth from the group of women watching the banquet (women did not dine with men on formal occasions). She opened an alabaster box filled with a precious unguent, whose beautiful scent immediately filled the room. She rubbed the unguent in Jesus' hair, then bathed his feet in it, to the indignation of many in the room. Judas Iscariot said to Andrew, "Why does the Master allow such behavior? Why was the unguent not sold and the money given to the poor?" The valuable unguent was worth a man's labor for a full year, and if sold could have fed five thousand.

Knowing their thoughts, Jesus spoke aloud, saying that they were wrong to blame Mary for her behavior. The poor were always with them, he said, while he would soon go to the Father. Mary was to be commended for her behavior because she alone believed him when he told them that he would soon be leaving them. She had purchased the unguent to anoint his body when he died, and to her credit she had used the precious oil while he still lived. In future times, when the gospel will be victorious, even then men will celebrate her deed.

Judas Iscariot took his words as a personal reproof and at that moment his subconscious resentments and rebellious thoughts entered his conscious mind. At that moment he decided to abandon the cause, when the moment was right.

ENTRANCE INTO JERUSALEM

On the next day in the early afternoon the group set forth for Jerusalem, two miles away, over the crest of the hill. Deciding on a public entry that would attract large crowds and prevent the Sanhedrin from acting against him, he also recalled verses in Zechariah, "Rejoice greatly, O daughter of Zion! Shout, O daughter of Jerusalem. Behold, your king comes to you. He is just and he brings salvation. He comes as the lowly one, riding upon an ass, upon a colt, the foal of an ass" (Zechariah 9:9). A conqueror entered a city riding on a mighty steed, but Jesus wished to impart a different message.

At Jesus' instructions, Peter and John went to the village of Bethpage (near Bethany), where a colt was tethered to a tree. When the owner asked what they were doing, they explained that the master required it. "If you mean Jesus of Nazareth, then take the colt," the man replied.

Jesus mounted the colt in the midst of several hundred followers. The word of his arrival had gone through the city and from its gates poured several thousand of the curious, interested to see this famous wonder-worker, perhaps even the Messiah as some said. In a carnival mood of exultant celebration the crowd pressed over the brow of the Mount of Olives below which spread the white city. Jesus halted the procession. Tears ran down his cheeks as he grieved over the Jews' refusal to accept his teaching, and over the impending destruction of this very city and the very mission of the Jews on this planet.

They descended the hill, entering through the eastern gate (see Map 5) and proceeding to the temple grounds, where Jesus dismounted. He entered the temple grounds but spoke no word. He wandered in the grounds, unmolested because the Sanhedrin greatly feared the mob. When the expectant apostles and followers saw that he made no move to proclaim the kingdom, a great depression fell over the crowd.

The resentment of Judas Iscariot flared into hatred when a prominent Sadducee, a friend of Judas' family, rushed up to him and in a spirit of gleeful ridicule slapped him on the back, saying, "Why so troubled of countenance, my good friend? Cheer up and join us all while we acclaim this Jesus of Nazareth the king of the Jews—as he rides through the gates of Jerusalem seated on an ass!"

CLEANSING OF THE TEMPLE

On Monday, the next day, Jesus returned with his apostles to the temple grounds. He went to the speaker's platform and was about to speak when he was interrupted by a loud altercation between a money-changer and a customer who had been cheated at the same time that a herd of cattle, driven from one pen to another, gave forth a huge bellow. All around in the great temple court was the swirl and babble of commercial transaction. The animals were kept in pens to be sold to pilgrims, who could offer for sacrifice only an animal without blemish according to Levitical law, and almost any animal could be rejected on technical grounds.

The animal had to be purchased only with silver coin minted during the reign of the Hasmoneans (166–63 BC), the House of Maccabeus. Only these coins could be used to pay the temple tax, one half shekel lodged against every male in Jewry (but not females, children, and slaves), or to fulfill a vow or for offerings to the temple. Because over twenty different kinds of currency were used in Palestine, money-changers had set up booths in the temple courtyard to exchange the various currencies for Hasmonean silver. They were permitted to charge around a forty percent commission on the exchange, some of which went to the temple treasury, but much went into the pockets of the priests. The meeting room of the Sanhedrin was just off the rivers of sacrificial blood, money-grubbing, and general chaos, but being greatly enriched by the practices just outside their door, they endured the situation.

In a fit of indignation Jesus suddenly stepped down from the speaker's dais, went up to the boy driving the cattle, took the knotted whip from his hand, and drove the cattle through the gate and out from the temple grounds. Then he returned to open every pen. He drove all the animals away. The excited mob, who deeply resented the practices of the corrupt priesthood, shouted Hosanna to the highest when they saw what he was doing, and rushed themselves to overturn the tables of the money-changers. By the time that Roman troops arrived on the scene, the mob had dispersed, the damage done.

During the rest of the day, returning to the dais, Jesus spoke to the crowds about the sonship of man in the kingdom of God.

JESUS AND THE PHARISEES

A commission of Pharisees pushed its way through the crowd listening to Jesus and interrupted him. "By whose authority do you teach?" one asked, a good question because all Jewry was based on a system of authority. Only the Sanhedrin had the right to ordain a scholar, in the presence of three others already ordained, and to make of him a "rabbi" (Hebrew "my master"). The question was especially pertinent directed to someone who had just committed violence against his own religion. After all the money-changers and the animal-sellers were under the priests' own direct license.

"I will answer your question if first you answer mine," Jesus said. "Concerning the baptism of John, where did it come from? By the authority of heaven or the authority of men?"

If they said by the authority of heaven, then he would say, "Then why didn't you believe him?" If they said by the authority of men, then the mob would be angered, for they considered John a great prophet.

"We do not know," they answered at last.

"Then neither will I tell you by whose authority I teach," Jesus answered.

He told the parable of the two sons:

A father told his younger son to go into the fields and tend the vineyard. At first the son would not go, but later he did. Then he told his older son to go into the fields. He said he would, but when the father was gone, he did not. Which of the sons did the father's will?

"The first son," the crowd agreed. "Even so. And the tax-gatherers and the fallen women, although at first they seem to refuse the call to repentance, later see their error and will go on into the kingdom of God before you who pretend to serve the Father in heaven while you refuse to do the works of the Father. It was not you, the Pharisees and scribes, who believed John, but the tax-gatherers and the fallen women. Nor do you believe my teaching, but the common people are very glad to hear my words."

He also told the parable of the absent landlord:

A certain rich man developed his property with fields and protective walls, then placed it in the hands of servants and went into a far land. When he sent messengers to collect the rent from his property, the wicked servants beat them and gave them nothing. Again and again he sent messengers, but they treated all in the same way, and even killed some. At last he sent his own son to collect the rent, but the wicked servants reasoned that if they killed the son, they could take his inheritance for themselves. And so they did. What do you think that the owner of the property will do when he learns what has happened to his son?

Had these learned men never read in the scriptures about the stone that the builders rejected? When the people found it, they made of it the cornerstone.

Later, Jesus told again the parable of the wedding feast, but added a new element:

Once the master of the wedding had gathered together many guests not at first invited, but summoned from far away lands, he noticed that one guest was not dressed in wedding clothing, although he had provided such clothing for all invited. Then the king said to his servants, "Cast out this thoughtless guest from my house to share the lot of all the others who have spurned my hospitality and rejected my call. I will have none here except those who delight to accept my invitation, and who do me the honor to wear those guest garments so freely provided for all."

When a believer in the audience cried out for a sign, proof that he was the Son of God as he claimed, Jesus said, "Only one sign will be given." Pointing to his body, he said, "Destroy this temple and in three days I will raise it up."

How could he possibly destroy a temple in three days that was fifty years in the building? Was he mad?

THE PROPERTY OF CAESAR

Jesus' behavior thoroughly confused the apostles. They felt that something tremendous was about to happen, but could not imagine what that was. That night they stayed at Bethany, but for the next night moved their camp close to the city on a piece of land in the gully that overlooked the public camping ground of Gethsemane at the foot of the Mount of Olives (see Map 5, page 229). Simon of Bethany owned this land and had given them permission.

Early on Tuesday morning Jesus gave a kind of farewell to Lazarus, cautioning him to prepare to leave Bethany at once, and to the women's corps, who met at the house of Simon of Bethany. He went down into the city with Andrew, Peter, John, and James, while the other apostles went to Gethsemane to set up camp.

The night before, the Sanhedrin had met again in emergency session. Jesus now threatened directly the basis of their worldly wealth. They simply could not arrest him while surrounded by adoring multitudes. Nearly in desperation, they chose several groups of learned doctors, who would through their wonderful cleverness and fine-honed wisdom discredit this so-called Messiah.

A group of students, whom the doctors rehearsed for their role, approached Jesus as he was about to speak in the temple grounds and asked him if, as a man of God, would he say that it was lawful or unlawful to pay tribute to the Romans. If he said it was lawful, then the crowd would turn from him in disgust. If he said it was unlawful, they could condemn him to the Roman authorities for sedition. Both Pontius Pilate, the Roman procurator who ordinarily resided at Caesarea on the coast, and Herod Antipas from Galilee were in the city at this time.

"Why do you tempt me?" Jesus said. "Show me the tribute money and I will answer you." They handed him a Roman coin, a denarius. "Whose image and inscription does this coin bear?" When they answered, "The image of Caesar," Jesus said, "Give to Caesar the things that are Caesar's and give to God the things that are God's."

CLEVER, VERY

Still before he could begin his teaching, a group of learned Sadducees pushed forward and one asked, "Master, Moses said that if a married man died, leaving no children, his brother should take the wife and raise up seed for the dead brother. Now there occurred a case where a certain man who had six brothers died childless. His next brother took his wife but also died, leaving no children. Likewise did the second brother take the wife, but he also died leaving no offspring. And so on until all six brothers had possessed her, and all six of them died without leaving children. Then the woman herself died. Now, we would like to ask you this: In the resurrection whose wife will she be since all seven of these brothers had her?"

It was the kind of question that the Pharisees liked to play with, bearing little relevance to anything in the real world. Furthermore, the practice of a brother conceiving children on his dead brother's wife was scarcely observed at this time.

"You are mistaken in asking such questions," Jesus answered, "because you know neither the scriptures nor the living power of God. You know that the children of this world marry and are given in marriage, but you do not seem to understand that they who are accounted worthy to attain the worlds to come, through the resurrection of the faithful, neither marry nor are given in marriage. Those who experience the resurrection from the dead are more like the angels of heaven, and they never die. These resurrected ones are eternally the sons of God. They are the children of light resurrected into the progress of eternal life. Even your father Moses understood this, for in connection with his experiences at the burning bush he heard the Father say, 'I *am* the God of Abraham, the God of Isaac, and the God of Jacob.' And so, along with Moses, I declare that my Father is not the God of the dead but of the living. In him you live, reproduce, and possess your mortal existence."

The Pharisees in the crowd enjoyed the discomfiture that Jesus' reply imposed on the Sadducees, in spite of their common goal to murder the Son of Man. Jesus appealed only to the law of Moses, because the Sadducees accepted only the first five books of the later assembled scriptures, the so-called books of Moses. They did not believe in resurrection of the dead, unlike the Pharisees, who believed in a literal resurrection of the body, providing of course that one were sufficiently steeped in the scriptures.

A Pharisee stepped forward and said, "Master, I am a lawyer and I would like to ask you which in your opinion, is the greatest commandment?" Jesus answered, "There is but one commandment, and that one is the greatest of all, and that commandment is, 'Hear O Israel, the Lord our God, the Lord is one, and you shall love the Lord your God with all your heart and with all your soul, with all your mind and with all your strength.' This is the first and great commandment. And the second commandment is like the first and springs directly from it, and it is, 'You shall love your neighbor as yourself.' There is no other commandment greater than these. On these two commandments hang all the law and the prophets."

When the Pharisee agreed with Jesus, who spoke in line with the highest thought in Jewish religion, the Pharisee added, "And this commandment is greater than all the burned offerings and the sacrifices." Jesus said to the Pharisee that he was close to the kingdom of God. That night the man went to the camp at Gethsemane and was baptized.

Jesus' skillful answer dissuaded other Pharisees from asking questions, so Jesus said he would like to ask them a question. "What do you think of the Deliverer? That is, whose son is he?" A scribe answered, "The Messiah is the son of David." "If the Deliverer is the son of David, how is it that in the psalm that the scribes accredit to David, David himself, speaking in the spirit, says, 'The Lord [that is, the Messiah] said to my lord [that is, David], sit on my right hand until I make your enemies the footstool of your feet.' If David calls him [the Messiah] Lord, how then can he be David's son?"

The scribes and priests could not answer this question. After the crucifixion they changed the standard interpretation of this psalm (110:1) to have "Lord" refer to Abraham and not to the Messiah. Others, to solve the problem, denied that David had written this so-called Messianic psalm.

JUDAS ABANDONS THE CAUSE

That evening at the home of *Joseph of Arimathea* Jesus spoke before a contingent of Greeks from Alexandria. Seeing that the historical mission of the Jews, begun when Machiventa Melchizedek appeared to Abraham, was coming to an end, he said to these gentiles, "He who believes this gospel believes not just in me but in him who sent me. When you look upon me, you see not only the Son of Man but also him who sent me. I am the light of the world, and whoever believes my teaching will no longer be in darkness. If you gentiles will hear me, you will receive the words of life and shall enter into the joyous liberty of the truth of sonship with God. If my fellow countrymen choose to reject me and to refuse my teachings, I will not sit in judgment on them, for I came not to judge the world but to offer it salvation.

"They who reject me and refuse to receive my teaching will be brought to judgment by my Father and those whom he has appointed to sit in judgment on those who reject the gift of mercy and the truths of salvation. To both Jew and gentile I declare the hour has come when the Son of Man will be glorified. You well know that, unless a grain of wheat falls into the earth and dies, it remains alone, but if it dies in good soil, it springs up again to life and bears much fruit. He who selfishly loves his life stands in danger of losing it, but he who is willing to lay down his life for my sake and the gospel's shall enjoy a more abundant existence on earth and in heaven, life eternal."

At that moment Jesus' Personalized Adjuster appeared before him, although no one else saw or heard when the Adjuster said, "I have glorified my name in your bestowals many times, and I will glorify it once more."

Seeing that something had happened, an onlooker in the room said: "An angel has spoken to him."

"All this has not happened for my sake but for yours," Jesus continued. "The Father will receive me and accept my mission in your behalf, but you must be encouraged and made ready for the fiery trial just ahead. Yet victory will eventually crown our united efforts to enlighten the world and liberate humankind.

"The old order is coming to judgment. I have cast down the Prince of this world. All men will become free by the light of the spirit that I will pour out on all flesh after I have ascended to my Father in heaven. I declare to you that if I be lifted up on earth and in your lives, I will draw all men to myself and into the fellowship of my Father. You have believed that the Deliverer would remain on earth forever, but I tell you that men will reject the Son of Man and that he will return to the Father. Only a little while will I be with you. Only a little time will the living light be among this darkened generation."

Then he led a solemn group back to temple, having declared his intention to deliver final words to his enemies. In the temple, still empty of money-changers and the sacrificial animals, Jesus spoke. He made a final offer of mercy to the leaders of the Jews, entrusted with enabling a close relationship with God. Then he denounced in vivid terms the priests, scribes, and rulers as being like whited sepulchers, nice to look at from outside, but inside filled with rotting corpses and dead men's bones. By first killing the prophets of olden times, then planning to kill himself, whose appearance the prophets foretold, the Jewish nation was to lose its special role on this planet. The ancient covenant was bankrupt and their destruction as a nation imminent.

In the course of his denunciation Judas Iscariot made his final decision to break with the group and return to the social and intellectual life of the Sadducees, from which he came.

In conversation that night with his apostles in the camp above Gethsemane, Jesus explained, to the bafflement of his hearers, that Jerusalem and its famous temple, for all their glory, would soon be utterly destroyed. He warned that when Roman armies invested the city (in AD 70) they must flee for their lives; and so in a few years they did, going mostly to Pella. He told them that although he was about to leave them, he would one day return, not in the weakness of the flesh but in the glory of the spirit. No statement of Jesus was so misunderstood. Generations of believers have expected his imminent and bodily return. In fact nothing whatever is known of the time or manner of Michael's reappearance on the planet of his seventh bestowal.

In reply to his denunciation, the Sanhedrin, now under the control of the Sadducees, met that night and before midnight on April 4, AD 30, voted unanimously to arrest Jesus before Thursday and to put him to death on trumped-up charges.

By custom the Apostles did not work on Wednesdays, a day of rest. This Wednesday

Jesus went alone into the hillside to commune with his Father, accompanied only by *John Mark*, a lad who had followed the camp since Jericho. John Mark held out a lunch basket to Jesus, who wished to be alone, then would not release it, begging to accompany him. And so he did, but the author of the earliest gospel never revealed what Jesus said to him on that day.

In the morning, too, Judas Iscariot slipped away to the house of the chief priest Caiaphas, where his cousin introduced him as a Judahite Sadducee who wished to break off his unfortunate association with this gang of ignorant Galileans. Jesus was, his cousin explained, evidently a well-wishing fanatic, perhaps not really sane, who had accomplished some remarkable healings through the agency of the prince of devils. Judas, he said, was willing to show the head of the temple guards where he could be arrested, away from the multitudes who supported him, and in secret. "And what shall be my reward?" Judas asked. "We will reward you appropriately once he is taken," sneered Caiaphas.

Judas arranged with the head of the guards to arrest Jesus on the next night, on Thursday, when Jesus was sure to be in their hidden camp above Gethsemane. Judas returned to the camp exultant in his heart that at last he would be able to even the score with Peter, James, and John, whom Jesus always preferred over him (although he was the treasurer and carried the bag). Once he had loved Jesus, but his love had gradually turned to hate. Had Jesus not allowed John the Baptist, whom Judas followed, to die like a dog in a hole? Had not his relatives in Jerusalem promised that a betrayal of Jesus would glorify his name among the children of Moses? Had he not joined the group that he might enjoy the power and fame and congratulations of being one of the leaders of the new order that the Messiah, the heir of David, would bring to the Jewish nation? Evidently there was to be no new order.

Judas was a coward, hence he saw Jesus' refusal to accept the kingship as cowardice. Judas was brought to his famous treachery through self-love, ambition for worldly glory, desire for revenge for petty slights, and eagerness to glorify his name in the old order, if there was to be no new. He was a pathetic man without a philosophy within which to accommodate disappointment. As a man he had failed to learn fundamental lessons about life, and he placed his self before knowledge of the spirit. He had never understood the message of Jesus.

In the meanwhile, David Zebedee, learning that the Sanhedrin were going to kill Jesus and that Jesus was not going to stop them, sent messengers to Bethsaida urging Jesus' family and his own mother to travel to Jerusalem at once. That night, when Jesus returned from the hills, he was remarkably cheerful and upbeat, only encouraging a deep depression in the camp and intensifying their terror of what was to come.

CHAPTER 35. THE LAST SUPPER

On the morning of the next day, his last as a free man, Jesus gave a farewell address to his apostles (except the absent Judas) and other followers gathered in the camp, including a group of devoted Greeks. Answering many questions, he advised them on their relationship with civil authorities and exhorted them to remember the essence of the gospel of the kingdom—the supreme desire to do the Father's will coupled with the supreme joy of the realization of sonship with God through faith. He bewildered them with his talk of the future of the gospel under governments that would one day come into being.

THE HOUSE OF ELIJAH MARK

Word came to the camp that Lazarus of Bethany had fled to Philadelphia. David Zebedee learned through his corps of messengers and spies that Judas, who had returned to camp, was in league with the Sanhedrin. This he reported to Jesus, but Jesus already knew. Taking aside John, Peter, and Philip, while David Zebedee distracted Judas who tried to overhear, Jesus told these apostles to go into the city. Inside the eastern gate they would find a man carrying a pot. The man would speak to them. They were to follow him.

The man led the three apostles to the house of Elijah Mark, father of John Mark, the boy who so assiduously followed Jesus and went alone with him to the hills behind Gethsemane. Elijah Mark took them to an upper room where Jesus and the apostles could dine in safety. Jesus had concocted this plan while alone with John Mark in the hills, to prevent Judas from revealing prematurely their whereabouts.

About half past three Jesus and twelve apostles set out for the city and that night's meal. Instead of crossing the crowded Kidron Valley to the east of the city, they went south towards Bethany along the west face of the Mount of Olives, which intersected with the Jerusalem road. At the brow of the hill they rested, waiting for dark. Jesus warned them not to interfere with what was about to happen and always in the future to continue their work. His Father's kingdom had many abodes, and one day they would inhabit them and be together with him in the spirit kingdoms of his Father. They were not to be afraid, although many hardships awaited them.

At dark they followed him into the city through one of the southern gates, passed through tangled streets, jostling past strangers who did not recognize them. John

Mark followed behind the group, then took a different route to reach his house so that he could greet their arrival.

LESSONS IN THE UPPER CHAMBER

The Jewish day ends at sunset, so the day of Passover really begins on Friday night, when the Passover meal was celebrated. The apostles were therefore befuddled when Jesus announced that they would dine on Thursday night, some thinking that he knew he would be arrested on the following day, and others that he simply wished to share a special meal with them. While he spoke with the parents of John Mark downstairs, the apostles were shown to a commodious upstairs room, fitted out for a traditional Jewish meal in an upper-class household. This was to be a private meal with no servants in attendance.

Inside the door on a table were basins, pitchers of water, and towels for washing the feet, but no servants to perform the duty. In the middle of the room was a U-shaped table. At the ends of the U were set bread, water, wine, and dishes of olive oil spiced with herbs. The divan of honor, raised higher than the other divans, where the host sat, was positioned near the end of one leg of the U, with a single couch to the right of the host. The other eleven couches were arranged to the left of the host around the table.

The apostles stood assessing the situation when Judas went to the couch of honor to the left of the host's couch. John Zebedee then angrily occupied the second seat of honor, the single couch to the right of the host. In a hullabaloo with angry words exchanged Simon Peter stalked all the way around the table, behind the hosts' couch, until he came opposite John Zebedee, and took possession of the couch of least honor. Peter thought that when Jesus saw where he sat, he might displace Judas or John, who arrogantly had taken for themselves the greatest honors. The other apostles occupied the other couches.

When Jesus entered the room he looked with some dismay on his longtime followers, who still argued over worldly honor and precedence. Without speaking he went to his place on the raised dais. He spoke to them.

He told them that his hour had come and that this would be their last meal together. They would not sit down again until entering the kingdom that his Father would give him, once he had completed that for which he had come into the world.

Taking a cup of wine now mixed with water, he said, "Take this cup and divide it among yourselves and, when you partake of it, realize that I shall not again drink with you the fruit of the vine because this is our last supper. When we sit down again in this manner, it will be in the kingdom to come."

It was the custom for the host to rise after the first drink of wine and wash his hands, but Jesus did not observe these customs. When Jesus got up from the couch, the apostles wondered what, then, was his intention. He went to the table with the basins and jugs of water and to the astonishment of all, as every man rose from his seat, he removed his cloak and wrapped a towel around his waist. He went to Peter in the couch of least honor and knelt before him, clearly intending to wash his feet, a task usually reserved for slaves. "You shall never wash my feet!" Peter said. "Then you shall not be part of what I am about to perform," said Jesus. "Then wash my hands, too, and my head!" cried the impulsive Simon Peter.

Jesus went around the room and washed the feet of all, including Judas Iscariot, who now was more sure than ever of the unworthiness of Jesus to lead the Jews to freedom and power and confirmed in his decision to betray him. When finished, Jesus put back on his cloak and again took his place at the table.

He spoke again, explaining how his behavior was meant to show that the secret of greatness in the spiritual kingdom is not like the methods of power in the material world. He reproved them for their behavior before he came into the room.

Two apostles now brought forth the second course—the bread, the wine, the dish of herbs, and a paste of fruits.

A TRAITOR IN OUR MIDST

Because of the Master's good cheer, conversation soon broke out and their mood heightened. Jesus spoke again, telling them that the forces of darkness had conspired to take the Son of Man and that he would not be with them at this time on the next day. "But it was not the Father's will that one of you betray me," he added.

Who? Is it I? The apostles asked among themselves, when they heard this.

"Already have I told you, him to whom I gave the sop," Jesus said, but it was so natural for the host to give a piece of bread soaked in oil to the diner on the left that none took notice that he had given one to Judas, although he had spoken plainly. But Judas did understand him and feared that others might know that he was the betrayer.

Peter, leaning toward John dining on the opposite side of the table, asked him to ask Jesus who he meant by this remark: Who was the betrayer? But Jesus spoke again. He said he hoped up to the last minute that a love of truth might triumph over evil. Then he turned to Judas and, within his hearing alone, said, "What you have decided to do, do quickly." Judas arose and left the room, on an errand, the others thought.

When the third cup of wine was passed, the "cup of remembrance," the remembrance of Jewish servitude in Egypt, Jesus stood up from his couch and said that this cup would be in remembrance of a new dispensation, an emblem of the Spirit of Truth that he would soon shed over the people. Before, the Passover celebrated the liberation of a people, but the new cup symbolized the liberation of the individual from the slavery of selfishness and ceremonialism. And they all drank from it.

He broke a loaf of bread and passed it around the room as the bread of remembrance, the symbol of the living word of truth incarnated in the likeness of mortal flesh. He said, "And as often as you do this, do it in remembrance of me. And when you do remember me, first look back on my life in the flesh, recall that once I was with you, and then, by faith, perceive that you shall all some time dine with me in the Father's eternal kingdom. This is the new Passover that I leave with you, the memory of my bestowal life, the word of eternal truth, and of my love for you, the outpouring of my Spirit of Truth upon all flesh."

They concluded the meal by singing the one hundredth and eighteenth psalm.

GOODBYE TO ALL THAT

All expected Jesus to rise and lead them from the room, but instead he spoke to them at great length as a farewell address. In his speech he imposed on them an alteration of the golden rule. They were not only to love their neighbors as themselves, but to love one another as he had loved them. He was soon to be gone, but one day they would follow him into the kingdom that his Father had given him. Not only had they chosen him, but he had chosen them.

"I am the true vine," he said, "and my Father is the husbandman. I am the vine, and you are the branches. And the Father requires of me only that you shall bear much fruit. The vine is pruned only to increase the fruitfulness of its branches. Every branch coming out of me which bears no fruit, the Father will take away. Every branch that bears fruit, the Father will cleanse that it may bear more fruit. Already are you clean through the word I have spoken, but you must continue to be clean. You must abide in me, and I in you. The branch will die if it is separated from the vine. As the branch cannot bear fruit except it abides in the vine, so neither can you yield the fruits of loving service except you abide in me."

Jesus was appealing to an old Jewish metaphor that said the Messiah would be a branch on the vine of David. A picture of a vine and cluster of grapes carved on Herod the Great's temple referred to this image.

Now they were to go forth into the world, Jesus said, which would hate them as it hated himself. In their suffering they should remember that he had suffered before them in service of the gospel of the kingdom. Most who hated him did not understand what they were doing, but some did understand. Such men, who knowingly reject the Son, reject the Father who sent him. They cannot survive. Although he would soon be gone, he would send the Spirit of Truth as he had promised so that through this spirit he would be present to all who sought to find him. That was the advantage of his departure from the flesh.

Thomas arose and said, "Master, we do not know where you are going, so of course we do not know the way. But we will follow you this very night if you will show us the way."

"Thomas, I am the way, the truth, and the life," Jesus said. "No man goes to the Father except through me. All who find the Father, first find me. If you know me, you know the way to the Father. And you do know me, for you have lived with me and you now see me … I am telling you these things while I am still with you so that you will be better prepared to endure those trials which are even now upon us. And when this new day comes, the Son as well as the Father will dwell in you. And these gifts of heaven will ever work the one with the other even as the Father and I have acted on earth and before your very eyes as one person, the Son of Man. And this spirit friend will bring to your remembrance everything I have taught you."

Although his gospel of the kingdom brings peace to the soul of the individual believer, he said, it will not bring peace on earth until humans believe his teaching wholeheartedly and establish the doing of the Father's will as the chief purpose in living the mortal life.

"In just a little while I will leave you for a short time," he said. "Afterward, when you again see me, I shall already be on my way to the Father so that even then you will not see me for long."

But the apostles could not grasp his meaning.

Jesus then went around the room and spoke to each of the eleven apostles personally, praising their better qualities and encouraging improvement where they fell short. When he at last came to Simon Peter, Peter in a typical exaggerated gesture declared his willingness to follow the Master through any trouble and even to die for him. "Peter, Peter, I say to you," Jesus answered, "this night the cock will not crow until you have denied me three or four times. And what you have failed to learn from peaceful association with me, you will learn through much trouble and many sorrows."

After singing a hymn, they returned to the camp at Gethsemane.

CHAPTER 36. THE ARREST AND TRIAL OF JESUS OF NAZARETH

THE BOY JOHN MARK HAD SLEPT DOWNSTAIRS while Jesus dined with the apostles, but when he heard them descend the stairs, he put on a linen coat and followed them back to the camp. From this moment on he lurked in the background and witnessed everything and heard much until the moment of the execution of Jesus of Nazareth.

The apostles were surprised that Judas had not returned, nor was in the camp when they returned around 10:00 PM. When they realized that Judas had betrayed them, furious conversation broke out interrupted by Jesus, who led them to a flat rock up the side of Olivet. There he prayed to the Father to assist them in what was to come. In his prayer he alluded to the betrayal by Judas.

Afterward he urged the excited apostles to go to their tents and rest but took Peter, James, and John with him into a nearby ravine to pray. Before the other eight apostles retired for the night, they followed Simon Zelotes to his tent where he handed out swords to all but Nathaniel and Andrew, who refused to believe that Jesus could not defend himself if he chose, by his own means. Had they all forgotten that the Master's kingdom was not of this world? In any event, no one expected trouble so late at night. The enemy would come on Friday morning, before the noontide preparations for the Passover, they thought.

ALONE IN GETHSEMANE

When Jesus and the three apostles reached the nearby ravine, Jesus went a short distance away and fell to the ground, his human heart oppressed by the terror of the events that lay ahead. Was there a way he could avoid drinking this cup? Was it really the will of the Father that he drink this cup?

An angel came to him, touched him, and strengthened him. Three times in the midst of his prayers and communion he returned to the attendant apostles, but every time he found them asleep, exhausted by recent events, although he wakened them and rebuked them for leaving him alone at this time of crisis. The assembled hosts of a vast creation hovered over this scene under the transient joint command of Gabriel and the Personalized Adjuster of Jesus, but the division commanders of these armies of heaven had repeatedly been warned not to interfere with transactions on earth unless Jesus himself gave the order.

By the time he returned to camp, Jesus was once again composed. Jesus urged Peter, James, and John to go to their tents and sleep. Just then two excited messengers arrived in the camp looking for David Zebedee. The other eight apostles slept, but the Greek followers encamped there had set a watch. When the messengers arrived, the Greeks came from their tents, armed and ready for trouble. Again Jesus urged Peter, James, and John to go to their tents and not to rouse the sleeping apostles, but they did not go.

ARREST

Jesus walked away from the group down through the grove to the olive press at the entrance to Gethsemane park. The Greeks and the three apostles were uncertain whether to follow, but John Mark did follow him, keeping out of sight, then hiding in a shed near the press so that he saw all that happened.

Jesus walked away because he did not want the apostles to be near him when the guards came to arrest him. He feared that the sight of Judas' betrayal would so infuriate them that they would resist, be themselves taken under arrest, and perish with himself. He knew that his imminent arrest was a plot arising among the Sanhedrin, but that the rebels of the realm Lucifer, Satan, and Caligastia, not yet confined, would enjoy the apostles' destruction. Jesus sat on the press and waited.

After Judas left the banquet in the upper chamber of the Mark home he went straight to his cousin's house. Together they went to the temple guard. Judas led them back to Mark's house, fully expecting to find Jesus there with only eleven other men, only two of them armed, Simon Zelotes and Simon Peter. When Judas learned that they had left fifteen minutes before, they and the temple guard returned to the temple grounds. Judas explained to the Sanhedrin that they would now need to arrest Jesus in Gethsemane, a risky proposition because three score of his followers were camped there and there were plenty of arms in the camp.

Fearing for his own safety, Judas requested that the Sanhedrin give him a cadre of forty men, but the elders did not have nearly so many. Judas and representatives of the Sanhedrin went therefore to the massive Antonia fortress, named after Marc Antony, a high-towered fort built at the northwest corner of the temple grounds. Pilate and his wife stayed there while in Jerusalem. The Roman troops were stationed in the fortress (see Map 5, page 229).

The first commander they spoke to referred the matter to his superior, and he to his own superior, until the request for troops came to the attention of Pontius Pilate, the Roman procurator for the province of Judea. His headquarters were in Caesarea on the coast, an artificial port built by Herod the Great, but he was in Jerusalem at this time of the Passover. His wife, who knew something about Jesus and was a kind of half believer, warned Pilate not to grant the Sanhedrin's request. Because the chief of the Sanhedrin had gone so far as to disturb him in his bed chamber with his wife, Pilate nonetheless thought it expedient to accede to the request. Any initial harm could always be undone later, Pilate thought.

Judas, now in company with around sixty men including Roman soldiers, the temple guard, and curious servants of the high priests, headed to the camp at Gethsemane around 11:30 PM. Judas was in a panic that he would be harmed once his former companions saw him betraying the Master. Therefore he went out ahead of the mass of men carrying torches in the night. He thought he could explain his presence by saying he had come to warn them of the soldiers' approach.

When Peter, James, and John saw the band of men carrying torches come around the hill to enter the camp, they hastened down to where Jesus sat on the wine press. Jesus stood up and went to the captain of the Romans and said, "Whom do you seek?" "Jesus of Nazareth." "I am he," he replied.

Although at this point there was no need for Judas to identify Jesus, Judas wished to establish his complicity so that the Sanhedrin might abundantly award his betrayal. He went to Jesus and kissed him on the brow and said, "Hail, master and teacher." Jesus said, "My friend, is it not enough to do this? Would you also betray the Son of Man with a kiss?"

Jesus pushed him away and spoke again to the captain of the Romans. "Whom do you seek?" "Jesus of Nazareth." "As I said, I am he. Let these others go and I will go with you."

The bodyguard of the high priest, in the cohort of the temple guard, stepped forward and bound Jesus' hands behind him, although the Roman captain had given no such instruction. When Peter saw this, he indignantly drew his sword and made to attack the bodyguard, but Jesus raised his hand and said, "Peter, put away your sword. They who take the sword will die by the sword. Do you not understand that it is the Father's will that I drink this cup? Do you not know that I could even now command more than twelve legions of angels and their associates, who would deliver me from the hands of these few men?"

Peter's threatening behavior alarmed the Roman captain and his men laid rough hands on Jesus and bound him securely. When the apostles and other followers heard the captain giving an order to arrest them too, they fled into the ravine.

John Mark ducked from the shed, hoping to follow the men taking Jesus into the city when a Roman, returning empty-handed from the ravine, saw him and caught hold of his cloak. Slipping out of it, John Mark ran naked into the night, up the ravine. He made his way to David Zebedee higher on the trail leading up to the camp. David sent immediately to the eight sleeping apostles to notify them that Jesus had been taken.

AT THE HOUSE OF ANNAS

Already at the wine press a dispute arose between the head of the Jewish guard and the Roman captain about where to take Jesus. The head of the Jewish guard wanted to take him to the house of the high priest Caiaphas in the city, but the Romans were accustomed to dealing with Annas, the father-in-law of Caiaphas, on matters of enforcing ecclesiastical law. Therefore they went toward the palace of Annas, built outside the city on the slopes of Olivet not far from where Jesus was arrested.

John Zebedee followed closely behind the mob. Peter followed much further behind. John Mark got a new cloak from the camp and, guessing they were taking Jesus to the house of Annas, the high priest emeritus, he arrived at the gate of the house in advance of the mob.

James Zebedee, separated from Simon Peter and John Zebedee, met with the other apostles at the wine press. Some thought they should make an armed assault to recover Jesus, but others reminded the group of Jesus' teaching of nonresistance, and James told them what Jesus said to Simon Peter when he drew his sword. Nathaniel reminded them that Jesus had warned them to take thought for their own safety. Five of the apostles therefore made their way to Bethany, while the others went into the city to hide. Jude, Jesus' brother, just then arrived at the nearly deserted camp ahead of his family, who had hastened down from Galilee on David Zebedee's instructions. David Zebedee sent Jude back up the road to take Jesus' family to the house of Mary and Martha in Bethany.

Jesus walked between the Jewish guard and the Roman captain. When John Zebedee came up close, the Jewish guard ordered that he too be bound as a follower of Jesus, but the Roman captain interfered. Roman law allowed an accused man to be accompanied by a friend throughout any trial proceeding: In this case Roman law would be taking precedence over Jewish intolerance, the captain made clear. Furthermore, the Roman captain added, this man, John Zebedee, was courageous and no rebel: He had not drawn against them in the garden, and he had the courage to follow Jesus into the maelstrom.

Hearing this, the base and cowardly Judas, whom no one addressed, shrank back from the crowd and arrived at Annas' house alone. There the Roman captain repeated his explicit instructions that John Zebedee was not to be molested and to remain with the prisoner. Thus John was present throughout the trial and execution of Jesus. In turning Jesus over to the Jewish guard the Roman captain sent his assistant to accompany Jesus and see that the Jews did not kill him without Pilate's consent. Jesus spoke no word.

Annas was a suave and astute politician, the most powerful man in Jewry. A distant relative of Salome, mother of John and James Zebedee, he had met Jesus twice, once when Jesus was a lad and a second time during his early ministry, when Jesus left Annas' house abruptly on seeing the coldness of Annas. Annas earned vast wealth from the temple incomes. The Romans trusted Annas, father-in-law of the current high priest.

Annas wanted the trial of Jesus to remain in the hands of the Sadducees because the only prominent Jews who had gone over to Jesus were Pharisees; you couldn't really trust them. Annas knew that Caiaphas had gathered together thirty strongly antiJesus Sanhedrists at his house in hopes of holding the trial there, but according to Jewish law a legal court could not be called before the morning sacrifices at the temple, which took place around 3:00 AM. Annas wished Jesus to pass the interim in his house, in accordance with the law. He did not want the brusqe and impolitic Caiaphas to deal with Jesus, and he entertained a twinge of hope that he might yet persuade Jesus to go away, to leave town; he regretted having to kill a good and just man. Yet the attack on the money-lenders had offended him deeply.

Annas received the prisoner in his capacious audience chamber. Annas sat on a throne and said, "Jesus, you know that we must do something about you, because you are disturbing the peace of the Jewish people." The majestic prisoner regarded him, but spoke no word. Annas realized that he could never persuade this man to leave the country or to abandon his teachings. Annas asked him what were the names of his followers, in addition to the agitator Simon Zelotes.

When Jesus still did not answer, he said, "Don't you care whether I am friendly to you or not? Do you have no regard for the power I have in determining the outcome of your coming trial?" Jesus said, "Annas, you know that you could have no power over me unless it were permitted by my Father. Some would destroy the Son of Man because they are ignorant. They know no better, but you, my friend, know what you are doing. How can you reject the light of God?"

Annas was puzzled by his friendly manner, but he was determined either to expel Jesus from Palestine or to kill him. He asked Jesus what it was that he taught, and who did he claim to be. Jesus replied that he had often spoken in public and that all Jerusalem knew what he taught, even if Annas did not. At this Annas' steward stepped forward and slapped Jesus across the face. How dare he speak to Annas in this fashion! Nor did Annas reprove the steward. "My friend," Jesus said to the steward, "if I have spoken evil, testify against the evil. But if I have spoken the truth, why would you strike me?"

Annas left the room in confusion and remained away for an hour. When he returned, he went up to Jesus' side and asked him if he claimed to be the Messiah. "So you have said," Jesus replied.

By now it was nearly dawn, and Annas decided to send Jesus to the court of the Sanhedrin assembled in the house of Caiaphas.

PETER IN THE COURTYARD OF ANNAS

Simon Peter had followed behind the crowd but came to Annas' house in time to see John Zebedee go in. Peter threw away his sword. Judas did not enter the house but veered off back to the city, toward the house of Caiaphas where the trial would be held. John saw Peter and spoke to the porteress, whom John happened to know because as a distant relative of Annas he had been to this house before.

The woman gatekeeper let Peter come into the courtyard. The night was cold and, frightened and ill at ease, Peter went over to the charcoal fire to warm himself, where several people were standing. The porteress came over to him and accused him of being a follower of Jesus, because obviously he knew John, who had just spoken to her about him. Peter said, No! She said she knew that he was because her sister had seen him in the company of Jesus. Then another servant came up and said, yes, he had seen Peter at the Gethsemane camp. He was definitely a follower. Peter vehemently denied that he followed Jesus or even knew anything about him.

Peter broke from the group and walked about the courtyard, but getting cold he returned to the fire. One of Annas' servants said that he certainly was a follower of Jesus because his accent was Galilean and Jesus was from Galilee. Swearing earnestly, profoundly afraid for his life, Peter denied it again.

He broke away again and stood in the porch. The gate-keeper and her sister came by and again accused him of being a follower of Jesus. Again Peter denied it, just as a cock crowed, greeting the morning on Olivet (poultry was not allowed within the city). With a shock of horror Peter remembered what Jesus had said. Yes, he had denied him repeatedly, and he had gone against explicit instructions not to endanger his life in the trouble to come. Peter had acted as a perfect fool. He was appalled by how badly he was behaving.

Just then the palace doors opened and Jesus was led from the house into the courtyard. In the glare of the flickering torches his eyes met Peter's, a look of love and pity that pierced Peter. Peter followed for a short distance after the group, then sat down by the road and bitterly wept. Later he arose and went to the camp above Gethsemane, now empty except for David Zebedee, who sent him to the house of Nicodemus in the city where Peter's brother Andrew was hiding.

AT THE HOUSE OF CAIAPHAS

Annas, Jesus, and the arresting party arrived at the house of Caiaphas around 3:30 AM, Friday morning. The Jewish authorities needed to move swiftly, because no work could be done after noon on the day before the Sabbath. According to Jewish law two witnesses must testify to every point in a capital case. To cover this problem the Sanhedrists had arranged for a dozen men to lie about things they had heard.

As Jesus stood silent before the Sanhedrist court, the suborned witnesses perjured themselves so clumsily that even the Sanhedrists were embarrassed. The witnesses seemed to agree on a single point, that Jesus had claimed in the temple grounds to be able "to destroy this temple built by hands and in three days to raise it up again without hands" (he had not actually said this). Annas observed that in making such claims Jesus *stirred up the people with ridiculous claims,* that he was a *revolutionary threatening the temple*, and that he *advocated magic*. Annas was attempting to come up with a charge that would be plausible to the Romans, for only they could issue the death warrant he needed. Then, thank God, he would be free of this Jesus problem forever.

When Jesus made no reply to any charge, the exasperated Caiaphas at last came down from the dais, stood before Jesus, shook his finger at him, and said, "I command you in the name of the living God that you tell us whether you are the Deliverer, the Son of God!" Jesus answered, "I am. Soon I go to the Father, and presently will the Son of Man be clothed with power and once more reign over the hosts of heaven."

When the high priest Caiaphas heard these words, he angrily tore his outer garments and exclaimed, "What further need do we have of witnesses? You all have heard this man's blasphemy. What should be done with this law-breaker and blasphemer?" The

Sanhedrists answered in unison, "He is worthy of death! Let him be crucified!" Caiaphas struck Jesus across the face and the Sanhedrists filed out, each spitting in Jesus' face or striking it with their hand.

According to Jewish law there had to be two court sessions in any capital case, separated by one full day during which the members of the court fasted and meditated. Because the Sanhedrists did not have a day, they allowed a single hour between the two court sessions. During this hour the guards and servants of the high priest abused and insulted Jesus, striking him, then asking, "Well, great Deliverer, prophesy to me—who struck you?" During this hour John Zebedee waited in an adjoining room, where Jesus had urged him to go, knowing that John would put himself in danger if he were to witness such abuse. During this time of torture Jesus spoke no word.

At 5:30 AM the court reconvened. Jesus was led into the adjoining room where John Zebedee waited. Now Judas was present in the room, although he gave no testimony. The session lasted one half hour. Annas knew that Pilate would have no interest in a charge of blasphemy, and so the court drew up three charges: (1) Jesus perverted the Jewish nation and incited them to rebellion; (2) he advocated not paying the tax to the Romans; (3) he was a rebel who proclaimed a new kingdom.

The Sanhedrists exalted Jewish law above all else, but their proceedings were a mockery of their own law. No two witnesses had agreed on any charge except the one about rebuilding the temple, and no witness for the defense had been heard. The only charge the court might have agreed on was that of blasphemy, but that rested solely on Jesus' own testimony, which Jewish law did not allow. No formal ballot was cast to condemn Jesus to death on this charge, nor was the accused present in the same room when the charges were formulated. So egregious were these and other irregularities that three of the Sanhedrists, in spite of their determination to destroy Jesus, left the room.

PONTIUS PILATE

At six on this Friday morning, April 7, AD 30, Jesus was led out of the palace of Caiaphas to appear before the Roman procurator, the only man with the power to confirm the sentence of death decreed by the Sanhedrist court. Pilate governed Judea, Samaria, and Idumea ("pertaining to Edom," the territory to the south of Judea), under the immediate supervision of the Roman legatus of Syria.

Pilate was a second-rate administrator, a religious skeptic, and wholly out of his depth in dealing with the troublesome Jews, whom he loathed, despised, and feared. His first big mistake came in a dispute about Roman legions entering the city with portraits of Caesar on them, forbidden idols to the Jews. Pilate blustered that if the rioters did not back off, he would put them to death. The leaders came forth and readily offered up their lives. When Pilate saw that they were really willing to die, he retreated from his threat. At that moment he lost his power over people like the bloodthirsty fanatics who had determined to put Jesus to death in the name of religion.

In later missteps Pontius Pilate used money from the temple treasury to expand the system of aqueducts to serve the holy festivals, thinking the people might like to have enough water to drink; the protestations were explosive. Then he hung up the emperor's shields on the palace of Herod in Jerusalem: The furor rose up even to the emperor Tiberius himself, who overrode Pilate's directive and ordered the shields removed.

The Jewish leaders could force Pilate to do what they wanted by threatening a public disturbance harmful to his professional career, already badly compromised. They could even force him to condemn a man whom Pilate himself knew was innocent and without fault. Later Pilate was recalled from Palestine after bloodily suppressing an uprising in Samaria, where a man arose who claimed to be the Messiah. Tiberius died while Pilate was en route to Rome. Being unable to interest Gaius (Caligula), Tiberius' successor, to reinstate him, Pilate retired to Lausanne (in modern Switzerland), wallowed in regret, and at last committed suicide. Pilate's wife Claudia Procula knew of Jesus' teaching through her Phoenician maid and after Pilate's death became active in the early Christian church.

AT THE ANTONIA FORTRESS

Pilate was expecting the group of around sixty men, including the Sanhedrists (mostly Sadduccean), Judas, the high priest Caiaphas, and the suborned witnesses. Annas did not come. Pilate greeted the delegation on the steps of the praetorium, a building added to the original Antonia Fortress.

"Of what do you accuse this man?" Pilate asked. The Jewish leaders wanted Pilate's confirmation of their sentence of death, not another trial. "If this man were not a doer of evil, we would not have brought him to you," a spokesman said, as Jesus was taken under guard inside the praetorium. Pilate replied that if they had not formulated any definite charges, they should judge him according to their own laws. The Jewish leaders explained that they wished to put him to death, which they could not do under Roman law.

"I will not sentence this man to death without a trial," Pilate said. "Nor will I examine him until you have presented your charges against him in writing."

The Sanhedrist clerk then presented to Pilate the three written charges that they agreed upon, although Jesus had not been legally convicted of any of these charges and in fact had never heard them. Pilate brought Jesus out in the open from the praetorium and had the charges read to him. When Jesus made no reply, Pilate ordered that he be taken back inside, where he could interview him.

Pilate was confused and alarmed. After he had given permission to use Roman soldiers in the arrest, he had learned a good deal about Jesus from his wife. Alone in a room with John and Jesus, Pilate invited them to sit. He sat beside them and questioned Jesus. When during their conversation Jesus said, "Every one who loves the truth hears my voice," Pilate cynically replied, "Truth ... What is truth? Who knows?"

Confident that Jesus was in no way guilty of these outlandish charges, but was instead some kind of harmless religious extremist of which there were all too many, Pilate went out to the Jewish officers in the court and told them he could find no guilt in the man and that he should be freed.

One of the Sanhedrists boldly stepped up, saying, "This man stirs up the people, beginning in Galilee and continuing throughout all Judea. He is a mischief-maker and an evildoer. You will long regret it if you let this wicked man go free."

"If he is a Galilean, then take him to Herod," Pilate snapped "Then report his findings to me." Galilee was after all the province of Herod Antipas, who was in Jerusalem for the Passover.

Therefore they took Jesus to Herod, resident in the huge Maccabean palace of Herod the Great built into the western wall of the city (see Map 5, page 229). As the guards led him away, Jesus told John that he could do no more and should go to his mother Mary at the house of Martha and Mary, sisters of Lazarus, in Bethany. John was to tell his mother Mary to come to see him before he died. And so John left.

Herod Antipas was curious about Jesus, who had caused him considerable worry in Galilee. He hardly remembered that as a young man Jesus had stood before him asking recompense for his father's death on a construction project, an honest request that Herod refused. Herod half hoped that Jesus would perform some miracle for him, but when he remained silent through fifteen minutes of taunting, Herod wrapped him in an old royal purple cloak (the color of royalty) and sent him back to Pilate. "I have no authority here in Judea," he said with relief. The Judeans would have to put Jesus to death for him. His hands still stank of the blood of John the Baptist, as everyone knew.

JESUS OR BARRABAS?

Pilate received Jesus on the porch at the head of the steps leading to the praetorium. The Jews could not come inside because it was against their law to enter a building that might contain leaven within twenty-four hours before the Passover. Had they done so, they would have had to undergo elaborate ceremonies of purification to participate in the Passover meal. Speaking to the Sanhedrists and the chief priests, Pilate said he could find no fault in the man, and neither could Herod Antipas, or he would not have sent him back. He was willing, however, to punish Jesus before releasing him, if that would satisfy their desire.

Just then a great mass of people came toward the praetorium to ask that Pilate release one prisoner in honor of the Passover, as had been the custom. Although the crowd shouted the name of *Barrabas*, a well-known agitator and murderer and thief caught on the Jericho road and condemned to death, Pilate suggested that they might take Jesus instead. But the chief priests shouted, No, let Barrabas go! When the people saw that the priests were determined to kill Jesus, they too shouted "Barrabas!" They were no longer impressed by a man who allowed himself to be taken by the priests and the Roman guards and placed on trial for his life.

"How can you prefer a murderer to a man whose greatest crime is that he spoke of himself figuratively as king of the Jews?" Pilate said, a stupid remark that inflamed the race-proud Jews, temporarily under the yoke of Rome, true, but soon to be freed that they might themselves rule over others once the Messiah, the anointed one, came in power and glory. How insulting to imply that this meek-mannered nonresistor call himself king of the Jews! They shouted all the louder for the release of Barrabas.

Just then a messenger interrupted Pilate. He bore a letter from his wife. Pilate stopped to read it. She had had a vivid dream and begged him to let Jesus go. While Pilate read, the priests circulated among the crowd, stirring up a claque to release Barrabas and execute Jesus.

When this irresolute and fear-ridden man returned to the crowd, he asked, "What should I do with this Jesus of Nazareth, king of the Jews?" "Crucify him!" they shouted with one voice. "Why? What has he done? Who can testify against him?" "Crucify him!" they shouted. "And Barrabas?" "Let him go!" "And Jesus?" "Crucify him!"

Terrified by the fury of the mob, Pilate made one last attempt to save Jesus, whom he knew to be innocent. He ordered that Jesus be taken into the open courtyard of the praetorium, tied to a post, and scourged with whips. His order was entirely illegal because according to Roman law only men sentenced to crucifixion could be scourged.

BEHOLD THE MAN

The enemies of Jesus outside the praetorium did not witness the scourging with knotted whips, but Pilate did. Before the Jewish guards and Roman soldiers were through, he ordered them to stop and bring Jesus to him. The tormentors again placed the purple cloak from Herod Antipas around his lacerated shoulders and now placed on his head a crown of thorns. One guard fell to his knees and said, "Hail to thee, king of the Jews," then spit upon him. A guard placed a stick as scepter in his hand, then snatched it away and struck him across the face with it.

Pilate brought this bleeding man again before the mob. He said, "Behold the man! I have whipped him. I find no fault in him and I would release him." "Crucify him!" they returned. "You wish him to die," he said, "but what is his crime? What has he done?" Caiaphas stepped forward and said, "We have a sacred law, and by that law this man must die because he made himself out to be the Son of God."

The high priest's words horrified Pilate as for the first time he wondered if Jesus really were a divine being disguised as a human, as often told in the Greek myths of his own education. Again he took Jesus inside the praetorium, sat down beside him, and asked him who he really was. But Jesus would not answer.

"Do you not realize that I have power to release you or to crucify you?" he said. Jesus answered, "You could have no power over me if it were not permitted from above. You could exercise no authority over the Son of Man unless the Father in heaven allowed it. But you are not so guilty, because you are ignorant of the gospel. He who betrayed me and he who delivered me to you, they have the greater sin."

Trapped between his fear of Jesus and his fear of the Jews, the supine Pilate again had Jesus presented to the mob. He proclaimed that this man was innocent and that he was going to release him. "If you do," said Caiaphas, stepping up, "you are no friend to Caesar, and I will tell him all."

"Behold the king of the Jews!" Pilate said. "Shall I crucify your king?" "Crucify him! We have no king but Caesar," the crowd roared.

Pilate called for a basin of water. He gave the order to free Barrabas, then washed his hands, saying that he was clean of the blood of an innocent man. He gave the order for Jesus' execution. The mob cheered and replied, "Let his blood be upon us and upon our children!"

CHAPTER 37. THE EXECUTION OF JESUS OF NAZARETH

JESUS WAS TURNED OVER TO THE ROMAN SOLDIERS about 8:30 Friday morning. The execution was delayed because the Roman commander decided to put to death two thieves also in his custody, rather than wait until after the Passover.

THE END OF JUDAS

At about the same time Judas arrived at the Sanhedrist court in the southeastern corner of the temple grounds. The guards and the Sanhedrin gathered to hear Caiaphas' report on the settling of the Jesus problem. Judas walked into the chamber, fully expecting to be greeted as a savior of his nation, to be extolled publicly and commended, to be richly rewarded for his patriotism and fellow religious feeling. Instead the Sanhedrin held him in thorough contempt. A servant of the high priests tapped him on the shoulder and gave him as reward thirty pieces of silver in a pouch, the price of a good, healthy slave. He disbarred Judas when the former apostle attempted to reenter the room.

Desperate in his darkness, Judas wandered the streets, then returned to the Sanhedrists' chamber, forced his way in, and loudly denounced his cowardice and his despicable act. He cast the thirty pieces of silver broadcast across the pavement. He then went into the valley of Hinnom south of the city (see Map 5, page 229). Taking his sash, he knotted one end to a tree and the other to his neck, then leaped from the cliff. The hastily tied knot slipped, and he fell to his death on the rocks below.

Judas was a self-centered individualist with many psychic, emotional, and spiritual troubles. As a personality, he was isolated. By disposition he was suspicious and vengeful, surly and vindictive. Emotionally, he was loveless and unforgiving. Socially he was unconfiding and self-contained. In spirit he was arrogant and selfishly ambitious. In life he ignored those who loved him, and in death he was without friend.

CRUCIFIXION

The Roman soldiers took Jesus back into the open courtyard of the praetorium. They removed Herod's purple cloak and dressed him in his own clothes. They mocked him but did not further torture him. The two thieves were brought into the court. One had heard Jesus speak at Pella and in the temple. They set forth for Golgotha ("place of the skull"), the usual place for executions, just before 9:00 AM (see Map 5, page 229). They went north, through the Damascus Gate, then turned west and south along the city walls. Beyond Golgotha were the villas of the rich and across the road many ancient and recent rock-cut tombs.

The condemned men were forced to carry their own crossbeams. The vertical members of the cross had already been fixed in place. As Jesus came out of the Damascus Gate he fell, exhausted and drained. The last time he had eaten was at the house of Elijah Mark, and he had had no sleep or water. He bled and was in pain from the torture. Furthermore he suffered from extreme mental anguish, acute spiritual tension, and human loneliness. When the captain of the guard, the same man who had arrested him, saw that he could not get up again, he ordered a passerby, one Simon of Cyrene, to carry the crossbeam. Jesus had met this man early in life during his travels around the Mediterranean and had helped him. Later Simon became a believer and worked in the early Christian church.

When they got to Golgotha shortly after 9:00 AM, they removed Jesus' clothes but covered his genitals with a cloth, a concession to Jewish objections to public nudity: Ordinarily a crucified man was naked. They bound his arms to the crossbeam, then nailed each hand to it and raised the beam to the notch in the upright member, where they nailed it in place. Jesus said, "Father forgive them, because they do not know what they do."

A post protruded from the upright beam to support the body's weight, so that the man did not die of asphyxiation but lingered in intolerable pain for two or sometimes three days. Then they drove a single long nail through both feet, about three feet above the ground. Jesus could hear those standing near him, and they could to some extent hear him. Near this very site, forty years later in 70 AD, thousands upon thousands of crucified Jews died in agony after the Roman sack and utter destruction of Jerusalem.

Above Jesus the captain of the guard nailed a whitened board with the words JESUS OF NAZARETH, KING OF THE JEWS, in Greek, Latin, and Aramaic. Pilate himself had written these words according to the custom of naming the executed man and giving the reason for his death. The head priests snarled and complained about his calling Jesus king of the Jews, but the ruined Pilate had in this way his little revenge against them.

John Zebedee had arrived at Golgotha with Mary, Jesus' youngest sister, Ruth, and his brother Jude just before Jesus was nailed to the beam. Once hoisted, he smiled down on them. Thirteen Roman soldiers were present, four assigned to each executed man and the captain. The four assigned to Jesus divided up his clothes between them, one taking the turban, another the sash, another the cloak, and another the sandals. Still the tunic remained, the garment closest to the flesh, and for this they threw dice.

At first a large crowd assembled. A local society opposed to crucifixion, a punishment learned from the Phoenicians and never imposed on Roman citizens, sent representatives to offer a sponge soaked in an opiate mixed in wine, to ease their suffering. But Jesus refused the sponge when it was extended to him at the end of a stick. He did not wish to alter the experience through pain-killing drugs. Later, when the guards broke out an early lunch at the foot of the cross, washing it down with sour wine, Jesus whispered, "I thirst." The captain of the guard extended the sponge stopper of his bottle of vinegar on the tip of his spear to moisten Jesus' lips.

John Zebedee had left the scene to go into the city and bring back his mother, Salome, and other women followers. While he was gone, the two thieves mocked Jesus, telling him to come down from his cross if he was who he said he was, until the thief who had heard Jesus speak defended him as an innocent man, unlike themselves, who in the name of resisting the political order had joined Barrabas in a life of rapine and murder. They deserved to die, he said, but Jesus did not.

This thief then cried out to Jesus, begging him to remember him in his kingdom. Jesus answered that yes, one day they would meet in Paradise.

John returned with his mother, Salome, and other members of the women's corps, including Mary Magdalene. As John stood near Mary, the mother of Jesus, Jesus said, "Woman, behold your son." Then he asked that John take Mary away. John led her to the house in the city where he was staying, then hurried back to Golgotha. On the next day Mary returned to Bethsaida, the harbor of Capernaum, where she lived in the house of John Zebedee until she died about one year later.

After Jesus' mother Mary left, the other women remained, but withdrew somewhat, waiting until Jesus was dead. Around noon an unseasonable sirocco came in from the Arabian desert, with its hot, stinging sand. The once numerous crowd quickly dwindled to around thirty—the thirteen Roman soldiers and the women followers. By 1:00 PM the sun was blotted out by the storm, and the onlookers crouched behind a rocky outcropping for protection from the biting sand. Only two men were present, except for the Romans—Jesus' brother Jude and John Zebedee, who returned from the city just before Jesus died.

As Jesus neared death, his human consciousness murmured passages from the psalms, including Psalm 22, which begins, "My God, my God, why have you forsaken me?" Mostly his words were unintelligible, but the captain of the guard heard these words. Standing by the cross as he died were John Zebedee, Mary Magdalene, Jude, and Rebecca, a woman of Sepphoris who had wished to marry Jesus.

Jesus said, "It is finished. Father, I commend my spirit into your hands."

The meaning of this remark is by no means clear because his Personalized Adjuster already led an entirely independent existence from his personality and was of course not affected in any way by the crucifixion. When the captain saw how Jesus died, he himself became a believer and later described to the gospel writer Luke what he remembered of this terrible day, including the conversion of one of the thieves. In this way these events entered the gospel account.

Because it was unseemly that men hang dying from crosses on the day preparatory to the Passover and Sabbath, the chief priests went to Pilate and asked him to break the legs of the crucified men, to dispatch them, and take them down and throw their bodies into the open burial pit at the south of the city (in the valley of Hinnom, where

Judas hanged himself). Pilate agreed and sent three soldiers to do the work. They arrived around 3:30 PM. They were surprised to find Jesus already dead after only five and a half hours on the cross. Just to be sure, one of the soldiers pierced his side with his spear. The man was certainly dead.

BURIAL

John sent the women to the home of Elijah Mark to pass the Sabbath while he remained at the cross until the Pharisees Joseph of Arimathea and Nicodemus arrived, both former members of the Sanhedrin but for long secret believers in Jesus. They brought with them a written order that they had obtained from Pilate allowing them to claim the body.

Once the sandstorm had abated representatives of the Sanhedrin came to prevent the followers of Jesus from interfering with the disposal of Jesus' body in the garbage pit, where it would be eaten by animals. Joseph of Arimathea presented to the Roman captain the written permission from Pilate, but the Sanhedrin seized hold of the corpse to drag it away. The Romans drew their swords on the Sanhedrin, drove them back from the corpse, and stood over it, then gave the body up to Joseph of Arimathea and Nicodemus to carry away.

No crucified man could be buried in a Jewish cemetery, but the family of the wealthy Nicodemus had a private burial site just across the road from Golgotha. Around 4:30 in the afternoon the four men—John Zebedee, Joseph of Arimathea, Nicodemus, and the Roman centurion—carried the body in a sheet, to the newly cut ten feet by ten feet burial chamber off the garden of Nicodemus.

While the women hung back, the men wrapped the body in linen soaked in aloes and myrrh that Joseph and Nicodemus had brought with them, a form of embalming. They placed a napkin over his face, then wrapped the whole body in a sheet and placed it on a shelf in the tomb. Roman soldiers rolled shut the large circular stone used to seal the tomb, and the men returned to the city to prepare for the Passover.

Mary Magdalene and three other women followers waited at a distance until dark came. The burial was hasty because of the impending Sabbath, and these women thought that Jesus had been improperly prepared for burial. The women left the tomb and went to the house of Joseph of Arimathea in the city. There they spent the Sabbath preparing oils and spices to embalm the corpse, which they planned to apply on Sunday, the next morning.

Only David Zebedee and Joseph of Arimathea really believed that Jesus would return from the dead on the third day, as he had promised. David now resided at the house of Joseph and was already preparing his messenger corps for the announcement that he fully expected. The Pharisees, who had more faith in this prophecy than did Jesus' own followers, asked Pilate for an armed guard to see that the tomb was not disturbed, at least for three days, to prevent Jesus' followers from stealing the body and claiming that Jesus had risen from the dead. A disgusted and impatient Pilate sent ten men to the tomb, along with ten men from the temple guard. These twenty men camped outside the tomb and rolled a second stone across the opening to which they applied the seal of Pilate.

JESUS IN MORONTIA FORM

Late Saturday night John Mark secretly summoned the apostles back to his house, where they assembled in the same room where they had dined with Jesus two nights earlier. Jesus' family now lodged at Bethany in the house of Martha and Mary.

The status of the personality of Jesus during the thirty-six hours between his death and 3:00 Sunday morning remains unknown, in spite of considerable speculation by celestial personalities. There is some evidence that Michael functioned both as universe sovereign and as Planetary Prince of Urantia during this period, but it is hard to know how to interpret the evidence. Certainly his now Personalized Adjuster commanded the various divine personalities on Urantia at this time. Probably his human personality slept, unconscious, like the sleep that mortals experience before awakening in the resurrection halls of the first mansion world of Satania.

Jesus' cruel death on the cross was not to effect man's reconciliation to a resentful God—such a notion, fundamental to Christian theology, contradicts everything that Jesus taught about the Father's infinite love for his children. Rather, his death was designed to stimulate man's realization of the Father's eternal love and his Son's unending mercy and to broadcast these universal truths to a whole universe, not just to Urantia. The crucifixion stirs deep emotions because it shows a man willing to lay down his life in loving service to his fellows, as much a revelation of man to God as Jesus' life was a revelation of God to man. The crucifixion makes it hard for us to complain about our troubles.

CHAPTER 38.
THE MORONTIA JESUS

THERE WERE ONE AND A HALF MILLION PEOPLE in Jerusalem at this time because Jews came from far and wide to celebrate the Passover. People were talking about the execution of Jesus. The many celestial personalities then present on Urantia discussed how they might help in the resurrection of Jesus, then realized that the creature does not resurrect the creator.

THE EMPTY TOMB

At about 2:45 AM on Sunday morning seven personalities from the Paradise incarnation committee arrived on Urantia and surrounded the tomb, from which arose an intense vibration. At 3:02 AM the morontia form of Jesus rose from the tomb and, hovering above it, communed with Gabriel, the Most Highs of Edentia, and with morontia personalities of the seven mansion worlds of Satania. Jesus, as in the ordinary human career, had entered the morontia realm.

The morontia Jesus signaled to his Personalized Adjuster, who dismissed all celestial personalities to return to their customary stations and duties. The stones in front of the tomb and the seals on the stones were still in place because the morontia form, which can be both visible to mortals and can touch the spiritual realm, passes through material objects.

The corpse within the tomb, through the intercession of archangels, was instantaneously disintegrated through a technique of speeding up time; it went through the natural process of decay, but without the usual duration of time. Secondary Midway Creatures then rolled away the largest of the two stones sealing the tomb, a large circular stone like a mill stone that rolled along a groove at the opening to the tomb.

The tomb now was empty, yes, but the physical material body of Jesus of Nazareth did not arise from the tomb. When the soldiers saw the stone rolled away apparently by itself, they fled in abject terror, the Romans to the Antonia Fortress and the Jews to the temple. When they informed the Sanhedrin, the panicked rulers bribed all twenty soldiers, both Romans and Jews, to say that Jesus' followers had come to the tomb and rolled away the stone.

About 3:30 PM Gabriel called the archangels together to inaugurate the dispensational resurrection of the sleeping survivors to accompany the termination of the Adamic bestowal. When the angelic hosts were assembled for this great undertaking, Michael in his morontia form spoke to them, saying, "As my Father has life in himself, so has he given it to the Son to have life in himself. Although I have not yet fully resumed the exercise of universe jurisdiction, this self-imposed limitation does not in any manner restrict the bestowal of life upon my sleeping sons. Begin the roll call of the planetary resurrection."

Under Gabriel's direction, the sleeping survivors appeared in the resurrection halls of the first mansion worlds. This was only the third complete dispensational resurrection roll call on Urantia, the first being at the time of the arrival of the Planetary Prince and the second during the time of Adam. Jesus' Personalized Adjuster now relinquished his authority, turning over the celestial hosts under his command to their various commanders. The Personalized Adjuster departed for Salvington, there to report to Immanuel the completion of the mortal transit of Michael. Gabriel remained on Urantia with the morontia Jesus.

FIRST MORONTIA APPEARANCES

At around 3:00 in morning, just as the first glimmer of sunlight edged the eastern horizon, Mary Magdalene and four other women of the women's corps went to the tomb bearing linen clothes and oils for Jesus' body. As they went out of the Damascus Gate they encountered the panicked guards running to the city. When they came to the garden, they were startled to find the large stone rolled away from the opening of the tomb, whose opening was oriented toward the east.

Although frightened, Mary Magdalene went into the tomb. In the faint light she could make out the grave clothes and napkin laid out on the niche. She cried out that someone had taken the body, but she could not understand why the grave clothes lay in their original position. She went back in to the garden and with the other four women sat on a rock. Someone was standing there, his back to the east so they could not see his face. He asked whom they sought, and they answered that someone had taken away the body of Jesus of Nazareth. Did he know anything about it?

"Did I not tell you that I would die and rise again, Mary?" Jesus said. Mary and the other women fell to his feet and Mary made to kiss them but the man said, "Do not touch me, Mary, for I am in a different form than when I was in the flesh." He told her to go tell the apostles, and especially Peter, that he had risen from the tomb, just as he promised.

The women ran to the house of Elijah Mark, but the apostles would not believe them. The highly depressed Thomas had left the group and was staying with a friend in the village of Bethphage near Bethany. When Mary mentioned Peter's name, as she had been instructed, Peter leaped up and burst from the chamber, followed closely by John Zebedee. Mary followed behind, but the other women went to the house of Joseph of Arimathea.

The men made their way quickly to the tomb. The grave clothes were as the women had said. Peter thought that the tomb had been rifled, but he could not understand the placement of the grave clothes. Mary Magdalene came in behind them. She stayed in the garden when the two men returned to the house of Elijah Mark.

Jesus then appeared to Mary a second time, again made visible in his morontia form through adjustments made by Midway Creatures and some morontia personalities that accompanied Jesus. That she had spoken to him the first time he appeared was a measure of her deep emotion and love, because as a woman with a past Mary Magdalene did not ordinarily accost men she did not know. Jesus told her to go back to the apostles and to tell them again that he had risen.

Jesus/Michael had decided to spend some time on Urantia as a morontia being, in this way undergoing an experience similar to that of ordinary mortals passing through the seven mansion worlds of Jerusem. Humans are not born with eternal life, but Jesus who is life made clear that believing that one was a son of God made eternal life a certainty. Jesus' morontia body was in every way similar to the morontia bodies that surviving mortals inhabit. Such bodies do not have circulating blood and do not ingest ordinary food. Nonetheless, they are quite real.

Only David Zebedee and Joseph of Arimathea believed the women saying that Jesus had risen from the dead when they had come to Joseph's house. They hastened to the tomb too, coming after Mary had returned into the city. They saw the grave clothes in their odd posture, the last to do so, for the Sanhedrin, hearing rumors that began to spread in the city, sent guards to clean out the tomb. The captain of the temple guard rolled up the grave wrappings in the linen sheet that had once covered the body and threw the bundle from a nearby cliff.

David Zebedee and Joseph of Arimathea joined the apostles in the upper room of the Mark home, but could not persuade them that Jesus had risen. David Zebedee upbraided them for forgetting what Jesus himself repeatedly told them, that he would die and rise on the third day. David then went to the house of Nicodemus where he assembled twenty-six of his best messengers, then sent them out to all cities of the Levant with the message that Jesus had risen from the dead. In this act of David Zebedee was sown the seed that became the Christian church, which from the beginning held as a central tenet the fact of the resurrection. This fact is something rather different from what Jesus taught, although in its affirmation of life over death through faith, it is more or less correct.

MORE MORONTIA APPEARANCES

Jesus appeared nineteen times in his transition from the morontia realm into the spiritual. The first time was to the five women in the garden. The second was to Mary Magdalene alone in the garden. The third was to Jesus' brother James in the garden of the house of Mary and Martha in Bethany, and the fourth to his family and friends gathered within their house.

In every case he appeared as a dim form, not recognizable until he spoke. After speaking words of comfort and advice, he would disappear. His fifth appearance was to the women's corps gathered in the house of Joseph of Arimathea, at about 4:15 PM, and the sixth to about thirty Greek believers assembled at the house of Flavius; some of these Greeks were in the camp near Gethsemane on the night of the arrest. Jesus emphasized to them that his teaching was for all humankind, not just for Jews.

At Emmaus, a village about seven miles west of Jerusalem, Jesus appeared for the seventh time to two brothers walking on the road, talking about the recent execution of Jesus of Nazareth. After speaking to them at length about the gospel of the kingdom, he went into their hut to eat, and as he broke the bread they recognized him. Then the form disappeared.

By now the entire city was abuzz with rumors about the resurrection of Jesus. Caiaphas, nearly in a panic, called a meeting of the Sanhedrin for that evening to discuss what to do about the followers of Jesus.

Peter remained in enormous psychological chaos, not sure whether to believe the rumors and wondering why Jesus would appear to the women and not to the apostles unless because of his own very shameful presence, he who had denied him outside the house of Annas. To relieve the others of his presence, who remained locked in the upper chamber in the house of Elijah Mark and afraid of arrest if they ventured forth, Peter slipped out of the house around 8:00 PM into the courtyard.

There Jesus appeared to him, his eighth morontia appearance, and spoke with Peter for several minutes. Peter rushed back into the upper room to announce that he had himself seen the Master, that he had in fact risen from the dead. His proclamation nearly removed the apostles' lingering doubt, until Andrew, Peter's brother, reminded the group of the time that Peter fancied in a dream that Jesus had walked upon the water.

Around 9:00 PM Jesus at last appeared to the apostles, his ninth morontia appearance, although the doors of the chamber were made fast and although Thomas was not with them but with a friend in Bethphage, where he suffered from grief and uncertainty. Jesus told the ten apostles that they were to go into Galilee, once all of them were gathered, and that he would see them there.

As participants in Michael's morontia-transition experience, there came to Urantia more than one million morontia directors and associates, together with transition mortals of various orders from the seven mansion worlds of Satania. On Monday Michael spent all day with these beings. On midnight of this day he passed into the second stage of morontia progression, and after intervals of several days advanced through all seven stages of the morontia progression, reaching the seventh on May 7. Even so the mortal progressor passes through the seven mansion worlds, acquiring ever finer morontia bodies as they go. After that the Most Highs of Edentia embraced him. At this moment Michael came to the end of his seventh bestowal in the world of material and morontia matter.

As Jesus progressed from one morontia level to the next, it became progressively harder to make him visible to mortals of the realm. In later appearances he spoke to a group of one hundred fifty followers of Abner and Lazarus in a meeting in

Philadelphia, then to the apostles again, a week after his first appearance to them, now including Thomas who had returned from Bethphage. Although his morontia form did not have the marks of the nails that Thomas hoped to see, the doubting apostle nonetheless fell to his feet and exclaimed, "I believe, my lord and master!"

On April 18 he appeared to eighty Greek and Jewish believers in Alexandria, Egypt, so that when David Zebedee's messenger arrived two days later to announce the Master's resurrection, they had already seen him.

The apostles stayed in Jerusalem a week. Mary the mother of Jesus spent much time with the women at the house of Joseph of Arimathea. Jesus had appointed Andrew as head of the apostolic corps, but released him from this role before his arrest. Peter emerged as the spokesman for the group, by common consent.

MORONTIA FORM IN GALILEE

The group departed for Galilee on Monday morning, April 17, telling many on the way of Jesus' resurrection. John Mark came with them. They arrived at the house of Zebedee in Bethsaida on Wednesday, and on Thursday evening decided to spend the night fishing. They caught nothing and when they pulled in close to shore around 6:00 AM Friday morning a man standing near a fire greeted them from the shore, as fish dealers from Tarichea down the coast often did (this was the Tarichea north of Tiberias). The man shouted that they should drop their net in a certain place, and when they did so the net was filled to overflow with fish. They recognized the man on the shore. Peter, with typical impetuosity, leaped into the water and swam toward shore with the boat close behind him, pulling the net filled with fish. This was the thirteenth morontia appearance of Jesus.

John Mark joined the group on the shore, but Simon Zelotes was not there: In a deep depression he had gone off to his own house in Bethsaida. They cooked fish over the fire and ate, while Jesus spoke with them. They remembered their former experiences here in peaceful Galilee, far from the treachery, hate, and murder of Jerusalem.

After an hour of general conversation, Jesus walked along the beach with the apostles two by two. In his conversations he advised them about personal faults of various kinds and about the need to go forth fearlessly to proclaim the truth of what he had taught.

To Peter he said, "Do you love me, Peter?" He said he did, and Jesus said, "Then feed my flock." To illustrate that Peter must follow him whatever others do, Jesus said as an example, "If I will that John should remain after you are gone, even until I come back, what is that to you? Only make sure that you follow me." This remark was taken to imply that Jesus would return before John's death, and the early promulgators of Jesus' gospel believed this to be true: that the Master would return at any moment to institute the kingdom of God.

Around 10:00 AM the morontia Jesus bid goodbye to the group but told them to meet him on the next day at the mount of ordination behind Capernaum. There on

Saturday, April 22, around noon he appeared to them, including now Simon Zelotes, who had rejoined the group. As they kneeled in a circle around him he spoke for an hour, repeating the terms of their ordination. Then he disappeared.

For the next week Jesus made no more appearances, preoccupied with morontia personalities and the transition through the seven stages of morontia life. On the following Sabbath, April 29, AD 30, Peter stood before about five hundred believers who had filtered into Bethsaida on reports of the Master's resurrection. He gave a speech of great power in which he proclaimed the resurrection of Jesus of Nazareth. Suddenly at his side appeared the morontia form of Jesus, his fifteenth appearance. Jesus wished peace upon the crowd, then disappeared.

Jesus had suggested to the apostles on the mount of ordination that he would appear to a large number of people, and that they were then to return to Jerusalem. Therefore on the next day, Sunday, April 30, they journeyed south, preaching along the way. They arrived in Jerusalem on May 3. They went to the house of Elijah Mark, who just then had died of a brain hemorrhage. The apostles sequestered themselves in the upper room and did not participate in the funeral of this generous man, but waited for something to happen.

On Thursday the group held a momentous meeting in which they determined to go forth into the city and proclaim the resurrection of Jesus. Only Thomas, Simon Zelotes, and the Alpheus twins did not agree to do this. At this moment, at the very beginning of the Christian church, Jesus' message of sonship with God and the brotherhood of man was turned into the proclamation of the resurrection of Jesus. Peter was the powerful spokesman for this view and by his eloquence persuaded the others, especially the women believers. Already the religion *of* Jesus was becoming a religion *about* Jesus.

The sixteenth morontia manifestation of Jesus took place on Friday, May 5, about 9:00 PM at the house of Nicodemus where were assembled the eleven apostles, the women's corps and associates, and about fifty other believers. In his talk to them Jesus emphasized that his resurrection verified his power but did not change his teaching.

One week later he appeared in Samaria at Jacob's well outside Sychar, where he had spoken to the Samaritan woman about the water of life. This Samaritan woman regularly led a group of other women to the well and on this day about seventy-five women were gathered there, discussing reports of the resurrection of Jesus. Suddenly he stood beside them and spoke.

The eighteenth appearance took place on May 16 about 9:00 in the evening before a group of believers in Tyre in Phoenicia. In his talk he emphasized that a child's faith may bring one into the kingdom, but then must the branch show forth rich fruit, for the branch without fruit will be cut away and discarded.

The Master's last appearance took place on May 18 when the apostles sat down to breakfast at the Mark household in Jerusalem. After speaking, Jesus led them forth up the side of Mount Olivet. They kneeled in a circle around him. After final words in which he promised to send over all flesh the Spirit of Truth, and soon, he disappeared. It was around 7:45 in the morning.

PENTECOST

Michael traveled to Edentia over Jerusem, where the Most Highs, under the watch of the Paradise Son, released him from the morontia state. Through the spirit channels of ascension Michael ascended to the status of Paradise sonship and supreme sovereignty on Salvington of the universe of Nebadon.

Peter called a meeting for that night at 10:30 at the Mark household of around one hundred twenty followers of Jesus present in Jerusalem. James, the brother of Jesus, also attended this meeting, as did Jesus' mother Mary, who had come back to Jerusalem in the company of John Zebedee; after Pentecost she returned to Bethsaida and lived with Salome until her death. Peter gave an electrifying speech describing the Master's ascension. Peter then explained they were going to choose by lot a successor to Judas, one Matthias, who played little part in later developments. Peter then urged all who were present to fall into prayer for the sending of the Spirit of Truth that Jesus had promised.

At around 1:00 AM the one hundred twenty believers felt a strange presence in the room. They were overcome by a feeling of joy and spiritual liberation. This was the Spirit of Truth come upon them, Peter said, and they should follow their inclination to go into the temple itself and proclaim the resurrection of Jesus of Nazareth (not the truth of what Jesus had taught them).

This day happened to be the day of the Jewish festival of Pentecost, Greek for "fiftieth" because it celebrated the handing down of the ten commandments fifty days after the exodus from Egypt. The city was thronged with people, many there just for the festival but even more who had stayed on after Passover. Around 2:00 PM Peter stood in the same place in the temple where Jesus had preached and in a vigorous speech proclaimed the truth of the risen Master. Peter, Andrew, James, John, Philip, and Matthew also spoke, speaking in Aramaic, Greek, and Hebrew, and in snippets of other languages they knew to the large polyglot crowd. Two thousand listeners followed Peter and some of the other apostles to the pool of Siloam south of the city, where the apostles baptized them in the name of Jesus Christ (that is, "Jesus the king"; baptism in the name of the Father, Son, and Holy Spirit came later).

Once again the Sadducees were alarmed, but the Pharisee Gamaliel, the greatest scholar of his day, advised them that if this teaching that they so feared were from man, it would perish, and if from God, it could not be stopped. Hence the apostles were allowed to continue preaching in the city and to baptize many.

THE SPIRIT OF TRUTH

The Spirit of Truth is really the spirit of the Father and the Son, living in the heart of believers and leading them to understand the truth in all things. For the truth makes you free. Once poured out, all normal human beings henceforth receive a Thought Adjuster at the time of the first moral decision.

The spirit relieves the loneliness that comes from knowing that Michael once lived among us in the flesh. Not just the apostles, but believers everywhere felt the presence of this spirit. The spirit does not reveal itself, but sponsors a close fellowship with Michael, the Son of God, and constant new understanding of what Jesus taught and the meaning of his life. Although the religion that arose from his life made the error of seeing Jesus as a ransom to an angry God lustful for blood, nonetheless much of his original teaching survived these distortions and will one day emerge as the true teaching of the incarnated Michael of Nebadon to transform all civilization on this planet. Jesus revealed the Father to man; the Spirit of Truth reveals the Creator Son to man.

The pouring out of the Spirit of Truth on all humankind, all races, both sexes, all castes and classes put an end forever to the control of religion by priests and sacred families. Before, women were scarcely allowed into religious life in any form, but now they are to be equal partners with the male; there were many women in the house of Mark when the spirit came.

The religion of Jesus does not promise a glorious afterlife, freed of the pain of this world, but brings to this world the enhanced joy of living in the spirit. The Spirit of Truth leads upward and onward in the hearts of progressive and aggressive believers. It is the force driving all humankind to a better future. The Spirit of Truth descended on the one hundred twenty believers in the house of Mark when they were just sitting there, independent of any ceremony, special place, or special behavior. Such traditional features of religious life are entirely unnecessary in finding spiritual truth.

BEGINNINGS OF THE CHRISTIAN CHURCH

The Christian church began at Pentecost, when many believed Peter's message of the risen Christ. Jesus lives; he died for humankind; he gave the Spirit; he is coming again. Jesus filled all the thoughts of earlier believers and determined their new concept of God. They were so excited over the new doctrine that "God is the Father of the Lord Jesus" that they were little concerned with Jesus' own message that "God is the loving Father of all humans," even of every single individual.

A new kind of community grew up whose members met regularly for a communal meal in remembrance of the last supper. Certain that the Christ would soon return and sit on the throne of the kingdom of heaven, they sold all their personal possessions and held all in common. Great was the suffering caused when Jesus did not return and the money ran out.

The break between the followers of Jesus and traditional Jewry came when a group of Greek believers came to Jerusalem from Alexandria. They converted numerous Greeks in Jerusalem to the gospel of Jesus, including one Stephen. The Jewish believers in Jesus appeared no longer to threaten the authorities, because they observed Jewish law and had no organization; they were simply the brotherhood of Jesus. But Stephen did not follow Jewish law or have much respect for it, and when he began to preach publicly something close to what Jesus had taught, the Sadducees and Pharisees took sharp notice. In the midst of a speech they picked up stones and killed him, the first Christian martyr.

It was obvious that the believers in Jesus needed to separate themselves from the unbelievers; the followers of Jesus could not continue as a Jewish sect. Within one month of the death of Stephen was established the Jerusalem church with Peter as its leader and James, the brother of Jesus, its titular head. Under relentless persecution by the Jewish authorities, many believers, especially Greek, left Palestine and went up the coast to Tyre, then to Antioch near where the Orontes River flows into the Mediterranean. There the term "Christian" was first used. From Antioch believers migrated over Asia Minor to Macedonia, Greece, and Rome. In this way promulgators of the new religion, even before Paul, were swiftly present in remote parts of the Roman Empire.

ENVOI: THE CHRISTIAN RELIGION

One may *preach* a religion *about* Jesus, but really one must *live* the religion *of* Jesus. In the enthusiasm of Pentecost, Peter without thinking inaugurated a new religion of the risen and glorified Christ. The Pharisee Paul of Tarsus, who never knew Jesus, later on transformed Peter's gospel into the religion of Christianity embodying Paul's own theological views and founded on Paul's personal religious experience on the Damascus road, when he saw a blinding light and heard a voice, "Saul, why do you persecute me?" (Acts 9:3).

The New Testament is almost entirely devoted, not to the religious life of Jesus, but to a discussion of Paul's religious experience and to a portrayal of his personal religious convictions. The notable exceptions, aside from parts of the gospels, are the Book of Hebrews and the Epistle of James. The New Testament is a superb Christian document, but it has little to do with the nature and meaning of the life of Jesus of Nazareth and it poorly reports his teachings.

Jesus taught the religion of personal experience in doing the will of God and serving human brotherhood; Paul founded a religion in which the glorified Jesus became the object of worship and the brotherhood consisted of fellow believers in the divine Christ. Paul's Christianity established the adoration of the divine Christ, but lost sight of the struggling and valiant human Jesus of Galilee, who, through a courageous personal religious faith and the heroism of his indwelling Adjuster ascended from the low levels of humanity to become one with divinity. Thus he became the new and living way by which all mortals may ascend from human to the divine.

The Christian religion is built around the person of Jesus of Nazareth, the incarnated Michael of Nebadon, known as the Christ, "the anointed one, the king." Christianity was spread throughout the Levant and the Western world by the followers of this Galilean, whose missionary zeal was as great as that of their illustrious predecessors, the Sethites and Salemites, as well as that of their Asiatic contemporaries, the Buddhist teachers. In the Christian religion, however, the original teachings of Jesus were Occidentalized and began to lose their potentially universal appeal. Christianity today is a religion well adapted to the social, economic, and political mores of the white races, but hardly the religion of Jesus. As a Urantian system of belief, Christianity arose through the compounding of influences from the following:

1. The Melchizedek teachings, a basic factor in all religions of Occident and Orient that have arisen in the last four thousand years.

2. Hebraic morality, theology, and belief in both providence and the great god Yahweh.

3. The Zoroastrian notion of the struggle between cosmic good and evil, which had already left an imprint on Judaism and Mithraism. In the prolonged struggle between Mithraism and Christianity, the doctrines of Zoroaster exerted strong influence on the dogmas and cosmology of the Hellenized and Latinized versions of the teachings of Jesus.

4. The mystery cults, especially Mithraism but also the worship of the Great Mother. The legends of the birth of Jesus were tainted with the Roman version of the miraculous birth of the Iranian savior-hero Mithras, whose advent on earth gift-bearing shepherds witnessed, informed by angels.

5. The historic fact of the human life of Joshua ben Joseph, the reality of Jesus of Nazareth, the Son of God.

6. The personal viewpoint of Paul of Tarsus, a city in which Mithraism was the dominant religion. Paul scarcely imagined that his well-intentioned letters to converts would one day be regarded as the "word of God."

7. The philosophic thought of the Hellenistic peoples, from Alexandria and Antioch through Greece to Syracuse in Sicily and Rome in Italy. Greek philosophy was more in harmony with Paul's Christianity than with any other current religious system and was an important force in the success of Christianity in the West. Greek philosophy, taken with Paul's theology, still forms the basis of European ethics.

*

It is nearly four thousand years since Machiventa Melchizedek, an emergency Son of Nebadon, bestowed himself on Urantia, and in that time the teachings of the "priest of El Elyon, the Most High God," have in some form penetrated to all races and peoples. Machiventa successfully achieved the purpose of his unusual bestowal. When Michael came to Urantia, the concept of the universal God existed in the hearts of men and women and still flames anew in the living spiritual experience of the many children of the Universal Father as they live their intriguing temporal lives on the whirling planets of space.

GLOSSARY

A

Abel. First child born to Adam and Eve after they left the first Garden of Eden; his half-brother Cain hated him and killed him.

Abner. Head of a colony of Nazarites near Engedi where John the Baptist was enrolled; Abner played an important role in the formation of the early Christian church; friend of John the Baptist; died at age 89, November 21, AD 74, in Philadelphia.

absolute reality. Level of reality where all things are forever, where there is no time and space, as on Paradise, the geographical center of the universe and model for all physical creation and source of the physical universe.

absolutum. Substance of which Paradise is composed, not found elsewhere in creation.

absonite reality. A subinfinite level of reality where things have no beginning and no end but are "eventuated," where time and space are transcended.

absoniters. Beings who always were; beings who live in absonite reality; orders of beings who are neither creatures nor creators; Transcendentalers.

Adam (and Eve). Material Sons of Urantia who lived in Mesopotamia around 38,000 years ago; not human but visible to mortal eyes; although they failed in their mission, they contributed DNA to the human species; reduced to human status after their default, presently members of the commission of twenty-four on Urantia affairs on Jerusem.

Adam. The original Material Son, begotten solely by the Creator Son; Adam's offspring when dispatched to evolutionary worlds bear his name.

Adamites. Progeny of Adam and Eve on Urantia, the violet race.

Adamson. The first born of the violet race of Urantia, being followed by his sister and Eveson, the second son of Adam and Eve.

Adamsonites. The descendants of Adamson and Ratta.

adjudication of nonexistence. A formal obliteration of the personality/consciousness of one who freely chooses evil; an everlasting fate.

Adjuster fusion. Where a personality merges with its Adjuster to become an immortal being.

Adjuster-fused mortals. Ascending order of children of God who began as mortals like ourselves and have attained Adjuster-fusion.

Adjutant Mind Spirits. The spirits of wisdom, worship, counsel, knowledge, courage, understanding, and intuition; functioning through the Creative Spirit to impart intelligence to evolving races; circuits of consciousness of the Universe Mother Spirit.

Amadon. One of the one hundred altered Andonites associated with the Caligastia one hundred and a follower of the loyal Van, approximately 500,000 years ago.

Amadonites. The eighty-eight modified Andonites who remained loyal to Van and Amadon during the Lucifer rebellion; built the garden home for Adam and Eve under the leadership of Van.

Amenemope. An Egyptian teacher who taught that God-consciousness determines conduct.

Ancients of Days. A category of Sons of the Trinity, begotten by the Paradise Trinity; powerful beings who reside on the capital of each Superuniverse, in this case Uversa, in groups of three; they supervise the Superuniverse courts; involved in organizing the Urantia revelation; preside over personality extinction but are not entitled to interfere directly in governance of Local Systems or Local Universes.

Andites. A highly able and advanced human race derived from an admixture of Nodite with Adamic life plasm.

Andon (and Fonta). The first humans, twin brother and sister who were born in 991484 BC; presently serve on the first mansion world where they welcome ascenders from their one-time home, Urantia; the first will creatures to evolve on Urantia; Andon was killed during an earthquake at age 42.

Andonic. Of or pertaining to the Andonites.

Andonites. Urantia aborigines; descendants of Andon and Fonta, the first humans; small black-eyed people with swarthy complexions who were mighty hunters and ate berries, fruits, and raw meat; closely related to modern Eskimos.

Andrew. The first apostle; an ardent follower of John who approached Jesus and offered to follow him as teacher; brother of Simon Peter; official head of the apostolic corps.

Andronover nebula. Spiral nebula from which our solar system originates.

angels. A downstepping of the Creative Spirit; invisible, nearly human creatures of time whose mission is to serve ascendant beings such as ourselves; experiential creatures, learning and progressing through direct experience with the world; see Ministering Spirits.

Annas. High priest emeritus in Jerusalem; distant relative of Salome, wife of Zebedee; father-in-law of Caiaphas; most powerful man in Jewry.

apostle (Greek, "one sent forth as messenger"). Name used for a member of the closed group of Jesus' followers.

Aramaic. Semitic dialect spoken in Palestine by the common people and by Jesus.

archangels. Descending order of the Sons of God; dual-origin offspring of the Creator Son and Universe Mother Spirit (angels are of single origin); highest type of spirit produced in large numbers in the Local Universe; take orders from the local Bright and Morning Star or his agents.

Archelaus. Son of Herod the Great who ruled Judea after Herod's death.

architectural spheres. Special constructions serving as headquarter spheres for the Superuniverse and its segments; channel force and energy to their domains.

ascendant beings. Those who have the potential to evolve and ascend to the Father, such as ourselves.

ascending sons. Creatures (like ourselves) working from the bottom levels of reality to the top levels.

Ascendington. Seventh sphere of the Circuit of the Father surrounding Paradise; holds the secrets of raising a material and mortal mind to a divine and spiritual entity; the rendezvous sphere for the ascendant creatures of time who are traversing Havona on the way to Paradise.

ascension career. After mortals survive death, they pass upward through the spheres of the System, Constellation, Local Universe and beyond as morontia beings and spirits gaining experience and knowledge of unimaginable kinds.

B

Bablot. Descendant of Nod who planned to erect a temple to memorialize the Nodite race, about 150,000 years ago.

Badonan. Great-great-grandson of Andon and Fonta whose tribe of descendants lived in approximately 991000 BC in northwestern India; six Sangik races sprang from his line.

Badonites. Members of the tribe of Badonan; superior strain of Andonites and parents of the Sangik family; only early humans who did not practice human sacrifice.

bestowal plan. Emanation from the Eternal Son whereby representatives from the Paradise Sons of God appear among the ascending races to reveal the love of the Father and mercy of the Son; a necessity to a developing planet and to the Creator Sons themselves.

Bestowal Son. Son of God born from a mortal mother who lives in the flesh and dies; purpose is to establish a new way of living and shed over the planet the Spirit of Truth; the Bestowal son becomes a teacher to mortal races about realities of the spirit.

Bethany. Village east of the ridge overlooking Jerusalem.

blue race. Sangik race with the brain power of the red race and the sentiment of the yellow race; received the most Adamic DNA of all the Sangik races; interbred with the Neanderthals in Europe; origin of the "white race."

body. Material or physical organism of man; living electrochemical mechanism of animal nature and origin.

Bright and Morning Star. First product of the union of the Creator Son and Creative Spirit; a brilliant and versatile being who possesses qualities of both parents, with authority to decide any matter in his Local Universe not pertaining to personality; the Bright and Morning Star in Nebadon is Gabriel.

Brilliant Evening Stars. Offspring of Creator Son and Creative Spirit who serve chiefly as liaison officers of Gabriel, the Local Universe chief executive; one or more function at the capital of every Constellation and System.

C

Caiaphas. High priest; son-in-law of Annas.

Cain. son of Eve and the Nodite Cano; Cain murdered Abel, his half-brother.

Caligastia one hundred. Morontia volunteers from Jerusem; former humans, each from a different planet, selected by Caligastia to accompany him to Urantia; Caligastia's corporeal staff, sex creatures implanted with Andonite life plasm; sixty joined the rebellion while forty remained loyal to Michael.

Caligastia. Lanonandek Son, Planetary Prince of Urantia from about 500,000 years ago; joined the Lucifer rebellion approximately 200,000 years ago; the "devil" of popular lore; currently imprisoned on a prison world near Jerusem.

Cano. Brilliant Nodite descended from the Andonite, red, and blue races and supporter of the Adamic mission; leader of a Nodite colony near the Garden of Eden; Eve's intercourse with him caused the Adamic default, approximately 37,800 years ago.

Capernaum. Town and Roman military post on the caravan route from inland Damascus (gateway to Mesopotamia) to the Mediterranean coast.

Celestial Recorders. Specialized beings who make one spiritual and one semimaterial copy of all records for use by high spirits of light and morontial beings; "living books" who report on every cosmic event since the origin of the Superuniverse.

chazen. Teacher in the synagogue schools.

cherubim. Order of angels that serves the seraphim in the Local Universes.

circles of attainment. Seven levels of spiritual status associated with qualities of intellect, socialization, and spiritual and cosmic insight.

Circuit of Spirit Gravity. Force centered in the Eternal Son that permits instantaneous communication between the Eternal Son and the Paradise Sons scattered throughout the universe of time; means through which eternity and divinity touch the material creation.

Conjoint Actor. Infinite Spirit as cocreator with the Universal Father and the Eternal Son.

Constellation Fathers. Order of Sons of God called Vorondadek Sons who are appointed by Gabriel, the Bright and Morning Star; the Most Highs.

Constellation. Grouping of 100 Systems, approximately 100,000 inhabitable planets; location of the legislative units in a Local Universe; local administrative center linking the Systems with the Local Universe center; ruled from a headquarters sphere by three Vorondadek Sons (the Most Highs); our Constellation is Norlatiadek whose capital is Edentia, center for a system of morontia training worlds.

corporeal staff. See the Caligastia one hundred.

Corps of the Finality. Father-fused ascenders, glorified mortals on Paradise who explore the infinity of the Universal Father, presumably in the levels of outer space; our aspiration as humans; members of the corps called finaliters.

cosmic mind. All-pervading mind emanating from the Third Source and Center and originating from the Seven Master Spirits; source of moral, scientific, and spiritual endowments in humans.

Creative Spirit. Proceeds from the Infinite Spirit; cocreator of a Local Universe with the Creator Son, hence called a Universe Mother Spirit, who is the personality of the local Creative Spirit; a ministering spirit of divine love and endless mercy to all intelligent creation; materializes the patterns that the Creator Son generates; source of myriad corps of angels; defines and pervades the space of a Local Universe; fosters evolutionary progression on the planets.

Creator Son. Descending dual-origin being, Paradise Son of God who embodies the Son aspect of the eternal and primordial Trinity; proceeds directly from the union of Son and Father on Paradise; belongs to the Order of Michaels; possesses all the powers and attributes of the Eternal Son; represents the Father in his created universe; the most powerful, versatile, and influential of all divine beings capable of direct administration of the evolutionary universes and of personality contact with immature beings; creates the patterns of life; God to the Local Universe; shares perfect divinity and creative power with the Universal Father and Eternal Son.

Cynics. Philosophical group who best preserved the Melchizedek teachings, after the Jews.

D

Dalamatia. The headquarters city of the Planetary Prince, situated on the Persian Gulf in Mesopotamia, approximately 500,000 years ago, its mission being to change the Urantia humans from hunters to herders; education and cultural center for the development of civilization concerned with philosophy, religion, morals, and higher intellectual and artistic achievements.

Daligastia. Lanonandek Son who served as the primary assistant to Caligastia, former Planetary Prince of Urantia; contrived with Caligastia to frustrate the Adamic plan and cause trouble in the Garden.

dark gravity bodies. Two belts of dark gigantic forms that surround outer Havona and hide it from the view of the Superuniverse worlds.

dawn mammal. Sprung from the North American lemur, lively, affectionate and curious; ancestor of the mid-mammal.

Decapolis. Ten Greco-Roman towns mostly to the east of the Jordan River.

Deity, Deities. The source of all that is divine; characterized by unity on all levels of reality; personalizable as God; prepersonal and superpersonal in ways not comprehensible by man; experienced as one although plural in manifestation.

descending Sons of God. Paradise Sons of God, including Local Universe Sons of God.

Dilmun. Ancient Nodite city founded after the submergence of Dalamatia; racial and cultural headquarters of the descendants of Nod.

Divine Counselors. Trinity-origin beings who advise the Ancients of Days on Superuniverse capitals; some are members of the revelatory commission sent from Uversa to Urantia to dictate early papers of *The Urantia Book.*

Divinington. First sacred sphere of the Father located in the first of the three sacred circuits surrounding Paradise; contains the secrets of Thought Adjusters, their origin and mission; the Thought Adjuster returns to Divinington by instantaneous transfer at a mortal's death.

doctrine of the atonement. Paul's notion that an angry Yahweh required blood sacrifice of an especially exalted kind to be at peace with his children.

E

Edentia. Architectural planet surrounded by seventy satellites; capital of Norlatiadek, our Constellation.

Elijah Mark. Father of John Mark; provided an upper room in his house in Jerusalem where Jesus and the apostles could dine in safety.

Elizabeth. Wife of Zacharias and mother of John the Baptist; cousin of Mary.

Engedi ("kid spring"). Location of an order of the Nazarites near the Dead Sea where John the Baptist was enrolled at the age of fourteen.

Enoch (1). Son of Cain and leader of the Nodites of Elam; (2) Hebrew patriarch; the first Urantian mortal to fuse with his Thought Adjuster while in the flesh.

Ensa. Our Minor Sector whose capital is U Minor the Third.

enseraphim. Technique of space travel using seraphim.

Eternal Son. Second Source and Center; the divine Word; enshrouds the personal presence of the Universal Father (divine thought); God as personality; the revelation of the spirit of God; center of spirit gravity in the Universe; the first completed thought of the Father; mercy; a cocreator; center of gravity of all spiritual reality; omnipresent spiritual force; the perfection of personality.

Eternals of Days. Personalities of Trinity origin; chief planetary administrators on spheres in Havona.

Eve. Material Daughter, mother of sixty-three children, thirty-two daughters and thirty-one sons, before the default and forty-two children after, in addition to offspring of joint parentage with mortal stock.

eventuate. To proceed directly from the Deities, to absonitize.

Eveson. Second-born son of Adam and Eve in the Garden of Eden, approximately 35900 BC.

evil. Deliberate turning away from God; alignment with the unreal; maladjustment to universe reality; conscious choice of what is known to be contrary to the will of the Universal Father.

evolution. Universal Father's plan of ascension; man becoming like God through experience and ascension; process whereby errors and disharmony of nature are brought into line with the perfection of the patterns on Paradise; the unfolding of the life force's urge to be better.

existential. Eternal without qualification.

extinction. Death of the personality who rejects God; takes place on the first mansion world.

F

faith. Belief without evidence; brings immunity to the corrupting influences of evil.

fandors. Extinct giant birds used for human transport on Urantia.

Father Melchizedek. Offspring of the Creator Son and Universe Mother Spirit who with both of them engenders the Melchizedek Sons; only one Father Melchizedek in each Local Universe.

Father-fused sons of God. Individuals who have merged with the Adjuster.

Feast of the Tabernacles. One of the three principal Jewish festivals when Jews traveled to the temple in Jerusalem; lasted seven days and participants lived in the open air in leafy temporary huts ("tabernacles") to remind the Jews of their days wandering the wilderness.

Fifth Epochal Revelation on Planet Earth. *The Urantia Book.*

finaliter. Member of the Corps of the Finality; divine being rich in experience in the worlds of time and space; primarily Father-fused beings but also includes Havona natives and Gravity Messengers.

finite reality. Level of reality where there is time and space and things have a beginning though not necessarily an end; where we live.

First Source and Center. Personalized as God the Universal Father, infinite and eternal, limited only by volition; the eternal I AM; unifier of all constituent elements of infinity and of the creation; creator of the universe.

Flavius. Greek Jew and "proselyte at the gate" at whose house Jesus often stayed.

Fonta. First female human being, whose initial expression of human will choice occurred when she was ten years old, in 991474 BC; Fonta was killed during an earthquake 32 years later at age 42 (see Andon).

Force Controllers. See Force Organizers.

Force Organizers. Intelligent beings who have no will and are involved in the organization of space energies and nonliving forces in the universe; proceed directly from the Infinite Spirit and reside on Superuniverse capitals; work with Creator Sons to control the lines of communication, circuits of energy, and lanes of power that bind creation into a single administrative unit; organize the formative energy of the space sectors; similar to Force Controllers, Power Controllers, Physical Controllers.

force-energy. Derived from self-existent Deity; imperishable and indestructible manifestation of First Source and Center.

G

Gabriel. Governs as regent for the universe of Nebadon when Michael is away; a Bright and Morning Star; head executive administrator of Nebadon; Michael's chief executive.

Garden of Eden. Named after the botanic glories of the constellation capital, Edentia; garden home located at first on a peninsula in the Mediterranean Sea; then in Mesopotamia between the Tigris and Euphrates rivers about 38,000 years ago; educational center devoted to practical arts, basic intellectual training, social culture, economic development, trade relations, and artistic achievements.

Gethsemane. Garden on the western slope of the Mount of Olives (Olivet) near Jerusalem where Jesus and his followers camped.

ghost cult. Religious behavior based on conviction that ghosts punish any change in the customs by which they themselves lived; gives force to customary behavior and traditional mores and establishes the reality of an invisible world; a burdensome scaffolding of the religions of fear.

ghost dream. Common dream when a relative or friend dreams vividly of someone who has died; origin of ghost cult and ghost fear.

God. Word symbol representing the reality of the uncreated first origin of all things, the oneness of all things; the creator; God is spirit, God is love; the infinite unity of all constituent elements of infinity and creation; the unlimited; a primal and infinite personality; the First Source and Center; the only unchanging thing in the universe.

Golgotha ("place of the skull"). Usual place for executions in Jerusalem where Jesus was crucified.

Gospel of John. Last of the four gospels, written in AD 101; speaks to Western peoples, especially the Christians in Alexandria who knew Philo's writings well; John Zebedee commissioned Nathan, a Greek Jew from Caesarea, to write it.

Gospel of Luke. Written by Luke in AD 82; pieced together from notes Luke wrote that derived from Paul whom he followed.

Gospel of Mark. First of the four gospels; written in AD 68 by John Mark; also considered the gospel of Peter because Mark worked with Peter to outline the narrative.

Gospel of Matthew. Written in AD 71 by Isador, a follower of Matthew, who attributed authorship to his master according to a common custom.

Grand Universe. Paradise, the one billion planets of Havona, and the seven Superuniverses, but not the outer space levels.

gravity circuits. Emanations from nether Paradise through which the Eternal Son controls physical and other features of the subordinate universes created by the Paradise Corps of the Creator Sons.

Gravity Messengers. Members of a unique messenger order able to function independently of space and time.

green race. Least capable of the six Sangik races; absorbed or annihilated.

guardian seraphim. Ascending angels who accompany, protect, and sponsor careers of individuals or groups of humans; an individual human who reaches the third circle of attainment receives a personal guardian seraphim.

H

Havona Pilgrim. Ascender who has entered the worlds of Havona.

Havona. Appeared simultaneously with the divinization of the Third Source and Center; the Central Universe; seven concentric bands containing one billion perfect spheres that encircle Paradise and its three sacred circuits of twenty-one satellites; the ideal and model for the universes of time; absonite reality.

Havoners. Lower Trinity beings; absonite native inhabitants of Havona; physical beings who conduct personal lives, advance in knowledge and wisdom, and serve as patterns for life forms within the Superuniverse; do not die or reproduce.

Hebrew: Language of Jewish learning spoken by priests and rabbis.

Herod (the Great). Titular ruler of Palestine in late first century BC; died 4 BC.

Herod Antipas. Son of Herod the Great; ruled Galilee and Perea during lives of John the Baptist and Jesus; died 39 AD.

Herodias. Herod the Great's unlawful wife; hated John the Baptist for his public denunciations of her.

Holy Spirit. Local Universe Mother Spirit; projection of the Infinite Spirit; perfect and unchanging pervasive ministering spirit.

human. Dual-natured being inheriting the emotions and appetites of the animal kingdom yet responsive to the pull of the spirit and capable of recognizing the creator.

I

indigo race. Last of the six Sangik races to emigrate from their homeland; absorbed the orange and green races in Egypt; later driven south into subSaharan Africa.

Infinite Spirit. Third Source and Center; agent of combined action and expression of Universal Father and Eternal Son; source of intelligence coming to the minds of material, morontial, and spiritual beings; Conjoint Actor; God of Action; member of Paradise Trinity; omnipresent divine mind; absolute mind; ancestor of cosmic mind; ministry; Circuit of the Mind.

iniquity. Open and persistent defiance of reality, a kind of cosmic insanity; the sign of vanishing personality control as the iniquitous personality ordinarily does not feel regret for misdeeds or accept forgiveness of sins.

Isle of Paradise. See Paradise.

isolated worlds. Worlds spiritually quarantined as a result of rebellion and cut off from all outside communications.

J

James Alpheus. Apostle; brother of apostle Judas Alpheus and son of the fisherman Alpheus; chosen by James Zebedee.

James Zebedee. Apostle; brother of John Zebedee and son of Zebedee and Salome.

Jerusem. Headquarters and administrative center of our Local System of Satania; architectural planet surrounded by seven architectural spheres each of which is encircled by seven subsatellites; heaven of popular lore.

Jesus of Nazareth. Joshua ben Joseph, son of Joseph and Mary; incarnated in 7 BC, executed in AD 30; a descending Creator Son; human incarnation of Michael of Nebadon, Creator Son of Nebadon.

Jews. Levantine people of mixed western and eastern heritage; Semitic-speaking people, ethnically close to the Phoenicians, Babylonians, and Rome's great enemy, the Carthaginians.

John Mark. Son of Elijah Mark; young boy who followed Jesus; author of the Gospel of Mark.

John the Baptist. Son of Zacharias and Elizabeth; cousin to Jesus; a Nazarite; baptized Jesus in the river Jordan; present head of the commission of twenty-four on Jerusem overseeing Urantia affairs.

John Zebedee. Apostle; brother of James and son of Zebedee and Salome; follower of Jesus; in AD 101 commissioned Nathan to write the Gospel of John; author of the Book of Revelation.

Joseph. Father of Jesus; Hebrew whose family line went back to Abraham and beyond to the Sumerians, the Nodites, the southern blue man, and to Andon and Fonta.

Joseph of Arimathea. Pharisee and former member of the Sanhedrin; longtime secret believer in Jesus.

Judaism of Jerusalem. Western or Hellenized branch of Judaism, deeply influenced by the *Septuagint.*

Judas Alpheus. Apostle selected by John Zebedee; son of fisherman; brother of apostle James Alpheus.

Judas Iscariot. Apostle chosen by Nathaniel; only Judaean apostle; son of prosperous family of Sadducees.

K

Kopet Dagh. Mountain range east of the southern end of the Caspian Sea within which a high Adamite civilization flourished for 7,000 years from the times of Adamson and Ratta (c. 36000–29000 BC).

L

Lanaforge. Current Sovereign of our System of Satania; a Lanonandek Son; presides over council of Planetary Princes and resident governors of isolated worlds; appeared on Sea of Glass on Jerusem two years after Lucifer rebellion broke out to replace Lucifer as System Sovereign.

Lanonandek Sons. Order of descending Local Universe Sons of God proceeding from the union of a Creator Son and his Creative Spirit; serve as heads of Local Systems and as Planetary Princes, rulers of inhabited worlds; have unusual powers of personal authority and freedom.

Laotta. Prominent Nodite woman, head of western Garden of Eden school; had intercourse with Adam following Eve's default with Cano; died giving birth to Sansa who was raised by Adam and Eve.

Lazarus. Son of Simon of Bethany and lifelong friend of Jesus; died at age 67 of the same disease that first carried him off at a much younger age, when Jesus "raised him from the dead."

Life Carriers. Order of the Sons of God who implant life on a planet; design life forms, carry them to planets, and remain on the planets to foster them until the appearance of mortal humans; offspring of Creator Son, Creative Spirit, and one of the three Ancients of Days; catalysts who execute the patterns of life, not creators; work to control aspects of the environment to favor the progress of evolution up to the moment when the first will creatures appear.

life-experiment world. Place where large deviations from general patterns are allowed; one inhabited planet in ten, including Urantia.

light and life. Perfected settlement in goodness, beauty, and truth; the goal for evolving planets.

Local Universe. Group of 100 constellations, about 10 million inhabitable planets; ruled by Creator Son from architectural sphere; our Local Universe is Nebadon whose capital is Salvington.

Lucifer: a Lanonandek Son created by the Creator Son Michael of Nebadon in conjunction with his local Creative Spirit; ruled on Jerusem with Satan, his first lieutenant; one of the most brilliant of his order; found in contempt of the Creator Son, imprisoned on a satellite of the seventh planet going around Jerusem.

Lucifer rebellion. Insurrection in the system of Satania 200,000 years ago; serious disturbance in the Constellation based on Lucifer's advocacy of a philosophy of "self assertion and liberty" and his disavowal of the authority of Michael of Nebadon, then vice-regent of the Universal Father.

Luke. Physician in Antioch in Pisidea (modern central Turkey); became follower of Paul in AD 47; wrote the Gospel of Luke in AD 82 and later the Acts of the Apostles; died in AD 90.

M

Machiventa Melchizedek. Emergency Son of Nebadon who bestowed himself on Urantia and taught of El Elyon, the Most High God; only member of the Melchizedek order to incarnate in human form; lived in human form in Salem (now Jerusalem) and was closely associated with Abraham, about 2000 BC; presently resides on Jerusem and serves as vice-regent of Urantia.

Magisterial Sons. Descending order of Paradise Sons of God; personalizations of the Universal Father and Infinite Spirit who judge sleeping survivors at the end of a dispensation; ministers who work closely with Creator Sons and perform judicial duties; may serve as incarnate material beings, teachers, in the likeness of a planetary race at critical junctures.

Major Sector. Group of 100 minor sectors, about 100 billion inhabitable worlds; ruled by personalities of Trinity origin; ours is Splandon whose capital is U-Major the Fifth.

mansion world students. All human personalities who repersonalize on the morontia worlds after mortal death.

mansion worlds. Seven subsatellites that surround the first transition world encircling Jerusem; training worlds at the morontia level where ascending mortals reside.

Mary. Mother of Jesus; her distinguished lineage traced back even to Eve.

Mary Magdalene. Woman fallen into prostitution through a series of misfortunes; the most effective recruiter of all the women's corps.

Master Architects of the Universe. Beings who dwell on Paradise and fashion and control the systems of energy and bodies in space until they acquire a certain level of gravity when control is turned over to Force Controllers.

Master Physical Controllers. Intelligent beings who originate in the Universal Mother Spirit and control the energy of the Local Systems; govern the level where nonteachable organisms respond to the environment.

Material Sons (and Daughters). Sons of God (such as Adam and Eve) resident on Jerusem where they are mentors and teachers to mortal ascenders; incarnate on evolving planets to enhance genetic code of the races; first pair fashioned in the Local Universe by the Creator Son working alone; subsequent reproduction is sexual; highest type of sex-reproducing beings; semimaterial beings easily visible to mortals on the evolving worlds; oversee shift from hunter-gatherer to agriculture and horticulture; their offspring are the Adamic or violet race.

matter. Basis of the universe; a form of energy that proceeds from and is controlled by the Universal Father; whatever responds to the gravity circuits of nether Paradise.

Matthew Levi. Apostle chosen by Andrew; customs agent who lived east of Capernaum.

Melchizedek receivers. Twelve Melchizedek Sons who came to govern Urantia from seven years after the Caligastia betrayal until the arrival of Adam and Eve, then returned after their default.

Melchizedek Sons. Descending order of Local Universe Sons; offspring of the Creator Son and the Universe Mother Spirit and a Father Melchizedek, himself created by union of the Creator Son and the Creative Spirit; a versatile order ready for nearly any universe assignment, especially educational, and the performance of emergency functions; number more than 10 million in the Local Universe; authored numerous papers in *The Urantia Book* on the Local Universe.

mercy. Love in action.

Mesopotamia. Cradle of civilization between the Tigris and Euphrates rivers; Dalamatia was at its southern edge, and the second Garden of Eden lay where the rivers come close together.

messiah ("anointed one" in Hebrew, translated by the Greek "Christos"). Deliverer, expected by the Jews, who would drive back the Romans and restore the ancient Davidic kingdom.

Michael of Nebadon. Creator Son of the order of Michaels who created Nebadon, our Local Universe, and incarnated on Urantia as Jesus of Nazareth in Palestine in 7 BC; represents the Father within his creation; proceeds directly from the Universal Father and Eternal Son; present titular Planetary Prince of Urantia, but dwells on Salvington, headquarters of Nebadon.

mid-mammals. Offspring of dawn-mammals; first evolutionary creatures on Urantia who walked almost upright, had opposing thumbs and little hair.

Midway Creatures (midwayers). Natives to individual planets; ministering spirits but not angels; semimaterial beings who have been here for millennia and except when translated as ascenders, do not leave; engendered on Urantia: (1) through union of members of the corporeal staff of the Planetary Prince (primary midwayers); and (2) through union between the sixteen invisible children of Adamson and Ratta; important to the advancement of culture and civilization; not ordinarily visible to the human eye; appear on most inhabited worlds, and always on life-experiment decimal worlds such as Urantia.

Mighty Messengers. Special messenger order; dictated several of the Urantia Papers.

Mind. Thinking, perceiving, and feeling mechanism of the human organism; the total conscious and unconscious experience; in personality intervenes between matter and spirit, a function of the Third Source and Center.

Ministering Spirits. Supernaphim, seconaphim, tertiaphim, seraphim, and Midway Creatures; beings that originate in the Third Source and Center (except Midway Creatures) and whose mission is to minister and serve.

Minor Sector. Group of 100 Local Universes, about one billion inhabitable planets; ruled by beings of Trinity origin; capital of our Minor Sector of U-Minor the Third is Ensa.

Mithraism. Roman "mystery cult" involving elaborate initiations and communal meals based on the story of a bull-conquering hero whose miraculous advent was witnessed by gift-bearing shepherds; later Mithras was associated with the Unconquered Sun, born on December 25.

Mithras. Persian god of resurrection.

morontia career. Educational journey of the surviving personality after mortal death through the Local Universe as preparation for the spirit life in the Superuniverse; education to eradicate permanently from mortal survivors such animal vestigial traits as procrastination, equivocation, insincerity, avoidance of problems, unfairness, and the seeking of ease; the transition life intervening between evolutionary material existence and high spirit attainment.

morontia food. Living energy unknown on the material worlds; has no residue and there is no excretion.

morontia form. Our next level of body after death; semimaterial body that reflects the creative mind that inhabits it.

morontia progressor. Survivor in the morontia schools who ascends through the spheres of the Local Universe gaining experience and knowledge.

morontia worlds. All fifty-six satellites surrounding Jerusem (seven main satellites each with its own seven subsatellites) as well as the seventy satellites surrounding Edentia, and the four hundred ninety satellites surrounding Salvington.

morontia. Phase of universe reality bridging the material and spiritual; realm of morontia progression and experience takes place in the Local Universe and ranges from the transition spheres orbiting Jerusem up through Salvington; morontia refers to material, mental, and spiritual realities, and to the soul.

mortal ascenders. Beings (such as ourselves) with the potential to reach Paradise.

mortal death. Escape from the material life in the flesh.

Most Highs. Three Vorondadek Sons who together govern a Constellation .

mota. Morontia system for organizing knowledge that is parallel to but higher than logic; regards cosmic meanings and universal interrelationships; learned on the third mansion world.

Mother Spirit. See Universe Mother Spirit.

Mount Seraph. Highest mountain on Jerusem, 15,000 feet, from which depart the transport seraphim.

mystery cults. Religious practices based on myths and ritual; had a strong influence on early Christianity.

Mystery Monitor. Another name for Thought Adjuster, emphasizing the Adjuster's role as the hidden monitor of mortal behavior.

N

Nalda. Samaritan woman of questionable moral character to whom Jesus made his first direct pronouncement of his divine nature and his sonship with God.

Nathan. Greek Jew from Caesarea commissioned by John Zebedee to write the Gospel of John in AD 101.

Nathaniel. Of Cana; friend of Philip of Bethsaida; an apostle.

Nazareth. Home of the young Jesus; a small town in the Galilean hills on a caravan route.

Nazarites (Hebrew "consecrated ones"). Ascetic order whose members took a solemn vow to abstain from all intoxicating drinks, to let the hair grow, and not to touch the dead; venerated by Jews.

Neanderthals. Mixture of Andonites and Badonites who emerged in NW India and widely emigrated; dominated the world for 500,000 years until the migrations of the six Sangik races.

Nebadon. Our Local Universe located on the outer edge of Orvonton, the seventh Superuniverse; a creation of the Son of God known as Michael and jointly governed by his Creative Spirit consort.

Nicodemus. Pharisee and powerful member of the Sanhedrin who secretly visited with Jesus at the house of Flavius.

Nod. One of the 100 corporeal members of Caligastia's staff; headed the commission on industry and trade and became leader of the sixty staff members who joined the rebellion; degraded to mortal status when they lost control of the tree of life.

Nodites. Descendants of the sixty disloyal corporeal staff, whose leader was Nod, and their human offspring; also a race and culture; freely mixed with Sangik races and left behind a biologically able progeny.

Norlatiadek. Our Constellation, one of 100 constellations of Nebadon, whose capital is Edentia.

O

omniaphim. Order of angels that serve on the Seven Sacred Spheres of the Spirit.

Onamonalonton. Great teacher of the red men from about 65,000 years ago, who lived in the redwood forest in California; taught of the Great Spirit who created the world.

orange race. One of the six Sangik races; a race of builders who emigrated to the lower Nile Valley and were destroyed or assimilated 100,000 years ago by the green and indigo races.

Order of Michaels. Paradise Sons who create Local Universes; see Creator Sons.

original sin. Theory of Paul of Tarsus that one is born in guilt and evil but can be redeemed.

Orvonton. Seventh of the seven Superuniverses that, in conjunction with the Paradise/Havona system, comprise the Grand Universe of time and space; divided into ten Major Sectors, each of which is divided into 100 Minor Sectors, each containing 1,000 Local Universes, each of which is comprised of 100 Constellations made up of 100 Local Systems governing a potential of 1,000 inhabited worlds.

outer space levels. Four zones, located beyond a quiet belt about 400,000 light years across, surrounding the belt of the Superuniverses; belts of cosmic energy wheeling around the inhabited Superuniverses.

P

Palestine. Southeastern end of the Mediterranean Sea, named after the Philistines, Greeks who migrated there in the second millennium BC; birthplace of Jesus of Nazareth; center for international trade enjoying peace and relative prosperity in the time of Jesus.

parable. Story that has one meaning, teaches a single truth.

Paradise. Eternal, nonpersonal master pattern for all material reality; motionless center for all personality activity and source of all force, energy, and power in the universe; geographical center of infinity; center of the universe and dwelling place of the persons of the Paradise Trinity, their coordinates and associates; source and goal of all spirit personalities, including ourselves; center of material gravity; absolute reality; original nonpersonal expression of the First Source and Center; perfect and unchanging, outside of time and space.

Paradise Citizens. Beings who always have lived and always will live on Paradise; Trinity-origin beings.

Paradise Sons. Same as Sons of God; a category of divine being produced by the Universal Father and the Eternal Son, of which Creator Sons are a subcategory.

Paul of Tarsus (died AD 65). Roman citizen, but a Hebrew of the Hebrews; spoke and wrote in Greek in his proclamations about a Jewish Messiah; founder of the Christian church (with Peter); established the Christian doctrine of atonement according to which the blood of Jesus washed away the sin of Adam to permit good relations between guilty man and an avenging God.

Pentecost. Jewish festival fifty days after Passover; on Pentecost in AD 30 the Spirit of Truth came to live in all normal human beings on Urantia where Thought Adjusters now are bestowed on all normal humans reaching an age of moral responsibility, spiritual choice.

Perfectors of Wisdom. Trinity-origin beings that advise the Ancients of Days on the Superuniverse capitals; some are members of the revelatory commission sent from Uversa to Urantia to dictate early papers of *The Urantia Book.*

personality. Bestowal of the Universal Father; the one unchanging reality in creature experience; makes moral behavior possible; self-consciousness and free will; the unification of the constituent elements of the individual mortal creature; neither body, mind, nor spirit, nor the soul, personality is the identity that survives in the morontia soul.

personality circuit. Universal circuit leading to the personal presence of the Universal Father, the source of all personality, approached through personality.

Personalized Thought Adjuster. Thought Adjuster who has merged with the personality of an ascending mortal; Adjuster who has achieved union with the personality it dwelt in; highest order of sonship with God; Thought Adjuster of Jesus of Nazareth was uniquely personalized directly by the Universal Father.

pervaded space. Horizontal extensions from Paradise wherein lie the central and surrounding universes; universe space.

Peter. See Simon Peter.

Pharisees. Learned Jews who awaited the coming of the Messiah who would destroy the Romans; many belonged to the small middle class and were mostly tradesmen.

Philip. Of Bethsaida; apostle who had known Jesus in Capernaum.

Philip. Son of Herod the Great and brother of Herod Antipas; ruler of the tetrarchy of Gaulinitis; a half-believer in the teachings of Jesus.

Philo of Alexandria (20 BC–AD 50). Contemporary of Jesus whose combination of Greek philosophy and Hebrew theology found broad agreement in Palestine; attempted to unify Greek philosophy and the Stoic doctrine of the *logos* with Hebrew traditional texts; one of the great teachers of the ancient world.

Physical Controllers. See Force Organizers.

Planetary Adams. Material Sons and Daughters; Adam and Eve.

Planetary Prince. Ruler of an inhabited planet; ordinarily a Lanonandek Son from Jerusem who oversees the orderly development of intelligent life on an inhabited world; once Caligastia, the current titular Planetary Prince of Urantia is Michael of Nebadon.

Planetary Sovereign. Title of the Planetary Prince when a world becomes settled in light and life; has expanded power and greater responsibility than the Planetary Prince.

Pontius Pilate. Roman procurator, the only man with the power to confirm the sentence of death decreed by the Sanhedrist court.

Power Directors. See Force Organizers; creators of architectural spheres; Power Controllers.

prepersonal. Without personality; Thought Adjusters are prepersonal specks of God.

primary midwayers. See Midway Creatures.

prison world. First satellite circling the seventh transition world circling Jerusem; the current home of Lucifer, Satan, and other leaders in the Lucifer rebellion.

proselyte of the gate. One who accepted the power of Yahweh but was not circumcised.

Q

quarantined world. World isolated as a result of rebellion; communications with the System capital are suspended; Urantia has been quarantined since the Lucifer rebellion.

R

Ratta. Pure-line descendent of Caligastia's corporeal staff who married Adamson, the first-born son of Adam and Eve, during the times of the second Garden; they had sixty-seven children and were grandparents to the secondary midwayers.

Rebecca. Daughter of a wealthy merchant and trader who fell in love with young Jesus; stood among the women at the foot of the cross.

red race. One of the six Sangik races; exceptional for intelligence and first organizers of tribal government.

Reflective Spirits. Trinity-origin beings of great power who dwell on the headquarters planets of the seven Superuniverses and enable instant communication with Paradise.

Reserve Corps of Destiny. Around 1,000 selected individuals trained unconsciously to preserve civilization in case of a catastrophe.

resurrection hall. Immense temple on the first subsatellite of the first transition world of Jerusem where surviving personalities are reassembled after mortal death.

S

Sacred Spheres of Paradise. Three circuits of seven immense spheres located between the Isle of Paradise and the billion planets of Havona; the innermost circuit is the Sphere of the Father, then the Son, then the Spirit, moving outward.

Sadducees. Jewish religious and political group installed under the Maccabees; social aristocracy who controlled the Jewish priesthood and reaped immense wealth from it.

Salem. Name of Jerusalem in the time of Abraham, about 2000 BC; where Machiventa Melchizedek incarnated in human form.

Salome. Wife of Zebedee and mother of John and James, who were to become apostles, and David, whose messenger corps made Jesus' missions possible; a distant cousin of Annas, the high priest emeritus in Jerusalem.

Salvington. Center of our Local Universe, Nebadon, and home of Michael who appeared on our planet as Jesus of Nazareth; architectural world surrounded by ten clusters of forty-nine planets each.

Samaritans. Peoples of mixed but nonJewish origins who lived in the hill country north of Jerusalem.

Sangik family. Badonite offspring who lived in northwest India 500,000 years ago, origin of six Sangik or colored races of Urantia.

Sanhedrin (from Greek "council"). Jewish ruling council in Jerusalem consisting of seventy-one members.

Satan. Lanonandek Son (offspring of Creator Son, Michael of Nebadon, and his Creative Spirit), who was first lieutenant of Lucifer; presented Lucifer's views on Urantia; now imprisoned on the prison world near Jerusem.

Satania. Our Local System, one of 10,000 Systems of inhabited worlds; located on the outer edge of the Local Universe of Nebadon; headquarters is the sphere Jerusem.

Sea of Glass. Receiving area for interplanetary travel on most architectural worlds; enormous circular crystal, one hundred miles in diameter and thirty miles thick that enables seraphic transport and, on Edentia, plays a role in the reception and distribution of cosmic energies.

seconaphim. Order of angels that serves in the Superuniverses.

secondary midwayers. See Midway Creatures.

Septuagint. Hebrew scriptures in Greek translation.

Serapatatia. Nodite leader descended from Caligastia's defaulted corporeal staff, one of Adam's most able and trusted lieutenants, manipulated by Caligastia to entrap Eve into the default of the Adamic mission by having intercourse with Cano; drowned himself in the great river of Eden.

seraphic transport. Ordinary means of travel to other spheres whereby seraphim wrap their spirit forms around nonmaterial personalities to carry them immense distances.

seraphim. Order of angels that serves in the Local Universe; offspring of the Universe Mother Spirit; some function as space traversers within the Superuniverse.

Seraphington. Sixth sphere in the Circuit of the Father that orbits the innermost circuit around Paradise; contains the secrets of seraphic transport.

Seth. Oldest son of Adam and Eve born in the second Garden; founded the Sethite priesthood whose ministry included religion, education, and health.

Seven Master Spirits. Source of cosmic mind; representations of the seven divine aspects of the Infinite Spirit, each one governing from one of the seven Sacred Spheres of the Infinite Spirit that go around Paradise and each one controlling one of the seven Superuniverses; together they control the circuits of the central eternal universe of Havona.

Simon (of Bethany). Father of Martha, Mary, and Lazarus.

Simon Peter. Apostle; brother of Andrew; formerly one of John the Baptist's supporters; Jesus changed his name to Peter which means "rock" in Greek, a firm foundation; became the spokesman for the apostles as a group.

Simon Zelotes. Apostle chosen by Peter; had been a member of the Zealots and before that a merchant.

sin. Conscious, deliberate doing of evil.

sleeping survivors. Individuals who in mortal life have not earned a guardian seraphim and upon mortal death enter an unconscious sleep where they remain until the next periodic judgment day.

Solitary Messengers. Corps of spirits who are the first begotten issue of the Infinite Spirit in time and space; act as ambassadors; with powers of interplanetary travel they are used for dispatch and service in situations where personality is essential to the achievement of the assignment; the only personalized beings able to synchronize with combined universal currents of the Grand Universe.

Solonia. Chief of the fifth order of angels who proclaimed Eve's default of the divine plan and requested the return of the Melchizedek receivers to Urantia; the "voice in the Garden"; responsible for four of the Urantia papers.

Sonarington. One of the seven sacred life spheres of the Father that orbit the innermost circuit around Paradise; personal receiving world of the Eternal Son and Paradise headquarters of the ascending and descending Sons of God; contains the secret of incarnation by members of the orders of the Sons of God.

Son-fused mortals. Individuals unable to attain Adjuster fusion who receive the gift of Spirit from the Eternal Son; permanent residents of Uversa.

Sons of God. Category of Paradise Sons, including Creator Sons, created by the Universal Father and the Eternal Son.

soul. Immortal morontial personality engendered by the tension between the mortal personality and the spirit hidden within; product of the seeking mind and the divine Thought Adjuster destined to begin the Paradise ascension.

Spirit of Truth. Spiritual energy from the Son that draws truth-seekers toward knowledge of the Universal Father through the person of the Son; once poured out, Thought Adjusters occupy the minds of all normal humans.

spirit. Soul of creation; water of life fulfilling those who receive it regardless of their material circumstance; that which endows one with a purpose and ideals that transcend our animal nature; the essence of existence.

Spirit-fused mortals. Individuals who do not fuse with their Adjusters but fuse with premind spirit on the mansion worlds; retain no memory of earthly life, awake as new creations and permanent residents of Nebadon.

spirit-gravity circuit. Pull of the Eternal Son; the motivating force behind the ascending career; the secret of mortal ascension and eternal life; energy of eternity and divinity touching material creation.

Splandon. Capital of our Major Sector, called U-Major the Fifth.

Stephen. First Christian martyr.

Stoics. Philosophy originated in Greece that taught that one must follow the laws of nature in response to the intelligence, or *logos*, that indwells all things.

superangels. Originate from the Third Source and Center; higher than angels.

supernaphim. Highest order of Central Universe ministering spirits; minister on Paradise and in Havona.

Superuniverse. Group of ten Major Sectors, about one trillion inhabitable planets; ruled by three Ancients of Days.

Supreme Being. God as a subabsolute being evolving with and experiencing the universe; cosmic reality; also called God the Supreme or the Almighty Supreme; complex and universal synthesis of the emerging phases of all finite realities; the great avenue through which the Father flows outward to all creation and through which finite creatures pass inward in their quest for the Father; sum total of all finite growth; Deity liberated from eternity.

Syria. Territory northeast of Palestine; center for international trade enjoying peace and relative prosperity in the time of Jesus.

System Sovereign. Administrator of a System, always a Lanonandek Son, an offspring of the Creator Son and the Universe Mother Spirit; has unusual powers of personal authority and freedom; carries out legislation passed at the Constellation level and juridical decrees made on Uversa.

System. Group of 1,000 inhabited or inhabitable planets in a Local Universe; our System is Satania whose capital is Jerusem.

T

Temple of Yahweh. Location of the great Jewish festivals in Jerusalem.

tertiaphim. Order of angels who serves the Creator Sons and the Ancients of Days in the Local Universes.

Third Source and Center. Infinite Spirit.

Thomas Didymus. Apostle; fisherman on the Sea of Galilee.

Thought Adjuster. Mystery Monitor, formless, prepersonal infinite spark of God sent by the Universal Father to indwell mortals; secret to human destiny; Father fragment; preserver of life memories of former mortal translated from material to morontia level; prepersonal "piece of God" that strives in secret ways to adjust one's thoughts to be loving, selfless, and worshipful; when fused with the morontia soul the new being becomes an ascending son of God.

Titus. Prominent man of Capernaum; traveled to Cana to beg Jesus to heal his dying son; later became important in the earliest Christian church and died in the persecution at Rome in AD 69.

Transcendentalers. Eventuated absonite beings who live on Paradise and Havona; perform myriad functions in the universe beyond the understanding of personal beings such as ourselves.

transition worlds. Seven satellites surrounding Jerusem.

translated midwayers. Ascending Midway Creatures who for various reasons are accepted as sons of God and traverse the same universe career as ascendant mortals.

tree of life. Shrub from Edentia that the Most Highs of Norlatiadek sent to Urantia at the time of Caligastia's arrival; fruit conferred continuous life on the Caligastia 100 and the 100 modified Andonites who served as their subordinates; located in the central courtyard of the temple of the unseen Father in Dalamatia; absorbed energies that defy aging in animal life; supported Van and Amadon on Urantia for 150,000 years.

Trinitized Teacher Sons. One of the three categories of Paradise Sons of God (along with Magisterial Sons and Creator Sons); divine beings generated by the Father, Son, and Infinite Spirit who roam the universe as teachers to divine and mortal personalities; special teacher corps that completes the journey of an inhabited planet in its evolutionary culmination and destination, the era of light and life.

Trinity. Universal Father, Eternal Son, and Infinite Spirit acting as one; the unity of Deity; coeternal with the I AM.

U

ultimaton. Indestructible particle that has Paradise as its nucleus; the building block of the electron, made up of 100 ultimatons; responsive to Paradise absolute gravity but not to linear gravity; first measurable form of energy, not yet described by Urantia scientists.

U-Major the Fifth. Capital of our Major Sector, Splandon.

U-Minor the Third. Architectural sphere that is the capital of our Minor Sector, Ensa.

Universal Absolute. Deity dominating nether Paradise; God without personality functioning in the universe.

Universal Censors. Trinity-origin beings that advise the Ancients of Days on the Superuniverse capital; some are members of the revelatory commission sent from Uversa to Urantia to dictate numerous early papers of *The Urantia Book.*

Universal Father. Personality of the First Source and Center; one aspect of the Triune creator; center of all things; source of the Isle of Paradise; center of the personality circuit; God as will; creator, controller, and upholder of all there is; an infinite, perfect, self-conscious being who shares in goodness every experience of every creature in the universe.

Universe Mother Spirit: See Creative Spirit.

universe of universes. Paradise, the Central Universe, the Superuniverses, and the outer space zones.

Univitatia. Permanent population of the Constellation headquarters worlds; an order of nonreproducing beings created by the Creator Son and Creative Spirit.

unpervaded space. Areas extending vertically above and below Paradise; reservoir space.

Urantia. Planet Earth, a life-experiment station; located at the outer edge of the Local System, Satania, in the Local Universe of Nebadon, in the Superuniverse Orvonton.

Uversa. Architectural world and capital of our Superuniverse Orvonton; surrounded by ten clusters of seven primary planets each with seven subsidiary worlds; the Superuniverse members of the Urantia revelatory commission came from Uversa.

V

Van. One of the one hundred members of the corporeal staff of Caligastia's staff; Van headed the committee on tribal coordination and racial cooperation; one of the forty members of the one hundred who did not join the rebellion; remained on the planet in cooperation with Amadon until the arrival of Adam and Eve, almost 150,000 years later, sustained by the tree of life; currently resides on Jerusem where he represents the interests of Urantia.

Vanites. Descendants of Van; ancestors of the Assyrians in northern Mesopotamia near Lake Van and Mount Ararat.

vice-regent. Ruler who does not possess full authority.

violet race. Planetary offspring of Adam and Eve; a monogamous superior race created to interbreed with natives of evolving planets.

Vorondadek Sons. Descending order of Local Universe Sons, offspring of a Creator Son and Creative Spirit; three Vorondadek Sons, called the Most Highs, govern each Constellation; 1,000,000 in number in Nebadon, also known as Constellation Fathers.

W

white races. Mostly descendants of the blue race, admixed with yellow and red, and upstepped by absorbing the greater portion of the violet race of Adam.

will creatures. Creatures capable of knowing God and worshiping him.

women's corps. Group of women who worked with Jesus; effective because they could speak freely to other women about the spirit of God that lived within them and could enter into bordellos and speak to the women there.

worship. Religious impulse; the conscious act of recognizing the intimate and personal relationship between the creators and the created; one of the seven Adjutant Mind Spirits.

Y

Yahweh. Volcano spirit of the Sinai; tribal god of the Hebrews, who thought he had chosen them from all other peoples to rule the earth.

yellow race. One of the six Sangik races; first Sangik race to abandon hunting for agriculture; endowed with superior ability to live in peace together; most successful of the Sangik races.

Z

Zacharias. Husband of Elizabeth and father of John the Baptist; member of the Jewish priesthood.

Zealots. Nationalist group determined to force change through political revolt against the Roman hegemony.

Zebedee. Boat builder from Bethsaida who had been a good friend of Jesus' father Joseph; husband of Salome; father of John and James, who became apostles, and David, whose messenger corps made Jesus' missions possible.